W9-DFU-348

Dynamic Models in Biology

DYNAMIC MODELS IN BIOLOGY

Stephen P. Ellner and John Guckenheimer

PRINCETON UNIVERSITY PRESS | PRINCETON AND OXFORD

©2006 by Princeton University Press

Published by Princeton University Press, 41 William Street, Princeton, New Jersey 08540

In the United Kingdom: Princeton University Press, 3 Market Place, Woodstock, Oxfordshire OX20 1SY

All Rights Reserved

Library of Congress Cataloging-in-Publication Data

Ellner, Stephen P., 1953–
Dynamic models in biology / Stephen P. Ellner and John Guckenheimer.
p. cm.
Includes bibliographical references and index.
ISBN-13: 978-0-691-11843-7 (alk. paper)
ISBN-10: 0-691-11843-4 (alk. paper)
ISBN-13: 978-0-691-12589-3 (pbk. : alk. paper)
ISBN-10: 0-691-12589-9 (pbk. : alk. paper)
1. Biology–Mathematical models. I. Guckenheimer, John. II. Title

QH323.5.E44 2006

570′.1′5118–dc22 2005048818

British Library Cataloging-in-Publication Data is available

This book has been composed in Stone and Stone Sans

Printed on acid-free paper. ∞

pup.princeton.edu

Printed in the United States of America

10 9 8 7 6 5 4 3 2 1

Contents

List of Figures

List of Tables

Preface

In 2001 the United Kingdom suffered a major outbreak of foot-and-mouth disease in livestock. The scientific experts assembled to advise the government included several groups of mathematical modelers. Within weeks of the first case report, modelers had developed systems of equations describing the farm-to-farm spread of infection, and were using them to help devise measures for controlling the outbreak (Ferguson et al. 2001a, b; Keeling et al. 2001). Adapting previous models, Ferguson et al. (2001) classified farms according to disease state: disease-free, infected, and vaccinated or slaughtered, with the infected class subdivided into several stages to represent the degree of infection and whether or not the disease had been detected. Equations were developed to describe how farms become infected, experience a local outbreak, and pass the disease to other farms, with model parameters estimated from the recorded dates and locations of all reported infections. Because the model is based on equations representing the process of disease transmission and spread, it could be modified to incorporate potential control measures and then used to predict their consequences. It was concluded that immediate slaughter of infected herds would slow disease spread but not halt it, and more aggressive measures were needed. Based in part on those analyses, the government instituted a policy of rapidly culling all farms in a ring surrounding a farm where infection is detected. These policies were successful: the disease soon started to decline and the outbreak collapsed within a few months. Detailed modeling and data analyses later confirmed that the control measures had greatly reduced the size of the outbreak, but also suggested that earlier application of the strict control measures that were eventually imposed could have reduced the outbreak by about half (Keeling et al. 2001).

The 2001 foot-and-mouth outbreak is a particularly dramatic example of how models, written in the language of mathematics and solved on the computer, are now used throughout biology. Experimental biology is reductionist, disclosing how life works one piece at a time. But at all levels life is *complex*, meaning that the properties of cells, organisms, populations, and ecosystems reflect the interactions among multiple components. In virtually all areas of biology, models provide the means for putting the pieces together, for example,

- to understand the consequences of multiple, interacting signal transduction pathways within cells and complex networks of interaction among genes;
- to explain the development of irregular heart behaviors such as fibrillation, and to improve methods for treating them;
- to predict which species are at greatest risk of extinction, to identify effective measures for their preservation, and to predict how ecosystems will respond to the changes in climate and massive loss of species now in progress.

In the past, biologists have generally been uninterested in mathematics, but that attitude is changing rapidly. With advances in many technologies, our intuition is no longer sufficient to analyze and interpret data. Quantitative models and methods are needed, sometimes because data are so extensive that automated methods are needed to examine them in detail (e.g., genome sequences) and sometimes because the pheomena are sufficiently complex that computer simulation is the only way to derive predictions from a model. We believe that biology students need to connect what they learn in a required calculus course to biology, and that mathematics students are greatly enriched when they see mathematics connected to real-world problems in a more tangible way than is taught in mathematics courses.

These connections come through models, but there are some complex issues that must be addressed to develop an appreciation for them. What are the principles used in developing a model? What simplifications are made to make a model tractable? What measurements are used to estimate model parameters and validate model predictions? How does mathematics help us analyze models, either by hand or through computer algorithms? How robust are the predictions of a model to uncertainties in the structure of the model, errors in our measurements, stochastic fluctuations of the environment, and the individual variability of living organisms? These are all difficult questions, and they deserve sustained attention as students learn the art of modeling biological systems.

This text is our attempt to address the lack of curricular materials designed to introduce undergraduate students to dynamic models of biological systems. It is based on a course that we have taught to a diverse audience of students majoring in the biological sciences, computer science, mathematics, engineering, and the physical sciences. We have adopted the following strategies in writing the text:

- We make no attempt to be comprehensive. Most chapters follow the "business school" model of focusing on a selective set of case studies that are treated in depth.

- Examples have been chosen to span the breadth of major areas within biology and to illustrate diverse types of mathematical models. Models of processes within cells, organisms, and populations are all based on the same principles. At this level, an integrative approach that does not divide biology along these boundaries seems appropriate. The models that we study are either ones that have stood the test of time or recent ones that are especially prominent now. Similarly, some of our case studies are classics, and others are very recent, but all of them are compelling to us and exemplify the reasons why we chose to work in this field.
- We restrict attention to dynamic models. This is a reflection of our personal interests and expertise. Ideally this course would be accompanied by a second semester devoted to genomics, bioinformatics, and structural biology where the mathematical theory is drawn primarily from combinatorics, probability, statistics, and optimization.
- We rely on high-level computer languages such as Matlab or R to implement models and perform calculations. This minimizes the user effort required to implement models, run simulations, and perform mathematical analysis, and allows us to quickly proceed to discussion of modeling results. The primary question we ask is what we learn about the biology from the models.
- We introduce the minimal amount of mathematics needed to understand the modeling results. We hope that students taking the course will be inspired to learn about the mathematics we introduce by studying the material in more depth in mathematics courses — but we make no pretense of discussing more than basic theoretical concepts. In contrast, biomathematics courses often select mathematics that is needed to study models in certain domains and teach this material to students who will engage in modeling within this domain. We think that approach is problematic in that students may need a deeper understanding of the mathematics than is typically covered in such courses.
- We discuss the modeling process itself. The first and last chapters of the book discuss the nature of models and give our advice about how to construct and use models. Intermediate chapters intersperse comments intended to place the models into a broad context, with citations to key references where the student can learn more.
- We expect students to read original papers from the biology literature. The published literature plays very different roles in biology and mathematics, and this influences our use of literature in the course. The mathematics literature records and systematizes theories that an individual is expected to personally reconstruct for full understanding. Biology literature describes experiments or observations and their outcomes, along with the hypotheses that the experiments were designed to test and the interpretation of the results. Critical evaluation of how experimental results are interpreted is an important part of the subject. Since the purpose of models is to address biological questions, we think that it is important for students—even at the level of this course—to engage in such critical thinking.

We require students to do a course project. Their assignment is to choose a published research paper using a dynamic model, implement the model, reproduce the modeling results in the paper, and then take one more step. By this we mean that they should formulate a question that was not fully addressed in the paper, modify or extend the model to do so, and then present their findings. This is a difficult assignment but we have been gratified by the papers that the students have written. Much more than the homework exercises, it prompts students to think creatively and critically about science in a setting in which answers are not absolutely right or wrong.

We think computer exercises are essential so that students will have personal experience as modelers and model-users. We use the first three or four computer lab sessions to teach the rudiments of programming and then add a bit each week as new types of models are encountered—how to compute eigenvalues, generate random numbers, solve differential equations, and so on. Our experience is that undergraduate biology students can and do learn how to do real scientific programming. Instructional materials and suggested exercises for computer labs can be downloaded at no cost from this book's web page at Princeton University Press (http://www.pupress.princeton.edu/titles/8124.html).

There is more in this book than can be covered in a single semester, allowing some flexibility as to the choice of topics and the mathematical level of the course. Here are a few suggestions:

1. Because we do not divide biology along disciplinary lines, premed students get a substantial dose of population biology before we get to neurotransmitters—and they have to pay attention, because we use population models to introduce some essential mathematics for everything that follows. Therefore we strongly recommend covering at least a good fraction of Chapter 1, so that students have a taste of the variety of applications to come.
2. Chapters 2 and 3 and Chapters 4 and 6 are paired in the sense that models of the same type are used in very different applications. The second of each pair can therefore be skipped without losing anything essential for later chapters. With a bit more effort only the second can be covered, with material from the first brought in as needed.
3. Chapters 5, 7, 8, and 9 are independent of each other and are not required for reading any other chapters—feel free to pick and choose, or to cover some of each (as we generally do).
4. Chapters 3 and 7 are the only ones where probability theory plays any significant role. A course on "deterministic models" can be constructed by omitting those chapters, and the sections in chapters 2 and 9 on stochastic models.

Acknowledgments

We would like to thank first our families for their support and patience while we wrote this book.

We are grateful to Cornell University for giving us the opportunity to teach our interdisciplinary class, for providing adequate teaching assistant support, and for a sabbatical leave to Steve Ellner while we completed the book. Thanks also to the Cornell Laboratory of Ornithology for hosting the sabbatical.

The decision to turn our class into a book was inspired by our colleague Rick Durrett's ability to write a book every few years while remaining prolific as a researcher, and by his explanation of how it's done ("do your lecture notes in TeX")—if only it were really that easy (but maybe it is for Rick). We also thank Sam Elworthy at Princeton University Press for encouraging us to start and finish the writing process and Jennifer Slater for her careful copyediting. Jonathan Rowell contributed notes on linear algebra that were the basis for parts of Chapter 2, and Section 6.6 on within-host dynamics of HIV is based on a guest lecture and lecture notes by Laura Jones. The National Science Foundation supported Carla Martin as a VIGRE Fellow to help collect materials for the course the first time we taught it, and a number of the exercises were written or improved by her co-TA Daniel Fink.

The students in the 2002 and 2004 renditions of our course were great. Their enthusiasm was infectious and their close reading of the notes helped us eliminate some mistakes. (Please tell us about any more that you discover!) We have also benefited greatly from comments on draft chapters by Hal Caswell, Leon Glass, Stefan Hames, M. Henry H. Stevens, and several anonymous reviewers. Special thanks to Carole Hom and to Mason Porter and Todd Stokes who field-tested draft chapters in courses at UC Davis and Georgia Tech, respectively, and gave us extensive feedback.

Textbooks do not emerge from the void. We owe a debt to previous authors and colleagues from whom we have learned (and probably "borrowed") a great deal, including Leah Edelstein-Keshet, Lee Segel, Simon Levin, Doug Nychka, Ron Gallant, Simon Wood, Harvey Gold, Charles Hall, Bertil Hille, James Murray, and Art Winfree.

Art and Lee, we miss you. Perhaps some day modeling will help to cure cancer.

References

Ferguson, N.M, C.A. Donnelly, and R.M. Anderson. 2001a. The Foot-and-Mouth epidemic in Great Britain: Pattern of spread and impact of interventions. Science 292: 1155–1160.

Ferguson, N.M., C.A. Donnelly, and R.M. Anderson. 2001b. Transmission intensity and impact of control policies on the foot and mouth epidemic in Great Britain. Nature 413: 542–548.

Keeling, M.J., M.E.J. Woolhouse, D.J. Shaw, L. Matthews, M. Chase-Topping, D.T. Haydon, S.J. Cornell, J. Kappey, J. Wilesmith, and B.T. Grenfell. 2001. Dynamics of the 2001 UK Foot and Mouth epidemic: Stochastic dispersal in a heterogeneous landscape. Science 294: 813–817.

1 What Are Dynamic Models?

Dynamic models are simplified representations of some real-world entity, in equations or computer code. They are intended to mimic some essential features of the study system while leaving out inessentials. The models are called dynamic because they describe how system properties change over time: a gene's expression level, the abundance of an endangered species, the mercury level in different organs within an individual, and so on. Mathematics, computation, and computer simulation are used to analyze models, producing results that say something about the study system. We study the process of formulating models, analyzing them, and drawing conclusions. The construction of successful models is constrained by what we can measure, either to estimate parameters that are part of the model formulation or to validate model predictions. Dynamic models are invaluable because they allow us to examine relationships that could not be sorted out by purely experimental methods, and to make forecasts that cannot be made strictly by extrapolating from data.

Dynamic models in biology are diverse in several different ways, including

- the area of biology being investigated (cellular physiology, disease prevalence, extinction of endangered species, and so on),
- the mathematical setting of the model (continuous or discrete time and model variables, finite- or infinite-dimensional model variables, deterministic or stochastic models, and so on),
- methods for studying the model (mathematical analysis, computer simulation, parameter estimation and validation from data, and so on),
- the purpose of the model ("fundamental" science, medicine, environmental management, and so on).

Throughout this book we use a wide-ranging set of case studies to illustrate different aspects of models and modeling. In this introductory chapter we describe and give examples of different types of models and their uses. We begin by describing the principles that we use to formulate dynamic models, and then give examples that illustrate the range of model types and applications. We do only a little bit of mathematics, some algebra showing how a model for enzyme kinetics can be replaced by a much simpler model which can be solved explicitly. We close with a general discussion of the final "dimension" listed above—the different purposes that dynamic models serve in biology. This chapter is intended to provide a basic conceptual framework for the study of dynamic models—what are they? how do they differ from other kinds of models? where do they come from?—and to flesh out the framework with some specific examples. Subsequent chapters are organized around important types of dynamic models that provide essential analytical tools, and examples of significant applications that use these methods. We explain just enough of the mathematics to understand and interpret how it is used to study the models, relying upon computers to do the heavy lifting of mathematical computations.

1.1 Descriptive versus Mechanistic Models

Salmon stocks in the Pacific Northwest have been in steady decline for several decades. In order to reverse this trend, we need to know what's causing it. Figure 1.1 shows data on salmon populations on the Thomson River in British Columbia (Bradford and Irvine 2000). Figure 1.2 presents the results from statistical analyses in which data from individual streams are used to examine how the rate of decline is affected by variables describing the impacts of humans on the surrounding habitat. Straight lines fitted to the data—called a linear regression model—provide a concise summary of the overall trends and quantify how strongly the rate of decline is affected by land use and road density. These are examples of *descriptive* models—a quantitative summary of the observed relationships among a set of measured variables.

Figure 1.2 provides very useful information, and descriptive models such as these are indispensable in biology, but it also has its limitations.

- It says nothing about *why* the variables are related the way they are. Based on the results we might decide that reducing road density would help the salmon, but maybe road density is just an indicator for something else that is the actual problem, such as fertilizer runoff from agriculture.
- We can only be confident that it applies to the river where the data come from. It might apply to other rivers in the same region, and even to other regions—but it might not. This is sometimes expressed as the "eleventh commandment for statisticians": Thou Shalt not Extrapolate Beyond the Range of Thy Data. The commandment is necessary because we often *want* to extrapolate beyond the range

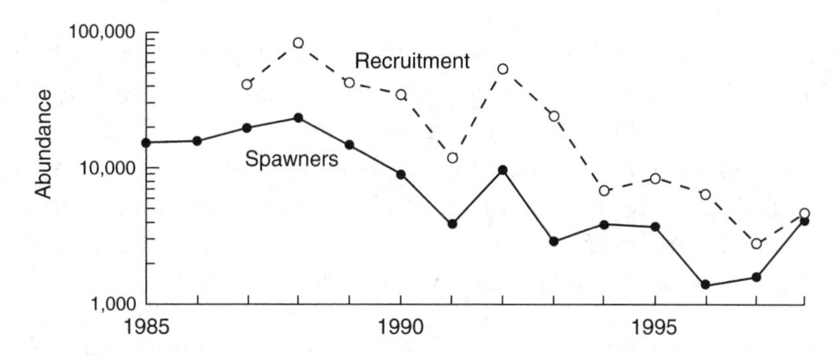

Figure 1.1 Decline in Coho salmon stocks on the Thomson River, BC (from Bradford and Irvine 2000).

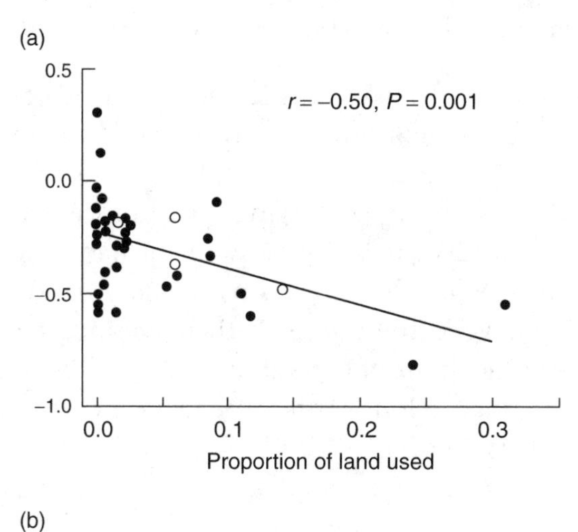

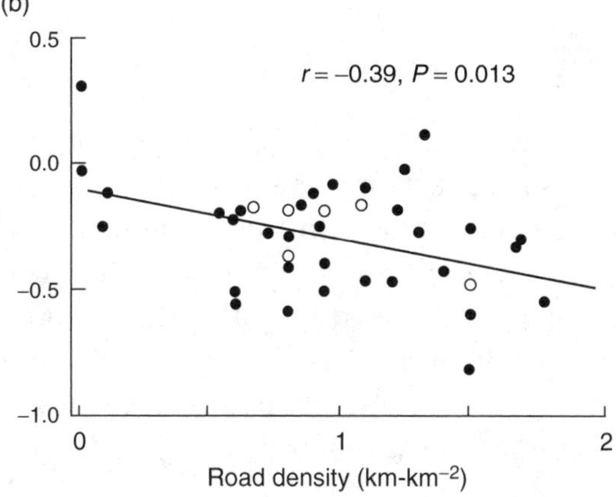

Figure 1.2 Rate of decline on individual streams related to habitat variables (from Bradford and Irvine 2000).

of our data in order to make useful predictions—for example, how salmon stocks might respond if road density were reduced.

The second limitation is related to the first. If we knew *why* the observed relationships hold in this particular location, we would have a basis for inferring where and when else they can be expected to hold.

In contrast, a *dynamic* model is mechanistic, meaning that it is built by explicitly considering the processes that produce our observations. Relationships between variables emerge from the model as the result of the underlying process. A dynamic model has two essential components:

- A short list of *state variables* that are taken to be sufficient for summarizing the properties of interest in the study system, and predicting how those properties will change over time. These are combined into a *state vector* X (a *vector* is an ordered list of numbers).
- The *dynamic equations*: a set of equations or rules specifying how the state variables change over time, as a function of the current and past values of the state variables.

A model's dynamic equations may also include a vector E of *exogenous variables* that describe the system's environment—attributes of the external world that change over time and affect the study system, but are not affected by it.

Because it is built up from the underlying causal processes, a dynamic model expands the Range of Thy Data to include any circumstances where the same processes can be presumed to operate. This is particularly important for projecting how a system will behave in the future. If there are long-term trends, the system may soon exceed the limits of current data. With a dynamic model, we still have a basis for predicting the long-term consequences of the processes currently operating in the study system.

1.2 Chinook Salmon

As a first example, here is a highly simplified dynamic model for the abundance of Chinook salmon stocks, based on some models involved in planning for conservation and management of stocks in the Columbia River basin (Oosterhout and Mundy 2001, Wilson 2003). Most fish return as adults to spawn in the stream where they were born; we will pretend for now that *all* of them do, so that we can model on a stream-by-stream basis. The state variable in the model for an individual stream is $S(t)$, the number of females spawning there in year t. The biology behind the simple models is as follows.

1. Fish return to spawn at age four or five, and then die. Of those that do return, roughly 28% return at age four, 72% at age five; this actually varies somewhat over time, but it turns out not to matter much for long-term population trends.

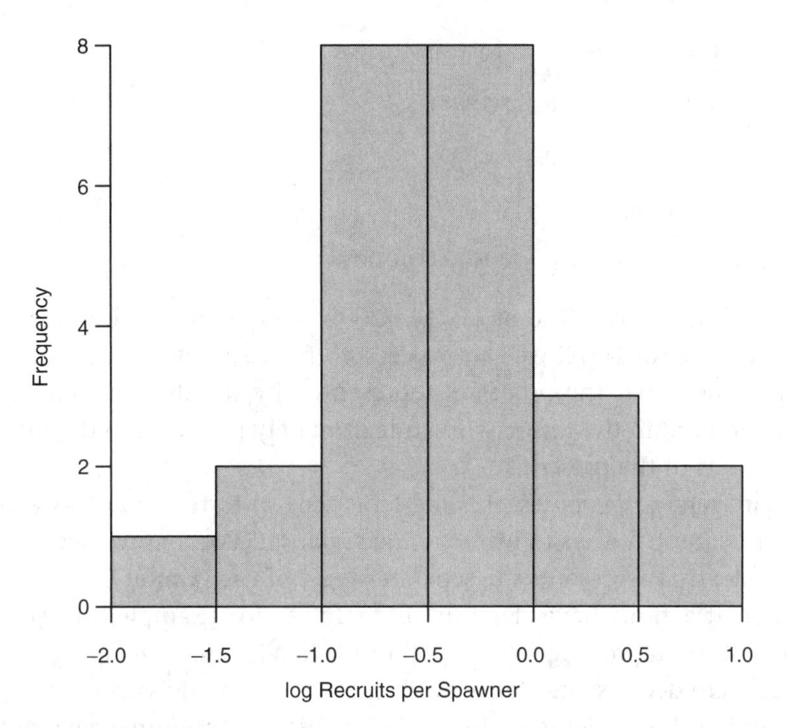

Figure 1.3 Frequency distribution of recruits per spawner for the Bear, Marsh, and Sulphur Creek stocks of spring/summer run Chinook salmon in "good" years.

2. Spawning success E is highly variable (Figure 1.3). In "bad years," E is essentially nil (about 0.05). About 18% of years (2/11 in the available data) are bad. In "good years," E can be large or small, and on average it is 0.72 (standard deviation 0.42). What makes years good versus bad may be related to El Niño (Levin et al. 2001).

For this simple illustration, we will ignore the fact that E varies over time and use instead a single "typical" value (in later chapters we discuss models that include variability over time). The median value of E is 0.56 and the average is 0.58, so for the moment let us take $E = 0.57$. The model is then

$$\begin{aligned} S(t) &= Ep_4 S(t-4) + E(1-p_4)S(t-5) \\ &= 0.16S(t-4) + 0.41S(t-5). \end{aligned} \qquad [1.1]$$

This is the dynamic equation that allows us to predict the future. Current spawners $S(t)$ come from two sources: those offspring of spawners four years ago who return at age four, and those offspring of spawners five years ago who return at age five. The model is nothing more than bookkeeping expressed in mathematical notation. So, to predict the number of spawners in 2002 from data in previous years, we would use the dynamic equation with $t = 2002$:

$$S(2002) = 0.16S(1998) + 0.41S(1997), \quad [1.2]$$

and we can then continue on forward,

$$\begin{aligned} S(2003) &= 0.16S(1999) + 0.41S(1998) \\ S(2004) &= 0.16S(2000) + 0.41S(1999) \\ S(2005) &= 0.16S(2001) + 0.41S(2000) \\ S(2006) &= 0.16S(2002) + 0.41S(2001). \end{aligned} \quad [1.3]$$

Notice that in the last line of [1.3], $S(2002)$ comes from the model. Having forecast a value for 2002, we can proceed to forecast beyond that time. Under the assumption that the processes represented by the dynamic equations will continue to hold in the future, we can legitimately project into the future based on conditions in the present.

For management purposes, this model is only a starting point—we also want to explore potential ways to improve the situation. For that we need a more detailed model that represents the separate stages of the salmon life cycle, because management actions have stage-specific effects—for example, protecting nests from predators, improving food availability when juveniles are growing, or modifying dams to decrease mortality during upstream and downstream migrations. By expanding the model to include relationships between management actions and stage-specific survival, it can then be used to predict which approaches are most likely to succeed (e.g., Wilson 2003; Ellner and Fieberg 2003). Such models and their use in conservation planning are a main topic of Chapter 2.

1.3 Bathtub Models

Our simple salmon model [1.1] was nothing more than bookkeeping—who were the parents of this year's spawners?—expressed in equations. The same is true of many models in biology, even complicated ones. A very important example is the *bathtub* with state variable $W(t)$ = amount of water in the tub (Figure 1.4). Bear with us, this is *really* true, and models with $W(t)$ = amount of protease inhibitor in the bloodstream are exactly the same in principle.

A dynamic equation for this model has to tell us how $W(t)$ changes over time. Time is a continuous variable in this case, but to derive the model we begin by considering a small interval of time from now (time t) until a little bit later (time $t + h$). We take h short enough that any changes in $I(t)$ and $O(t)$ over the time interval are small enough to be ignored. Then

$$\begin{aligned} W(t+h) &= W(t) + \text{Inflow rate} \times \text{time elapsed} - \text{Outflow rate} \times \text{time elapsed} \\ &= W(t) + I(t) \times h - O(t) \times h. \end{aligned} \quad [1.4]$$

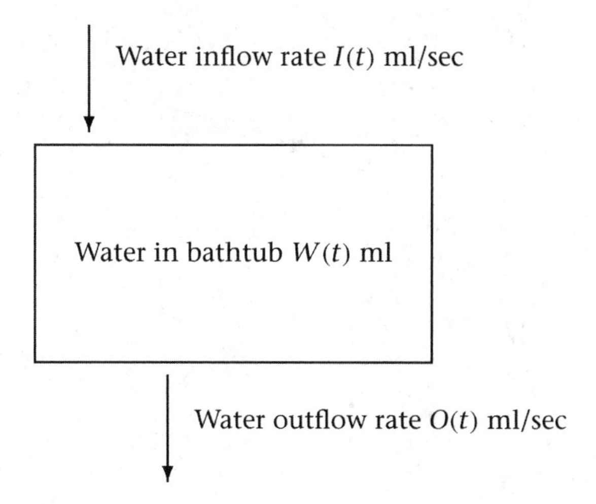

Figure 1.4 Compartment diagram of the bathtub model. The rectangle denotes the state variable—the amount of water in the tub—and the arrows denote flows.

We want to let $h \to 0$. To do that, we rearrange [1.4] into

$$\frac{W(t+h) - W(t)}{h} = I(t) - O(t). \qquad [1.5]$$

The left-hand side of [1.5] is a *difference quotient* for the derivative of W with respect to time. So we can now let $h \to 0$ and we get

$$\frac{dW}{dt} = I(t) - O(t). \qquad [1.6]$$

In words: Change in W = amount coming in − amount going out, with all amounts being *amounts per unit time*.

1.4 Many Bathtubs: Compartment Models

If we connect several bathtubs, so that the water (or whatever) flowing out of one tub can then flow into one or more other tubs, we get a *compartment model*. The state variables in a compartment model are the amount of some stuff in a number of distinct locations or categories ("compartments") within the system. Despite their simplicity, or perhaps because of it, compartment models are very widely used in biology.

The "stuff" in the model can be essentially anything. Sometimes it really is the amount of some material in a particular location—the level of lead in blood, liver, brain, and so on, or the amount of nitrogen in different layers of soil. It can also be amounts or numbers in a particular state or category: gene copies in the active versus inactive state; infected versus uninfected T-cells in the immune system; ion channels in the open versus closed state in a neuron; small, medium, and large sea turtles. The key assumption, as in our simple bathtub, is that items

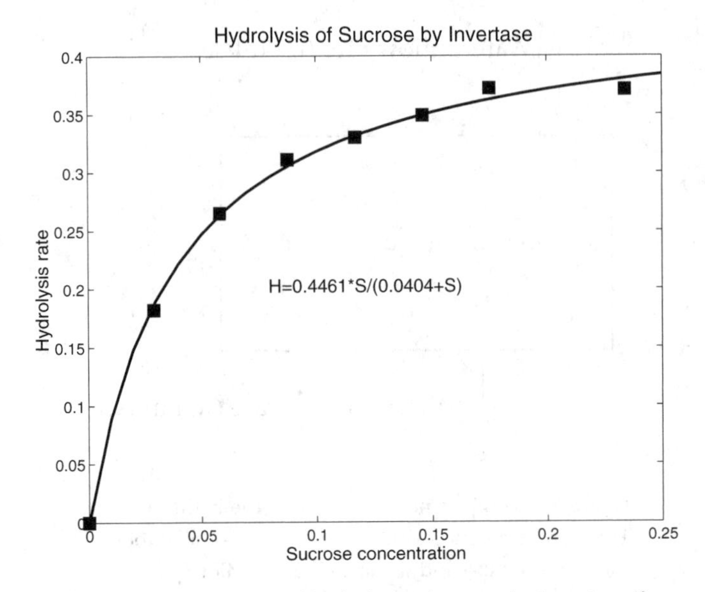

Figure 1.5 Data on the rate of hydrolysis of sucrose by invertase (from dickson 1998), and the fitted Michaelis-Menten equation with parameters estimated by least squares.

within a compartment are indistinguishable from each other—this is expressed by saying that compartments are "well mixed." Thus we only need to keep track of the quantity of material in each compartment—how much water is in the bathtub, but not the time when each molecule arrived or which other bathtub it came from.

1.4.1 Enzyme Kinetics

A simple example of a compartment model is an enzyme-mediated biochemical reaction. This is a classical example (Michaelis and Menten 1913). The reaction diagram, as chemists write it, is

$$S + E \underset{k_{-1}}{\overset{k_1}{\rightleftharpoons}} SE, \quad SE \overset{k_2}{\rightarrow} P + E \tag{1.7}$$

where $S =$ substrate, $E =$ enzyme, and $P =$ reaction product.

Figure 1.5 shows some data with sucrose as the substrate and invertase as the enzyme. Without a model we can partially understand what's going on. At low concentrations of sucrose the reaction is substrate limited: if we double the amount of sucrose, the hydrolysis rate doubles. At high concentrations of sucrose the reaction is enzyme limited: double the amount of sucrose and the hydrolysis rate stays the same. Assuming a smooth transition between these regimes, we can deduce the qualitative shape of the dependence—but nothing quantitative, or how the curve is determined by the underlying reaction rate constants.

To make a quantitative connection between process (the reactions) and pattern (the plotted data), we convert the reaction diagram into a compartment model. Apart from the reaction diagram, the only assumption is the *Law of Mass Action*: The rate of a chemical reaction is proportional to the product of the concentrations of the reactants. The k's in the diagram are the constants of proportionality.

To express these assumptions as a dynamic model, let s, e, c, p denote the concentrations of S, E, SE, P. The diagram then says that

- S and E combine to form SE at rate $k_1 se$
- SE separate to $S+E$ at rate $k_{-1}c$, and to $P+E$ at rate $k_2 c$

We need dynamic equations for each of the four "bathtubs" s, e, c, p. An S is lost whenever S and E combine to form an SE (which is like water flowing out of the bathtub), and gained whenever an SE separates into $S+E$. The principle is still the same:

rate of change = inflow rate − outflow rate.

We can construct a compartment diagram of this system by looking at the reaction diagram to find all the processes that create or destroy each chemical species, and then represent those as inputs to, and outputs from, each of the bathtubs (Figure 1.6). From the compartment diagram we can read off the rest of the dynamic equations:

$$\begin{aligned} ds/dt &= k_{-1}c - k_1 es \\ de/dt &= (k_{-1} + k_2)c - k_1 es \\ dc/dt &= k_1 es - (k_{-1} + k_2)c \\ dp/dt &= k_2 c. \end{aligned} \tag{1.8}$$

The starting point for the process is to have only some substrate and enzyme present at concentrations s_0 and e_0, respectively:

$$s(0) = s_0, \quad e(0) = e_0, \quad c(0) = p(0) = 0.$$

Our interest is in the rate of product formation, so we do not really need the equation for p: the rate of product formation is $k_2 c(t)$. We can get rid of another dynamic equation by noting that

$$dc/dt + de/dt = 0.$$

This implies that $c(t) + e(t) = c(0) + e(0) = e_0$, so

$$e(t) = e_0 - c(t). \tag{1.9}$$

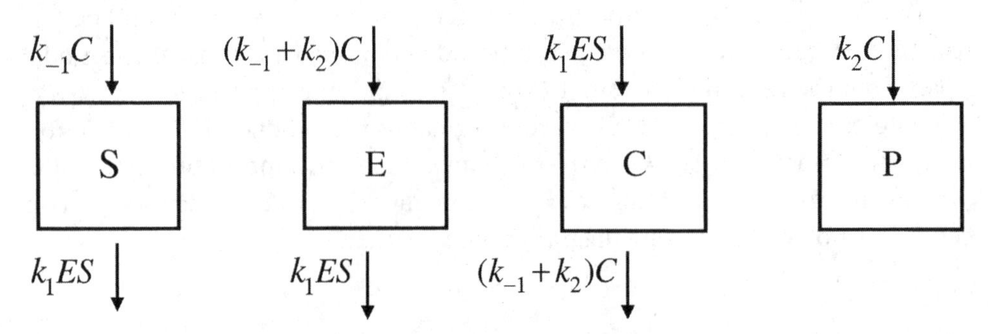

Figure 1.6 Compartment diagram for the enzyme-mediated biochemical reaction. The compartments are S = unbound substrate, E = unbound enzyme, C = bound substrate-enzyme complex, and P = reaction product. A single reaction can create or destroy molecules of several different chemical species, so an arrow in the reaction diagram can imply several different arrows in the compartment diagram.

This makes physical sense, because e and c are the unbound and bound forms of the enzyme, which is not created or destroyed in this reaction scheme (whereas substrate is irreversibly converted to product).

We can now substitute $e_0 - c$ for e into the equations for s and c, leaving us with a system of only two equations:

$$\begin{aligned} ds/dt &= k_{-1}c - k_1(e_o - c)s \\ dc/dt &= k_1(e_0 - c)s - (k_{-1} + k_2)c. \end{aligned} \qquad [1.10]$$

We cannot solve these (except numerically on the computer) but we can get the information we want by a simple approximation. The key property of enzymes is that they are effective at very low concentrations, only a few percent or less of the substrate concentration. So we can assume that e_0 is *small*. Because [1.9] implies that $c(t) \le e_0$, we infer that $c(t)$ is also small.

Taking advantage of this assumption requires a change of variables. Let

$$v = c/e_0 \quad \Rightarrow c = e_0 v.$$

Then

$$\begin{aligned} dv/dt = (1/e_0)dc/dt &= k_1(1 - c/e_0)s - (k_{-1} + k_2)c/e_0 \\ &= k_1(1 - v)s - (k_{-1} + k_2)v. \end{aligned} \qquad [1.11]$$

More interesting is what happens to the other equation:

$$\begin{aligned} ds/dt &= k_{-1}e_0 v - k_1(e_0 - e_0 v)s \\ &= e_0(k_{-1}v - k_1(1 - v)s). \end{aligned} \qquad [1.12]$$

Comparing [1.11] and [1.12], we see that s changes much more slowly because e_0 is small—so in [1.11] we can proceed *as if* s were constant. This brings us down to the single equation

$$dv/dt = k_1 s - (k_{-1} + k_2 + k_1 s)v. \quad [1.13]$$

It is easy to see graphically how [1.13] behaves: dv/dt is positive at $v = 0$ and decreases linearly with increasing v (Figure 1.7). So the behavior of solutions is that $v(t)$ approaches the value v^* at which $dv/dt = 0$:

$$k_1 s = (k_1 s + k_{-1} + k_2)v^* = (k_1 s + k_{-1} + k_2)c^*/e_0$$

$$k_1 e_0 s = (k_1 s + k_{-1} + k_2)c^* \quad [1.14]$$

$$c^* = \frac{k_1 e_0 s}{k_1 s + k_{-1} + k_2} = \frac{e_0 s}{K + s}, \text{ where } K = (k_{-1} + k_2)/k_1.$$

c^* is called a *stable equilibrium point*: the system tends toward it, and then stays there.

Now we bring back the equation for p, which says that the rate of product formation is $k_2 c^*$. So defining $V_{\max} = k_2 e_0$, the rate of product formation is

$$\frac{V_{\max} s}{K + s}. \quad [1.15]$$

Equation [1.15] is important enough to have a name, the Michaelis-Menten equation, named for the originators of the theory presented above. The importance comes from the very general prediction for the rate at which an enzyme-mediated reaction yields its final product.

In addition we have learned something important about the reaction system itself—the c equation is "fast." One value of models is that the things you know may turn out to imply other things that you did not realize before. The c equation describes how enzyme moves between bound and unbound states, as a function of how much substrate is around. It says that the fractions of bound versus unbound enzyme quickly reach an equilibrium value determined by the current amount of substrate (Figure 1.7). The amount of unbound enzyme then determines the rate at which substrate is converted to product via the intermediate step of binding to enzyme.

1.4.2 The Modeling Process

The development of our enzyme kinetics model illustrates the process of dynamic modeling.

1. A *conceptual model* representing our ideas about how the system works ...
2. ... is expressed visually in a *model diagram*, typically involving boxes (state variables) and arrows (material flows or causal effects).
3. Equations are developed for the rates of each process and are combined to form a *mathematical model* consisting of dynamic equations for each state variable.

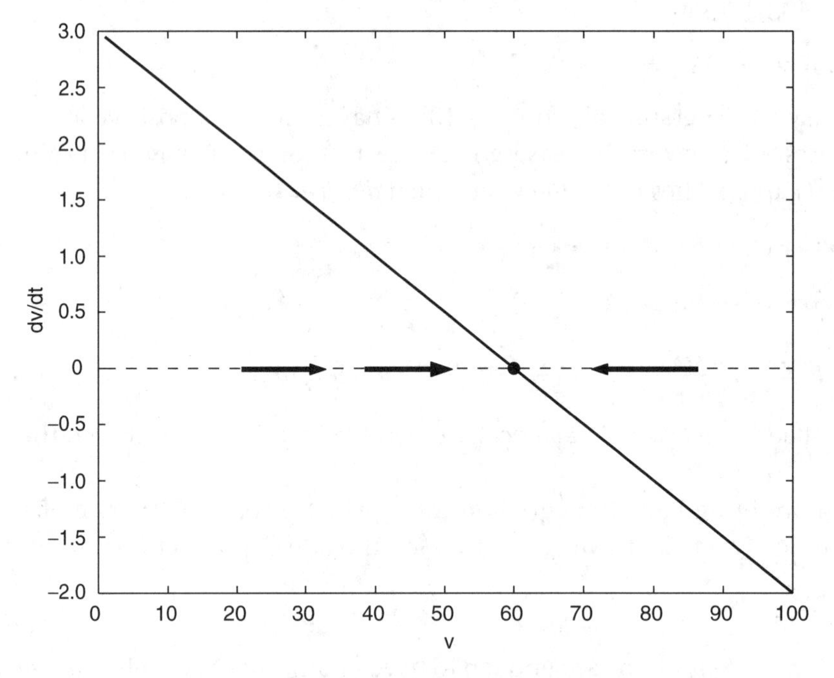

Figure 1.7 Graphical analysis of $dv/dt = k_1 s - (k_{-1} + k_2 + k_1 s)v$.

4. The dynamic equations can then be studied mathematically or translated into computer code to obtain *numerical solutions* for state variable trajectories.

An important part of the conceptual model is identifying which state variables and processes are important for your purposes and need to be included in the model. The conceptual model may be derived from known properties of the system. It may also be a set of hypotheses suggested by what is currently known about the system, and the goal of the model is to test the hypotheses by comparing model behavior against data. In this latter situation, a single model is less useful than a set of models representing different hypotheses. Instead of having to decide if your model fits the data, which is hard and subjective, you can see which model fits the data best, which is easier and more objective.

Rate equations can be based on various sorts of knowledge.

1. *First principles:* Fundamental laws of physics or chemistry that apply to the situation at hand. This is rarely enough; more often you need
2. *First principles + data:* The form of the equation is known from first principles or prior experience, but parameter values are not known. For example, if a Michaelis-Menten form is known to be appropriate, the two parameters in the model can be estimated from a small number of experimental measurements.
3. *Process rate data:* Curves are fitted to experimental measurements of how a specific process rate changes as a function of the model's state and exogenous variables,

using statistical methods that make few assumptions about the form of the equation.

4. *Previous models:* If it worked for somebody else, it might work for you, too. Then again ...
5. *System dynamics data:* Rate equations are deduced by trying to make the model produce state variable trajectories similar to experimental observations. This "inverse method" is technically difficult and used only as a last resort: a state variable is generally affected by many processes, and it is often not possible to disentangle one from another.

We will return to the modeling process in more detail in Chapter 9, and in particular we will describe methods for developing rate equations and fitting them to data.

1.4.3 Pharmacokinetic Models

More contemporary and practical examples of compartment models are models for the absorption, redistribution, and transformation of drugs or other ingested substances within the body, a set of processes collectively referred to as "pharmacokinetics." By appropriately modifying parameter values in a model for these processes, data on one species (such as rats) can be used to make predictions for another (such as humans), for example about how much of the substance reaches any particular organ. Similarly, effects observed at low doses can be extrapolated to potential effects at higher doses.

Figure 1.8 shows a relatively simple example from Easterling et al. (2002), a model for arsenic transport and metabolism in rat hepatocytes. Arsenic is a natural contaminant of drinking water, occurring in many areas at levels high enough to increase the risk of adverse effects including cancers of the skin, lung, liver, kidney, and bladder. This model is simple in part because it represents only one component in the development of a full-body model for arsenic.

The model was developed and tested with experimental data on individual rat hepatocytes incubated in media with various concentrations of arsenic. Each variable name in the model diagram (Figure 1.8) corresponds to a compartment in the model, arsenic (iAS) or one of its metabolites either in the cell or in the surrounding medium (MMA = methylarsonous acid, DMA = dimethylarsinic acid, p-iAS = protein-bound arsenic). Arrows represent transport and metabolic transformations that result in transfers between compartments, and the dotted line represents the inhibitory effect of arsenic on the rate of formation of DMA from MMA. The p's and k's are rate constants for these processes.

This simple model was already useful for understanding intracellular arsenic metabolism. In order to adequately model the data, it was necessary to assume that transformation from DMA to MMA was inhibited by arsenic. Three models

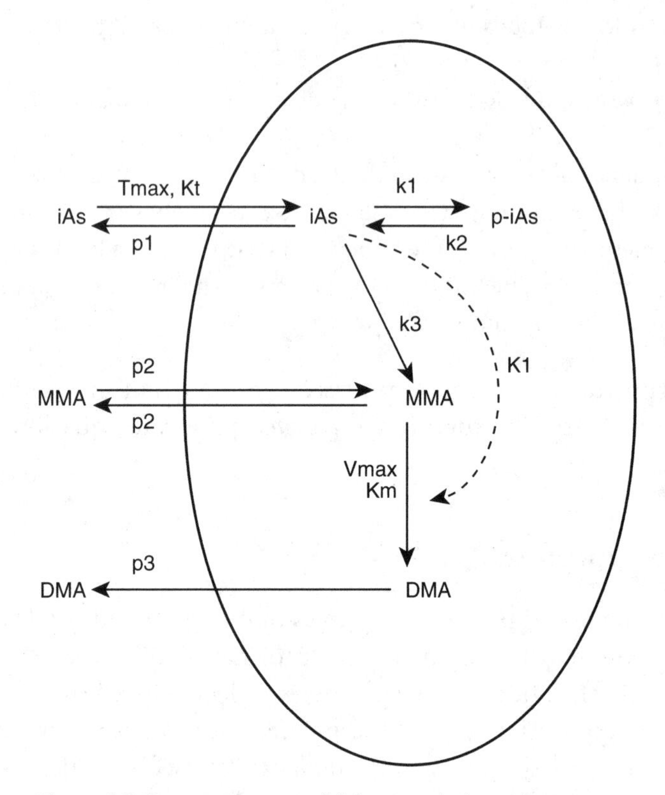

Figure 1.8 Compartment model for arsenic in rat hepatocytes (from Easterling et al. 2002).

for inhibition were compared, and one of them was found to fit better than the others. Simulations of the three models were then used to identify experiments that could distinguish decisively between them, in particular measurements of MMA within the cell 46 hours after initial exposure to arsenic at 1.4 μM.

A much more complicated pharmacokinetic model, because it is complete and intended for practical use, is the Integrated Exposure Uptake Biokinetic Model for Lead in Children (IEUBK) developed and distributed by the U.S. Environmental Protection Agency (EPA) Technical Review Workgroup for Lead (EPA 2002; White et al. 1998). Figure 1.9 shows a diagram of the model.

Despite the elimination of leaded gasoline and lead-based paints in the United Stated, lead exposure remains a problem. A recent front-page headline in our local paper (Bishop 2002) concerned the cost overrun for cleaning up lead-contaminated soil surrounding a local recreation site used for fishing and swimming. "An environmental cleanup near Ithaca Falls that initially was projected to cost $2 million and last about a year is now expected to cost twice as much and last twice as long." The contamination remained from previous manufacturing in the area. To bring lead exposure down to an acceptable level, a layer of soil was literally being vacuumed up off the ground and trucked off to a landfill.

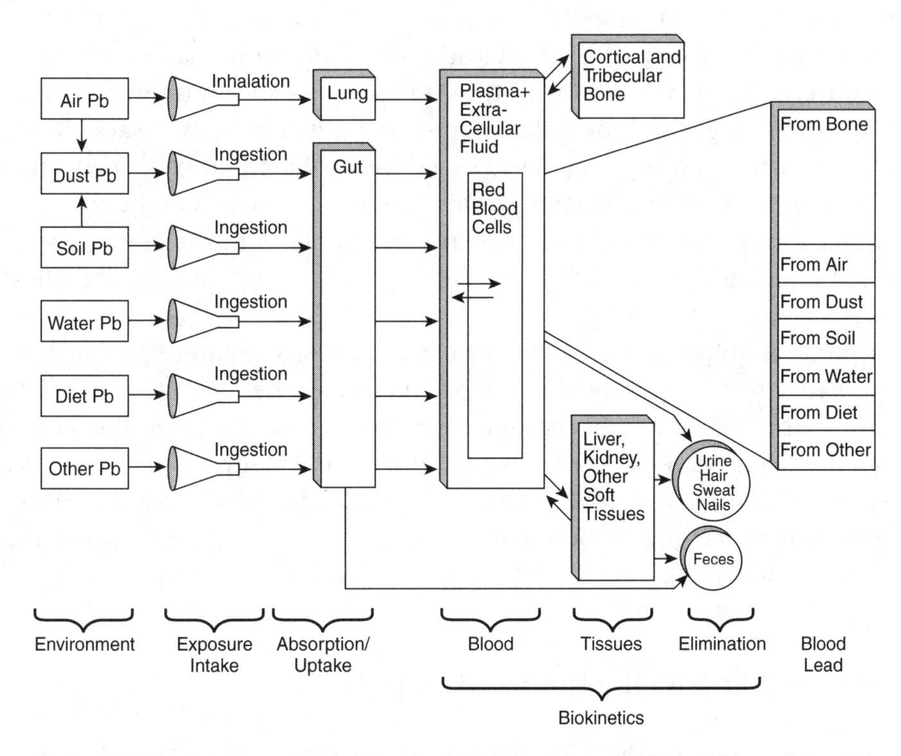

Figure 1.9 Compartment diagram for the IEUBK model for lead in children (from EPA 1994a). Plain rectangles represent environmental levels of lead, shaded rectangles are model compartments (forms of lead in the body that are distinguished in the model), and circles are losses of lead from the body.

And what should be an "acceptable level" of lead in the soil near a family swimming hole, based on the expected health impacts? The IEUBK model was developed to answer such questions. Producing version 0.99 required over a decade of effort by several different EPA programs (EPA 1994b). It was developed as an alternative to descriptive statistical models, because those models were difficult to generalize beyond the specific situations for which data had been collected. This is an example of the "curse of dimensionality" for descriptive models: the impact of lead exposure on a child depends on so many variables that it is infeasible to collect data spanning the full range of possibilities for all of them.

The full IEUBK model has three components: Exposure, Uptake, and Biokinetics. The exposure and uptake components are descriptive: a static set of equations that predict the amount of lead entering the child's body and transferring into the bloodstream, as a function of how much lead the child encounters in various daily activities but without reference to the biological processes involved. The biokinetic component is a dynamic compartment model. It starts by calculating the volumes and weights of model compartments (e.g., the volume of blood plasma) as a function of the child's age and the initial amounts of lead in each compartment. The model equations specify the rates of lead transfer between compartments and the rate of lead loss through elimination—the rates of flow between the "bathtubs"—and the equations are solved numerically, from birth to age 84 months. Body fluid volumes and organ weights are also changed as the child ages, but this is modeled descriptively, based on published data on child growth (EPA 1994b). Putting all this together, lead levels in the soil around Fall Creek can be translated into the expected lead level in the blood (given an assumed frequency and duration of visits) and then into the expected health impacts.

The model is quite complex—a listing of its equations requires 22 printed pages (EPA 1994b), and there are roughly 100 parameters. Models intended for practical use are often complex, because accuracy is more important than aesthetics so the model grows to match the available data. However, it is important to remember that the model is only complex because it has many parts. All of its pieces are simple bathtubs obeying the basic balance equation: rate of change = (total input rate) − (total loss rate).

1.5 Physics Models: Running and Hopping

In compartment models the underlying physical law is conservation of mass. In this section we introduce a model based on the physical laws of mechanics. The variables in the model are positions and velocities, and the dynamic equations come from Newton's laws of motion.

Many animals use legs to move. The number of legs and the gaits they employ differ markedly. To study locomotion, *biomechanical* models that represent the animal as a rigid skeleton connected by joints are a starting point. Gravity and muscles exert forces on the skeleton, causing the animal to move. By measuring the physical properties of the animal—its shape, mass distribution, and muscle strength—we want to be able to analyze and predict the motion. With humans, our purpose may be to improve athletic performance or to restore function following an injury.

Different gaits have different mechanical characteristics. When humans walk, our feet never lose contact with the ground and we alternate between having both feet on the ground and a swing phase in which one foot is on the ground and the second leg swings like a pendulum. When we run, we alternate between a flight phase in which both feet are off the ground and a stance phase in which one foot is on the ground. Kangaroos hop with a flight phase alternating with a stance phase in which both feet are on the ground simultaneously.

A pogo stick with a single leg provides a simple model for running and hopping (Figure 1.10). The animal is regarded as a "point mass" with a springy leg attached to the mass. As a first approximation, you can think of the spring as the tendons

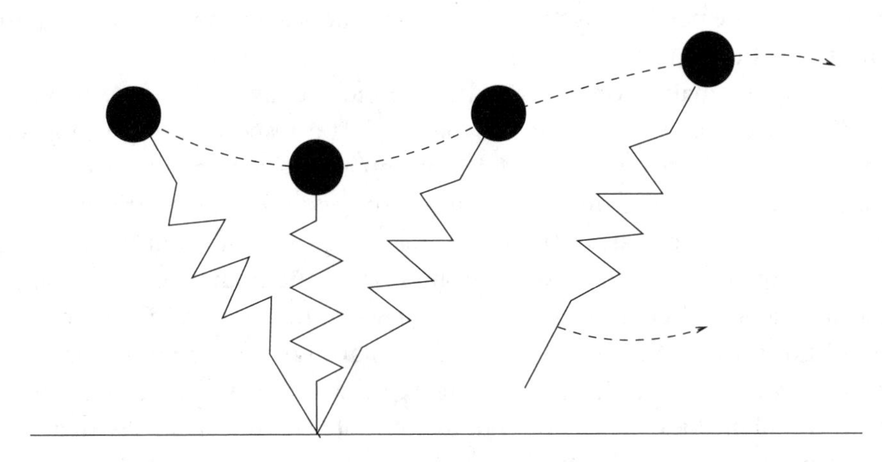

Figure 1.10 A cartoon of the pogo stick or monopode.

in the leg. By contracting muscles, the animal changes the force of the leg spring, enabling it to bounce off the ground. When in flight, we assume that the animal is able to swing the leg so that it will point in a new direction when the animal lands on the ground. At landing, the leg shortens, compressing the spring. The compressed spring exerts a vertical upward force that together with additional force exerted by the muscles propels the animal into its next flight phase.

We construct a system of equations that embodies this verbal description of a bouncing "monopode." For simplicity, we consider only the motion of the monopode in a vertical plane. There are two components to the virtual animal: a body that is assumed to be a point mass with mass m, and a massless leg whose (variable) length is denoted by r. Euclidean coordinates (x, y) will be used to label the position of the body: we write $(x(t), y(t))$ for the position of the body at time t. The leg is attached to the body by a "hip" joint that is free to rotate, so we label the angle of the leg with the downward vertical as θ. The position of the foot relative to the body is $(r\sin(\theta), -r\cos(\theta))$ so its position in "world coordinates" is $(x + r\sin(\theta), y - r\cos(\theta))$. The ground is assumed to be the x-axis $y = 0$.

When the foot of the monopode is above the ground, we say that the monopode is in "flight." During flight, we ignore friction and assume that gravity is the only force acting on the body, so Newton's laws of motion ($F = ma$) are just those for a particle moving under the influence of gravity:

$$\begin{aligned} m\ddot{x} &= 0 \\ m\ddot{y} &= -mg. \end{aligned} \qquad [1.16]$$

Here $\ddot{x}$ and $\ddot{y}$ denote the acceleration (second derivative with respect to time) of the monopode's body in the x and y directions, and g is the gravitational force constant 9.8 m/s^2. The motion of the body in flight is along a parabola determined by the position and velocity of the body at the instant when the flight begins. The position of the leg has no influence on the body motion until the foot hits the ground.

When the foot makes contact with the ground, we say that touchdown occurs and the stance phase of the motion begins. The response of the monopode to the impact is determined by additional physical properties that have to be specified to complete the model. We make the assumption that the foot sticks at the point of impact, rather than slipping along the ground or bouncing into the air. At touchdown, the downward motion of the body begins to compress the springy leg, so there is an additional force on the body exerted through the spring. Thus the equations of motion for the stance phase are more complicated than those for the flight phase. They are expressed most simply in a coordinate system in which the origin is placed at the foot (Figure 1.11). We denote the position of the body relative to the foot by (u, v). The force exerted on the body by the spring is proportional to the deviation between the body's location and its "resting location" (u_0, v_0)—where it would be if the spring were at its resting length r_0. The force vector is therefore

$$\vec{F} = f(r)(u_0 - u, v_0 - v) \qquad [1.17]$$

where $f(r)$ is the "spring constant" and $(u_0, v_0) = (r_0/r)(u, v)$. For a linear spring $f(r)$ is constant, but in general it can depend on how far the spring has been compressed or stretched. The equations of motion for the body are therefore

$$\begin{aligned} m\ddot{u} &= f(r)(u_0 - u) \\ m\ddot{v} &= f(r)(v_0 - v) - mg. \end{aligned} \qquad [1.18]$$

We assume that the motion of the body reaches a positive lowest height during stance: the spring force becomes larger than the force of gravity, slows the vertical motion of the body, and then propels the body upward. If the body hits the ground, then we regard the monopode as having fallen, and stop. We also assume that there is a maximal leg length l and that when the leg reaches this length, the foot comes off the ground and a new flight phase begins.

There is still one additional item that must be specified to obtain a complete model. Since the leg is assumed to be massless, there is no energetic cost in moving the leg relative to the body during flight. Changes in the angle θ do not affect the flight phase of the body, but θ affects foot position and therefore influences when touchdown occurs. We assume that our monopode can point the leg during flight as one of two things it can do to control its motion. It does not matter when or how this happens as long as the foot does not make contact with the ground during repointing, so we assume that the repointing occurs instantaneously when the body reaches its apex (highest point) during the flight phase. The second way in which we assume that the monopode can control its motion is to change the spring constant $f(r)$ during stance. This corresponds to using muscles to push along the leg or to make the leg stiffer.

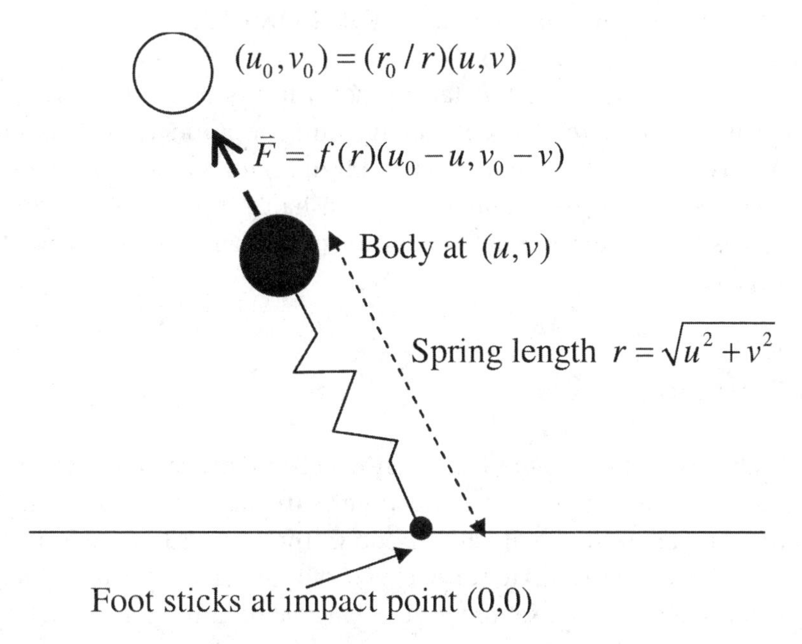

Figure 1.11 The monopode in stance phase. The dynamic equations are written in coordinates (u, v) giving the position of the body relative to the point where the foot meets the ground. The open circle shows the body's "resting position" corresponding to the leg at resting length r_0, and the bold dashed arrow indicates the resulting force vector generated by the spring.

There are many questions that we can ask about this model of running animals. First, we can try to fit observational data to such a model. Blickhan and Full (1993) show that the model works well for animals as "diverse as cockroaches, quail and kangaroos." Even though these animals are mechanically far more complicated than the monopode, it is possible that the behaviors that they actually use to control their motion do not fully exploit this complexity, and are similar to what would occur if the organisms were simple monopodes.

Second, we can ask what kind of motion results from different *control strategies*—the rules used for adjusting position and movements on the basis of the current situation. A goal that we may want to achieve is to produce a regular stable running motion in which each stride is the same as the previous one, except for being further along in the direction of motion. Koditschek and Buehler (1991) showed that the model does not always do this. A fixed control strategy can give rise to irregular, "chaotic" motions in which apex height continues to fluctuate from one hop to the next. Carver (2003) investigated "deadbeat" control strategies for which regular running motion at a desired speed can be obtained from a large region of initial positions after a single stride. Third, we can ask whether the model might help with the design of better legged robots. Raibert (1986) utilized models like this one in the design of monopode, biped, and quadruped robots. Similar principles are being utilized by Buehler and co-workers in advancing the performance of hexapod robots (Saranli et al. 2001, or see http://www.rhex.org for movies of the robot in action).

The monopode model is just a starting point in understanding animal locomotion. More elaborate models with a large number of independently movable body parts have more fidelity to the actual biomechanics of animal locomotion, but producing simulations that are indistinguishable from "motion capture" of actual animals requires a better understanding of neuromuscular control than we currently have.

1.6 Optimization Models

Biologists also use *optimization* models, in which the modeler assumes that the study organisms are trying to achieve some particular goal. For example, we might posit that an animal will gather food in the habitat that gives it the best chance of getting enough food to survive, without itself getting eaten by its predators. Or we might ask how a set of neurons *ought to* be connected, and how they should affect each other, in order to best achieve some important task like identifying the direction a sound is coming from. An optimization model does not worry about how the goal is achieved in practice—how the animal evaluates different habitats, or how neurons accomplish their growth and connection patterns during development. We assume that natural selection will find a way, and see if the actual behavior matches the predictions based on the assumed goal.

Optimization models are especially informative when different assumptions lead to different predictions that can be compared with observations. For example, Segrè et al. (2002) predicted the changes in cell metabolism in *Escherichia coli* caused by a gene deletion ("knockout") that eliminated a particular metabolic pathway, based on two different assumed goals. The first was that certain parameters of the metabolic reaction network had values such that the net growth rate of knockout mutant cells was the highest possible given the deletion. The second was that network parameters minimized the disruption in the flux rates on the remaining pathways—that is, that the cell has evolved to be highly robust against mutations. Although the outcome was not uniform, overall the second model was far more successful at predicting the effects of several knockout mutations. Clutton-Brock et al. (1996) compared predictions about size-dependent reproductive decisions in Soay sheep—whether to have zero, one, or two offspring in a given year—based on two different assumptions about how well females can predict changes in population density and their effect on mortality rates. Again, neither model was exactly right, but the model assuming that females could make reliable short-term predictions did far worse than one assuming that females could not. In these studies the models were not *really* making predictions—the data were already in hand. But by asking what observable features of the study system would result from different assumptions about unobservable features, it was possible to make inferences about properties that could not be observed directly.

1.7 Why Bother?

A dynamic model is built up from equations representing the processes thought to account for the patterns observed in the data, whereas a descriptive model only has to represent the patterns themselves. That difference typically means that constructing a dynamic model is a lot more work than constructing a descriptive statistical model. So why would anyone bother?

1. ***Scientific understanding.*** A model embodies a hypothesis about the study system, and lets you compare that hypothesis with data.

- Can processes A and B account for pattern C? Having observed some pattern in our study system, a model lets us ask whether it is *logically* possible for our hypothesized causes to have produced it, by creating (on the computer) an artificial world in which A and B are the only things happening, and seeing if and when C can be the outcome (e.g., Harris-Warrick et al. 1995). Hypotheses that fail this test can be rejected without taking the trouble to do experiments.
- "Biological detective" work: Which of several contending sets of assumptions is best able to account for the data? Scientists are often clever enough to invent several sets of causes (A_i, B_i) that are logically capable of producing something like observed pattern C. One way to decide between them, or at least narrow the field of possible explanations, is to implement each hypothesis as a quantitative model, and see which of them can match the available empirical data (Hilborn and Mangel 1997; Kendall et al. 1999).
- Given that processes A and B occur, what consequences do we expect to observe? Patterns may not be seen until they have been predicted to occur.
- Where are the holes in our understanding? Trying to build a model is a great way to see if you really have all the pieces.

Paradoxically, when a dynamic model is being used as a research tool to increase our scientific understanding, it is often most useful when it *fails to fit the data*, because that says that some of your ideas about the study system are wrong.

2. ***Using our scientific understanding to manage the world***

- Forecasting disease or pest outbreaks
- Designing man-made systems, for example, biological pest control, bioengineering

- Managing existing systems such as agriculture or fisheries
- Optimizing medical treatments or improving athletic performance

In this book, and already in this chapter, there are many examples where models have been put to practical uses like these. Forecasting and management can be done, and often are done, using descriptive models. But a lot of effort also goes into building mechanistic models for forecasting, design, and management. Why is this?

3. *Experiments are small, the world is big.* Reviewing studies of plant competition, David Tilman (Tilman 1989) found

- Nearly half were on a spatial scale of $\leq 1\ m^2$
- About 75% on a scale $\leq 10 \times 10\ m^2$
- About 85% ran for ≤ 3 years

When large-scale (in space or time) experiments are infeasible, dynamic models can be built using the processes studied by small-scale experiments and then used to derive their larger-scale implications.

4. *There are experiments that you would rather not do*

- Endangered species management by trial and error
- Setting dosages for clinical trials of new drugs on humans
- Setting "safe" limits for exposure to toxic substances

A model can never be a complete substitute for experiments, but if the experiment is infeasible or undesirable, a model based on current scientific knowledge is far better than guesswork. Models can also be used to help plan risky experiments to increase the odds of success, such as drug trials or recovery plans for an endangered species.

5. *The curse of dimensionality.* Sometimes a purely experimental approach is not feasible because of the *curse of dimensionality*: the data requirements for estimating a statistical model with many variables grow rapidly in the number of variables.

As an example, one of our projects concerned the design of fish-farming systems (Krom et al. 1995; Ellner et al. 1996). The goal was to raise marine fish in a limited area of artifical ponds, allowing very little pollution due to nutrient export, which would have gone into the sea near a coral reef. The limiting factor in the system was the buildup of ammonia (NH_4), which reaches toxic levels if too many fish are excreting into a limited volume of water. One way to limit ammonia buildup is to have water flow through the system, but this releases a lot of nutrient-

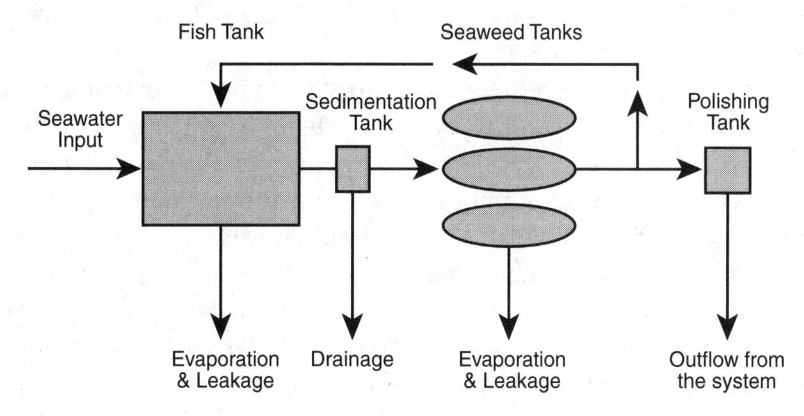

Figure 1.12 Diagram of the experimental fishpond system showing water flows between compartments.

rich pollution. So we designed a system that circulated water between ponds containing the fish and tanks containing seaweed that could be sold as a second crop, followed by a final "polishing" tank with seaweed before the effluent is released (Figure 1.12). Variables under the designer's control include

- Volumes of the fish ponds, seaweed tanks, and polishing tank
- Stocking densities (number of fish, weight of seaweed) in each tank
- Fish feeding schedule (how much, and when)
- Rate of water input
- Rate of water recirculation
- Degree of shading on fish and seaweed tanks to control phytoplankton (phytoplankton consume ammonia—which is good—but compete with seaweed, which is not)

How can we find the optimal combination of values for each of these twelve design parameters? If we want to determine experimentally where a function of one variable takes its maximum value, we need enough data points that one will be close to the maximum. So if we're *really* lucky, three data points will be enough. With two variables, to find the maximum with the same accuracy requires a 3×3 array of data points; hence nine experiments. But with twelve variables under our control and three values for each of these, that would be $3^{12} =$ 531,141 experiments. This crude estimate turns out to be qualitatively correct, in the following sense. If we assume only that the function is differentiable, then the average estimation error ϵ using optimal methods scales asymptotically as $\epsilon \sim N^{-1/(2+n)}$ (Stone 1982), where n is the number of variables and N the number of data points. This inverts roughly to $N \sim C^{2+n}$ data points to achieve a given estimation error, where $C = 1/\epsilon$.

So instead of a purely experimental approach, we built a dynamic model for the flow of inorganic nitrogen through the system. We quantified the processes causing N to be enter and leave each component of the system, resulting in a model that could predict ammonia levels and the nutrient export rate as a function of all the control variables. We could do this with fewer experiments because each process in the model depends on only a few variables—often only one. Also, many experiments could be done at the lab bench, such as measuring how fish size affects the ammonia excretion rate. That made it possible to get data spanning any values that might occur in the real system. Moreover, for many processes it was reasonable to assume on mechanistic grounds that the equation describing the process rate was a line with zero intercept, which can be estimated from a single data point. A dynamic model was feasible within our time and budget constraints, whereas a purely statistical approach would not have been. And in subsequent trials, the resulting system design used far less water than a system without recirculation.

For these kinds of practical applications the model needs to be consistent with all available data, so that you have some grounds for trusting its predictions about situations where data are not available. Therefore, an interaction between modeling and experiments is useful:

- Given a model, you can do lots of experiments quickly and cheaply on the computer.
- Those results can be used to identify potentially good designs or management options.
- Model predictions can be checked by doing a few well-chosen experiments.

1.8 Theoretical versus Practical Models

Although dynamic models are used for many purpose, we can put them under two broad headings: *theoretical understanding* of how the system operates, and *practical applications* where model predictions will play a role in deciding between different possible courses of action (Table 1.1). It is important to keep this distinction in mind because it affects how we build and evaluate models.

A theoretical model—such as the monopode model—has to be simple enough that we can understand why it is doing what it does. The relationship between hypotheses (model assumptions) and conclusions (properties of model solutions) is what provides understanding of the biological system. Replacing a complex system that we don't understand with a complex model that we also don't un-

Practical Models	Theoretical Models
Main goals are management, design, and prediction	Main goals are theoretical understanding and theory development
Numerical accuracy is desirable, even at the expense of simplicity	Numerical accuracy is not essential; the model should be as simple as possible
Processes and details can be ignored only if they are numerically unimportant	Processes and details can be ignored if they are conceptually irrelevant to the theoretical issues
Assumptions are quantitative representations of system processes	Assumptions may be qualitative representations of hypotheses about the system, adopted conditionally in order to work out their consequences
System and question specific	Applies to a range of similar systems

Table 1.1 Classification of models by objectives

derstand has not increased our understanding of the system. Theoretical models are often expressed as a few dynamic equations (e.g., matrix or differential equations) representing the most relevant processes for the particular phenomena of interest.

A practical model—such as the IEUBK model—sacrifices simplicity in order to make more accurate predictions for a particular system. There may be a lot of equations with terms based on quantitative data. As a result, practical models are usually too complicated for anything but computer simulation. However, theoretical models are nowadays typically studied by a combination of mathematical analysis, approximations, and computer simulation. Even simple-looking nonlinear models may be beyond the reach of totally rigorous mathematical analysis. More often, a model is explored by simulation and we then try to use mathematics to understand what the computer has shown us.

The goal-centered nature of practical models makes it conceptually easy to evaluate them: a good model is one that lets you make better decisions. "Theoretical understanding" is not something we can quantify, so model evaluation is more difficult. A quail is not a monopode, but a monopode model can describe its gait well: good model or bad? (Answer: That's the wrong question).

The model types described above and in Table 1.1 are extremes of a continuum, and many models fall somewhere in between. Constraints on time and effort, data limitations, and the need to communicate with stakeholders place limits on the complexity of practical models. Theoretical models are often challenged by comparisons with experimental data. For that purpose models often remain theoretical in the sense of including only a few "essential" processes, but the rate equations are allowed to become more complicated, so that a model with the right basic structure should survive a quantitative comparison with data.

1.9 What's Next?

This chapter has tried to introduce the distinctive features of dynamic models, some of the considerations involved in creating them, and some of the ways in which they are useful. What's next is the process of taking you, gradually, from being a spectator to acting as a participant in dynamic modeling. The remaining chapters in this book will introduce you to some important types of dynamic models that have stood the test of time and are likely to be around for a long time to come. New ideas are hard to come by, but fortunately a lot of new problems can be solved using old ideas. Cutting-edge science is still being done with centuries-old modeling frameworks, augmented by modern computers. Along the way, we will introduce mathematical and computational methods for working with dynamic models and understanding what they do. The final chapter then comes back to look in more detail at some of the issues we have touched on here about the process of formulating and testing dynamic models.

Exercise 1.1. Find a scientific research paper published within the last five years that uses a dynamic model in your area of biology (your major, concentration, or any area of particular interest to you), and read the paper to answer the following questions.

(a) Give the complete citation for the paper: authors, date, journal, pages.

(b) What was the purpose of the model—that is, what was accomplished by building and using the model?

(c) What are the state variables of the model?

(d) Identify one of the model's simplifying assumptions—some known aspect of the real world that the model omits or simplifies.

Exercise 1.2. This exercise refers to the dynamic model for salmon stocks. Table 1.2 gives some estimated spawner counts by age (the estimates are not integers because they are derived by scaling up from sample data). Use these data to estimate

(a) the value of $p_4(t)$ for as many years as possible

(b) the value of $E(t)$ for as many years as possible

Note that it may be possible to estimate $p_4(t)$ and/or $E(t)$ for years other than those appearing in the leftmost column of the table. Recall that for fish spawning in year t, $E(t)$ is the total number of offspring per spawner that will return to spawn, at either

Year	Four-year-old spawners	Five-year-old spawners
1985	84.2	210.4
1986	72.6	151.5
1987	116.3	340.0
1988	26.0	1083.3
1989	27.8	63.1
1990	72.4	112.7
1991	23.7	157.0
1992	47.2	125.8
1993	44.7	664.4

Table 1.2 Estimated spawner counts by age for the Bear Valley stock of Chinook Salmon on the Snake River

age four or age five; $p_4(t)$ is the fraction of those that return at age four. *Hint*: The first line in the table says that in 1985 there were $S = (84.2 + 210.4)$ spawners. Which entries in the table tell you how many offspring of those spawners returned eventually to spawn?

Exercise 1.3. Find the differential equations for the diagrams in Figure 1.13 representing continuous-time compartment models. This exercise uses some notational conventions for compartment models. $\rho_{i,j}$ is the flow rate (amount/time) *from* compartment j *to* compartment i. If an outflow rate is proportional to the amount in the compartment of origin, this is called *linear donor control* and the constant of proportionality is $a_{i,j}$. That is, $\rho_{i,j} = a_{i,j}q_j$ where q_j = amount in jth compartment.

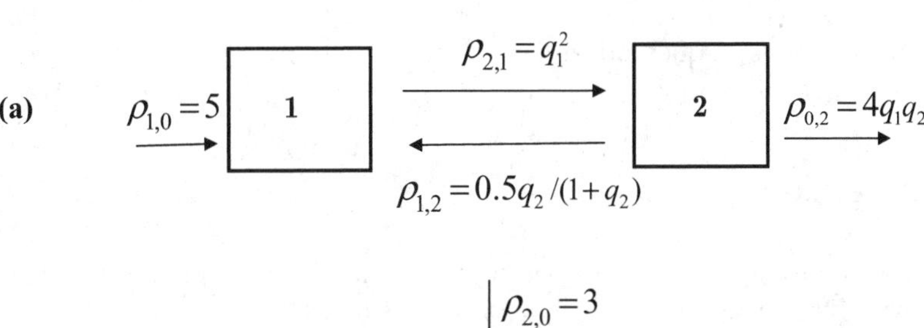

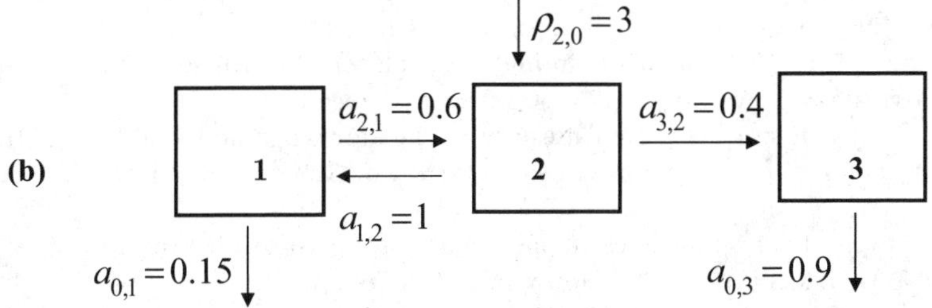

Figure 1.13 Compartment diagrams for Exercise 1.3.

Exercise 1.4. Another convention is that production within a compartment is represented as an input from outside, and any internal losses (consumption, degradation, etc.) are represented as a loss to the outside. For example, in a diagram of the Ellner et al. (1996) fishpond model, excretion of ammonia-N by fish into the fishpond water column would be drawn like $\rho_{2,0}$ in Figure 1.13(b), pointing into the box for the state variable "ammonia-N in the fishpond."

(a) Using this convention, draw a compartment diagram for the Ellner et al. (1996) fishpond model, omitting the polishing tank so that there are six state variables (ammonia-N and ToxN in the fish tank, sedimentation tank, and seaweed tanks). Label each arrow with a ρ (subscripted appropriately), and below the diagram list what each flow is. So for example, if the fishpond is compartment 1, then your list could start with

$\rho_{1,0}$ = ammonia-N excretion by fish in the fishtank

(b) Which of the flows that you listed above are linear donor controlled in the model?

Exercise 1.5. Draw compartment diagrams (with flows labeled as in Figure 1.13) that correspond to the following dynamic equations.

(a) $$\frac{dq_1}{dt} = 1 - q_1, \quad \frac{dq_2}{dt} = q_1 - 2q_2,$$

(b) $$\frac{dq_1}{dt} = -q_1, \quad \frac{dq_2}{dt} = q_1 - q_2q_3 + q_2/(1 + 5q_2), \quad \frac{dq_3}{dt} = 3 - q_2/(1 + 5q_2).$$

Exercise 1.6. We wrote *A quail is not a monopode, but a monopode model can describe its gait well: good model or bad? (Answer: That's the wrong question).*

(a) Why is that the wrong question?

(b) What was the right question, and why?

1.10 References

Bishop, L. 2002. Cleanup estimates to double. Ithaca Falls project will run over in time, money. *Ithaca Journal* (http://www.ithacajournal.com), October 4, 2002.

Blickhan, R., and R. Full. 1993. Similarity in multilegged locomotion: Bouncing like a monopode. Journal of Comparative Physiology A 173: 509–517.

Bradford, M. J., and J. R. Irvine. 2000. Land use, fishing, climate change, and the decline of Thompson River, British Columbia, coho salmon. Canadian Journal of Fisheries and Aquatic Science 57: 13–16.

Carver, S. 2003. Control of a Spring-Mass Hopper. Ph.D. Thesis, Cornell University.

Clutton-Brock, T. H., I. R. Stevenson, P. Marrow, A. D. MacColl, A. I. Houston, and J. M. McNamara. 1996. Population fluctuations, reproductive costs, and life-history tactics in female Soay sheep. Journal of Animal Ecology 65: 675–689.

Dickson, T. R. 1968. *The Computer and Chemistry; An Introduction to Programming and Numerical Methods*. W. H. Freeman, San Francisco.

Easterling, M. R., M. Styblo, M. V. Evans, and E. M. Kenyon. 2002. Pharmacokinetic modeling of arsenite uptake and metabolism in hepatocytes—mechanistic insights and implications for further experiments. Journal of Pharmacokinetics and Pharmacodynamics 29: 207–234.

Ellner, S., A. Neori, M. D. Krom, K. Tsai, and M. R. Easterling. 1996. Simulation model of recirculating mariculture with seaweed biofilter: Development and experimental tests of the model. Aquaculture 143: 167–184.

Ellner, S. P., and J. Fieberg. 2003. Using PVA for management in light of uncertainty: Effects of habitat, hatcheries, and harvest on salmon viability. Ecology 84: 1359–1369.

EPA (U.S. Environmental Protection Agency). 1994a. Guidance Manual for the Integrated Exposure Uptake Biokinetic Model for Lead in Children. Publication number 9285.7-15-1, EPA 540-R-93-081, PB93-963510. Office of Solid Waste and Emergency Response, U.S. Environmental Protection Agency Washington, DC 20460. Online at http://www.epa.gov/superfund/programs/lead/products.htm

EPA (U.S. Environmental Protection Agency). 1994b. Technical Support Document: Parameters and Equations Used in the Integrated Exposure Uptake Biokinetic (IEUBK) Model for Lead in Children (v 0.99d). Publication number 9285.7-22, EPA 540/R-94/040, PB94-963505. Online at http://www.epa.gov/superfund/programs/lead/products.htm

EPA (U.S. Environmental Protection Agency). 2002. User's Guide for the Integrated Exposure Uptake Biokinetic Model for Lead in Children (IEUBK). Windows version, 32 bit version. Publication number EPA 540-K-01-005. Office of Solid Waste and Emergency Response, U.S. Environmental Protection Agency, Washington, DC 20460. Online at http://www.epa.gov/superfund/programs/lead/products.htm

Harris-Warrick, R., L. Coniglio, N. Barazangi, J. Guckenheimer, and S. Gueron. 1995. Dopamine modulation of transient potassium current evokes phase shifts in a central pattern generator network. J. Neuroscience 15: 342–358.

Hilborn, R. and M. Mangel. 1997. *The Ecological Detective: Confronting Models with Data*. Princeton University Press, Princeton, NJ.

Kendall, B. E., C. J. Briggs, W. W. Murdoch, P. Turchin, S. P. Ellner, E. McCauley, R. M. Nisbet, and S. N. Wood. 1999. Inferring the causes of population cycles: A synthesis of statistical and mechanistic modeling approaches. Ecology 80: 1789–1805.

Koditschek, D. and M. Buehler. 1991. Analysis of a simplified hopping robot. International Journal of Robotics Research 10: 587–605.

Krom, M. D., S. Ellner, J. van Rijn, and A. Neori. 1995. Nitrogen and phosphorus cycling in a prototype "non-polluting" integrated mariculture system, Eilat, Israel. Marine Ecology Progress Series 118: 25–36.

Levin P. S., R. W. Zabel, and J. G. Williams. 2001. The road to extinction is paved with good intentions: Negative association of fish hatcheries with threatened salmon. Proceedings of the Royal Society of London Series B 268: 1153–1158.

Michaelis, L., and M. L. Menten. 1913. Die Kinetik der Invertinwerkung. Biochemische Zeitschrift 49: 333–369.

Oosterhout, G. R., and P. R. Mundy. 2001. The Doomsday Clock 2001: An Update on the Status and Projected Time to Extinction for Snake River Wild Spring/Summer Chinook Stocks. Report prepared for Trout Unlimited. Online at http://www.decisionmatrix.net/models.htm and http://www.tu.org/pdf/newsstand/library/doomsday_clock_2001_report.pdf

Raibert, M. 1986. *Legged Robots that Balance*. MIT Press, Cambridge, MA.

Saranli, U., M. Buehler, and D. E. Koditschek. 2001. RHex: A simple and highly mobile hexapod robot. International Journal of Robotics 20: 616–631.

Segrè, D., D. Vitkup, and G. M. Church. 2002. Analysis of optimality in natural and perturbed metabolic networks. Proceedings of the National Academy of Sciences USA 99: 15112–15117.

Stone, C. J. 1982. Optimal rates of convergence for nonparametric regression. Annals of Statistics 10: 1050–1053.

Tilman, D. 1989. Ecological experimentation: Strengths and conceptual problems. pages 136–157 in G. E. Likens (ed.), *Long-Term Studies in Ecology: Approaches and Alternatives*. Springer-Verlag, New York.

White, P., P. Van Leeuwen, B. Davis, M. Maddaloni, K. Hogan, A. Marcus, and R. Elias. 1998. U.S. Environmental Protection Agency: The conceptual structure of the integrated exposure uptake biokinetic model for lead in children. Environmental Health Perspectives 106 (Suppl. 6): 1513.

Wilson, P. H. 2003. Using population projection matrices to evaluate recovery strategies for Snake River spring and summer Chinook salmon. Conservation Biology 17: 782–794.

2 Matrix Models and Structured Population Dynamics

If a population or species is in decline and at risk of extinction, it is clear what we need to do: increase the birth rate, decrease the death rate, or both. But exactly *whose* birth or death rate are we talking about? For example, given limited human and financial resources, would it be more effective to create additional good nesting sites so more adult birds can breed each year, to shield eggs from predators so that each nest produces more offspring, or to augment food supply so that juveniles have a better shot at surviving to breeding age?

Trying to answer these questions in advance, rather than by trial and error, is an increasing part of the conservation biologist's job. This chapter focuses on a type of model that is widely used for this task (Morris and Doak 2002): matrix models for the dynamics of structured populations. "Structured" means that the model incorporates differences among individuals. Models based on differences in age were developed centuries ago for the study of human populations—one basic result is credited to Euler (1707–1783)—and we will start by considering that case. To understand structured population models we need some aspects of matrix theory. We review this from the ground up, beginning by defining matrices and reviewing basic matrix algebra. We then consider some applications, including conservation planning, and finally some generalizations of the model.

This chapter and the next introduce our *modus operandi* of using a small number of in-depth case studies as the vehicle for presenting different types of models and for motivating the study of their properties. To indicate the range of applicability for matrix models, we will not give you a quick survey of examples from other areas of biology, or even a list of books and articles with other applications. Instead, we follow this chapter with one where matrix models (and other kinds of model) are applied in a totally different biological context. Within our own field of biology we feel confident about choosing good "role models" for you to study. For the rest, you'll have to explore the literature and decide for yourself.

2.1 The Population Balance Law

The starting point for population modeling is the fundamental balance law

$$N(t+1) = N(t) + \text{births} + \text{immigration} - \text{deaths} - \text{emigration} \qquad [2.1]$$

where $N(t)$ is the number of individuals in the population or the population density (number per unit area) at time t. The balance law becomes a complete model when we specify formulas for the quantities on the right-hand side. The simplest model is to assume a closed population without immigration or emigration, and that the per capita (i.e., per individual) birth and death rates are constant:

$$\text{Births} = b \times N(t), \quad \text{Deaths} = d \times N(t).$$

The balance law then becomes

$$N(t+1) = N(t) + bN(t) - dN(t) = \lambda N(t) \qquad [2.2]$$

where $\lambda = 1 + b - d$. This model is simple enough that we can solve it. Starting from any initial population size $N(0)$ we get

$$N(1) = \lambda N(0)$$

$$N(2) = \lambda N(1) = \lambda(\lambda N(0)) = \lambda^2 N(0)$$

$$N(3) = \lambda N(2) = \lambda(\lambda^2 N(0)) = \lambda^3 N(0),$$

and so on, leading to the general solution

$$N(t) = \lambda^t N(0). \qquad [2.3]$$

This is *exponential population growth*: defining $r = log\,\lambda$, we then have $\lambda^t = (e^r)^t = e^{rt}$ and so

$$N(t) = e^{rt} N(0).$$

(Note that "log" means the natural (base-e) logarithm; we will use log_{10} to indicate base-10 logarithms.)

Growth cannot go on forever, so [2.3] cannot be valid forever if $\lambda > 1$. This kind of limitation bothers biologists much more than it bothers physical scientists, who are used to the idea that different models for a given system may be valid in different circumstances. Anderson and May (1992, p. 9) compare simple biological models to Newton's first law of motion:

> A body remains in its state of rest or uniform motion in a straight line, unless acted on by external forces.

Exponential population growth has the same character—it tells us what happens if current conditions persist without change:

> A closed population of self-reproducing entities—such as viruses, cells, animals, or plants—will grow or decay exponentially at a constant rate, unless a change in conditions alters the per entity birth or death rate.

Therefore, more general models are derived by considering the factors that can alter the average per entity birth and death rates.

2.2 Age-Structured Models

The biological theme of this chapter is that per entity birth and death rates are affected systematically by differences among individuals, such as their age. To take effects of age into account we need to describe the population by a state vector listing the numbers of individuals of each age. It is natural (but not necessary) to use years as the time unit; the state variables are then $n_a(t)$, the number of a-year-old individuals in year t, with a running from 0 to the maximum possible age A.

For now we continue to assume a closed population without immigration or emigration. The model's dynamic equations are bookkeeping expressed in mathematical symbols, as in the salmon model in Chapter 1. Consider one of the authors: 50 years old in January 2004. In order to reach that state, he must have been 49 years old in January 2003 and survived the next year. Consequently

$$n_{50}(2004) = p_{49}n_{49}(2003) \tag{2.4}$$

where p_{49} is the probability that a 49-year-old survives to age 50. In general, this line of reasoning tells us that

$$n_a(t+1) = p_{a-1}n_{a-1}(t) \text{ for } a > 0 \tag{2.5}$$

where p_x is the probability that an x-year-old individual survives to be age $x+1$.

To complete the model we need to specify the number of births each year. If we assume that per individual birth rates are only a function of age, we can define f_a to be the average number of newborns next year, per age-a female this year. Then we have

$$\begin{aligned} n_0(t+1) &= f_0n_0(t) + f_1n_1(t) + \cdots + f_An_A(t) \\ &= \sum_{a=0}^{A} f_an_a(t). \end{aligned} \tag{2.6}$$

Two conventions can be used: count everybody, or only count females—in which case f_a **only** includes female offspring. The females-only convention is by far the more common, and we will always use it here.

Exercise 2.1. Find at least two important assumptions that are necessary for [2.5] to be true.

Exercise 2.2. Define $l_0 = 1$, $l_a = p_0 p_1 p_2 \cdots p_{a-1}$, the probability of an individual surviving from birth to age a. Explain in words why

$$n_0(t+1) = \sum_{a=0}^{A} f_a l_a n_0(t-a). \tag{2.7}$$

Exercise 2.3. The general theory developed later in this chapter tells us that in the long run a population governed by the age-structured model typically grows exponentially, as in [2.3]. In particular, $n_0(t) = c\lambda^t$ will hold (with greater and greater accuracy as time goes on), for some λ and c. By substituting this approximation for $n_0(t)$ into [2.7], show that the long-term population growth rate λ satisfies the equation

$$\sum_{a=0}^{A} \lambda^{-(a+1)} l_a f_a = 1. \tag{2.8}$$

This is called the Euler, Lotka, or Euler-Lotka equation.

Exercise 2.4. (a) Show that the left-hand side of [2.8] is a decreasing function of λ by computing its derivative with respect to λ. (b) Compute the values of the left-hand side in the limits $\lambda \to 0$ and $\lambda \to \infty$. (c) Explain why (a) and (b) imply that [2.8] has one and only one positive real solution.

2.2.1 The Leslie Matrix

It is convenient and informative to express the age-structured model in matrix notation. In this form it is called the Leslie matrix model, after British ecologist P. H. Leslie who popularized age-structured models for animal populations in the mid-twentieth century.

First we need to review a bit about matrices. A matrix is a rectangular array of numbers. A matrix $\mathbf{A}$ with entries a_{ij} is said to have size $m \times n$ if it has m (horizontal) rows and n (vertical) columns. Thus the row index i takes the values $1, 2, \ldots, m$ (1 indicating the top row) and the column index j takes the values $1, 2, \ldots, n$ (1 indicating the leftmost column). A matrix with one column is called a column vector, and a matrix with a single row is called a row vector.

Matrix algebra was invented for studying systems of linear equations in several unknowns, such as

$$\begin{aligned} 3x_1 + 5x_2 &= 1 \\ x_1 - 2x_2 &= 0. \end{aligned} \tag{2.9}$$

Solving one such equation in one unknown is a snap:

$$\begin{aligned} 3x &= 1 \\ (1/3) \times (3x) &= (1/3) \times (1) \\ x &= 1/3. \end{aligned} \tag{2.10}$$

Matrix algebra lets us make [2.9] look like [2.10], so that we can solve it in the same way. We put the coefficients in [2.9] into a matrix

$$\mathbf{A} = \begin{bmatrix} 3 & 5 \\ 1 & -2 \end{bmatrix} \qquad [2.11]$$

and put the variables and right-hand side into vectors $\mathbf{x} = \begin{bmatrix} x_1 \\ x_2 \end{bmatrix}$ and $\mathbf{b} = \begin{bmatrix} 1 \\ 0 \end{bmatrix}$.

Matrix-vector multiplication is then defined so that [2.9] is equivalent to the single matrix equation:

$$\overset{\mathbf{A}}{\begin{bmatrix} 3 & 5 \\ 1 & -2 \end{bmatrix}} \overset{\mathbf{x}}{\begin{bmatrix} x_1 \\ x_2 \end{bmatrix}} \overset{=}{=} \overset{\mathbf{b}}{\begin{bmatrix} 1 \\ 0 \end{bmatrix}}. \qquad [2.12]$$

The definition that makes [2.9] and [2.12] mean the same thing is the following: if $\mathbf{x} = (x_1, x_2, \ldots, x_n)$ is a column vector and $\mathbf{A}$ is a matrix with n columns, then $\mathbf{Ax}$ is the column vector

$$\begin{aligned} &(x_1 \times \text{first column of } \mathbf{A}) \\ &+(x_2 \times \text{second column of } \mathbf{A}) \\ &\quad\vdots \\ &+(x_n \times \text{last column of } \mathbf{A}). \end{aligned} \qquad [2.13]$$

Algebraically, that works out to the following formula: for a matrix $\mathbf{A}$ with n columns and a column vector $\mathbf{x}$ of length n, $\mathbf{Ax}$ is the vector whose ith element is

$$(\mathbf{Ax})_i = \sum_{j=1}^{n} A_{ij} x_j \qquad [2.14]$$

where A_{ij} is the number in the ith row and jth column of the matrix $\mathbf{A}$. Equation [2.14] says that the ith element of $\mathbf{Ax}$ is the *inner product* of the ith row of $\mathbf{A}$ with $\mathbf{x}$, where the inner product of two vectors $\mathbf{v}$ and $\mathbf{x}$ of length n is defined as

$$\mathbf{v} \cdot \mathbf{x} = v_1 x_1 + v_2 x_2 + \cdots + v_n x_n. \qquad [2.15]$$

This expression is sometimes called the *dot product*, and the alternate notations $\langle \mathbf{v}, \mathbf{x} \rangle$ or $(\mathbf{v}, \mathbf{x})$ are also used.

The inner-product interpretation of [2.14] is how people are usually taught to do matrix-vector multiplication by hand. For example (and make sure that you understand this example!),

$$\begin{bmatrix} 1 & 2 \\ 3 & 4 \end{bmatrix} \begin{bmatrix} 5 \\ 6 \end{bmatrix} = \begin{bmatrix} (1,2)\cdot(5,6) \\ (3,4)\cdot(5,6) \end{bmatrix} = \begin{bmatrix} 1 \times 5 + 2 \times 6 \\ 3 \times 5 + 4 \times 6 \end{bmatrix} = \begin{bmatrix} 17 \\ 39 \end{bmatrix}. \qquad [2.16]$$

However, the conceptual definition [2.13] is essential for understanding the biological meaning of matrix models.

So what good does this do us? Suppose that we could find a multiplicative inverse to **A**—a matrix $\mathbf{A}^{-1}$ such that $\mathbf{A}^{-1}(\mathbf{Ax}) = \mathbf{x}$ for any vector **x**. Then we could solve [2.12] the same way we solved [2.10]: just multiply both sides of equation [2.12] by the inverse of **A** to get the solution $\mathbf{x} = \mathbf{A}^{-1}\mathbf{b}$. Figuring out when such inverses exist and how to compute them was one of the major accomplishments of nineteenth-century mathematics.

As is often the case in mathematics, a tool invented for one purpose turns out to be useful for many others. Notice that the equation [2.6] for births in the age-structured model has the same form as [2.14]. The survival equation [2.5] is also a sum, with only one term. So we can express these in matrix notation by putting the survival and birth rates in the right places:

$$\begin{bmatrix} n_0(t+1) \\ n_1(t+1) \\ \vdots \\ \\ n_A(t+1) \end{bmatrix} = \begin{bmatrix} f_0 & f_1 & f_2 & \cdots & f_A \\ p_0 & 0 & 0 & \cdots & 0 \\ 0 & p_1 & 0 & \cdots & 0 \\ \vdots & \vdots & \ddots & & \vdots \\ 0 & 0 & & p_{A-1} & 0 \end{bmatrix} \begin{bmatrix} n_0(t) \\ n_1(t) \\ \vdots \\ \\ n_A(t) \end{bmatrix} \quad [2.17]$$

or simply

$$\mathbf{n}(t+1) = \mathbf{Ln}(t) \quad [2.18]$$

where **L** is the matrix in [2.17], and $\mathbf{n}(t)$ is the population vector $(n_0(t), n_1(t), \ldots, n_A(t))$. The top row of the matrix contains the births, and the other nonzero entries are survival.

For example, consider a plant with a maximum age of 2—this might be a plant that flowers once, at either age 1 or age 2, and then dies. Suppose that newborn offspring (age 0) have a 50% chance of surviving to age 1; age-1 plants produce f_1 offspring each on average, and have a 25% chance of surviving to age 2, and age-2 individuals have f_2 offspring each on average. The Leslie matrix is then

$$\mathbf{L} = \begin{bmatrix} 0 & f_1 & f_2 \\ 0.5 & 0 & 0 \\ 0 & 0.25 & 0 \end{bmatrix}. \quad [2.19]$$

Definition [2.13] tells us how to "read" a matrix like [2.19]. Since $n_i(t)$ multiplies the ith column of the Leslie matrix, the ith column of **L** gives the individuals of each age "next year" resulting from a single age-i individual "this year," as a consequence of their survival and fecundity. So the first column says that for each age-0 individual this year, there will be (on average) half a 1 year old next year—the ones that survive. The second says that for each age-1 individual this year, the population next year will have f_1 age-0 individuals (the offspring of age-1 individuals) and 0.25 age-2 individuals (the age-1 individuals who survive to next year). Age-2 individuals all die, so their only contribution to next year's population is their offspring (f_2 per 2 year old). In the same way, the jth column of the general Leslie matrix [2.17] says that for each j year old "this year," next year's population will have f_j offspring (age 0) and p_j survivors (age $j+1$).

Note that it does not matter how survivorship and breeding at a given age are related to each other. For example, it could be the case in [2.19] that half the age-1 individuals reproduce and then die (having $2f_1$ offspring each, on average) while those that do not reproduce have a 50% chance of surviving to age 2. Or it could be the case that all age-1 individuals reproduce, and all have a 25% chance of surviving to age 2. Either way the matrix is the same.

Exercise 2.5. Verify that the following are correct in three different ways: using [2.13], using the inner-product method illustrated in Equation [2.16], and by writing a script to do the calculations on the computer.

$$\text{(a)} \begin{bmatrix} 1 & 3 \\ 2 & 4 \end{bmatrix} \begin{bmatrix} -1 \\ 2 \end{bmatrix} = \begin{bmatrix} 5 \\ 6 \end{bmatrix}, \quad \text{(b)} \begin{bmatrix} 1 & 0 \\ -1 & 2 \end{bmatrix} \begin{bmatrix} 2 \\ 3 \end{bmatrix} = \begin{bmatrix} 2 \\ 4 \end{bmatrix}$$

$$\text{(c)} \begin{bmatrix} 0 & 0.5F & F \\ 0.5 & 0 & 0 \\ 0 & 0.25 & 0 \end{bmatrix} \begin{bmatrix} 10 \\ 20 \\ 10 \end{bmatrix} = \begin{bmatrix} 20F \\ 5 \\ 5 \end{bmatrix}.$$

Exercise 2.6. Write a script file to run simulations of the biennial plant model [2.19] with $f_1 = 1$, $f_2 = 5$, starting from a single age-1 individual at time 0. Have the script plot as functions of time (1) the log of the total population size $N(t) = n_0(t) + n_1(t) + n_2(t)$, and (2) the fraction of individuals of each age, $w_i(t) = n_i(t)/N(t)$, $\quad i = 1, 2, 3$, for $t = 1$ to 50. What long-term properties of the population do you see in your simulation results?

Exercise 2.7. Write down the Euler-Lotka equation [2.8] for the biennial plant model [2.19] with $f_1 = 1, f_2 = 5$, and numerically solve it for the value of λ. How does this compare to the rate of population growth that you saw in your simulations? [You can find λ approximately by having your script compute the left-hand side of [2.8] at a finely spaced set of λ values, and finding one at which the sum is closest to 1. Or if you're adventurous, find a function in your scripting language that finds the roots of univariate functions, e.g., `uniroot` in **R** or `fzero` in MATLAB.]

2.2.2 Warning: Prebreeding versus Postbreeding Models

The interpretation of the f's depends on when births occur relative to the annual census time. Suppose we census the population on January 1 each year. In humans, births occur year round, so f_a should be the average number of births over the coming calendar year, to an individual whose age was between a and $a+1$ on January 1, but only counting offspring that survive until January 1 of next year. So f_a is the sum over all such individuals of

(average number of births on Jan 1) × (survival from this Jan 1 to next Jan 1)

+(average number of births on Jan 2) × (survival from this Jan 2 to next Jan 1)

+(average number of births on Jan 3) × (survival from this Jan 3 to next Jan 1)

and so on.

In other cases it is more accurate to assume a once-per-year seasonal "pulse" of births, as if all offspring for the year were born at once. Let m_a be the number of offspring that an a year old has in the current birth pulse. If we census the population immediately after the pulse (*postbreeding* census) then

$$f_a = p_a \text{ [survival to next year]}$$
$$\times m_{a+1} \text{ [\# offspring in next year's birth pulse]}.$$

But if we census just before the pulse (*prebreeding* census), then

$$f_a = m_a \text{ [\# offspring now—but not counted until next year]}$$
$$\times p_0 \text{ [fraction of offspring who survive to be counted]}.$$

Both of these are valid under their assumptions about census timing, and both are used. As a result, formulas for things like life expectancy, population growth rate, and so on, exist in two different versions for prebreeding and postbreeding models. An additional complication is that some authors (e.g., Caswell 2001) number age-classes starting at 1 rather than 0, so that their $n_1(t)$ is equivalent to our $n_0(t)$. Even experts get confused by all these options, and many books and papers include a mix of formulas based on incompatible assumptions. So when you see f_4 in a book or paper, it's important to check what the author intends it to mean.

2.3 Matrix Models Based on Stage Classes

In most applications to nonhuman organisms, the oldest age A really consists of individuals aged *A or older*, due to lack of data. The meaning of "extreme old age" is that most individuals die before they get there, so there always are relatively few observations of what happens to extremely old individuals. For example, suppose during your study of a lizard population, there is one hardy 4 year old who lives to be 5, then lives to be 6, and then dies, while all other individuals die before they reach the age of 4. So would you then take $p_4 = 1, p_5 = 1, p_6 = 0$? A better option is to assume that all individuals above some age are identical, so that you can get a reasonable estimate of their average survival probability.

Collapsing all ages above some cutoff into one "age-class" is our first example of the tradeoff between *model error* and *parameter error*. Model error means errors due to incorrect assumptions, where the model simplifies or omits known aspects of reality. Combining all individuals of age 3 or above is likely to create model error, because we have no grounds for believing that there really are no systematic differences between a 3 year old and a 5 year old. Parameter error means errors due to parameters being estimated inexactly from a limited set of data. By combining all 3+ year olds, we avoid the parameter errors that would result from estimating p_4, p_5, and p_6 from a sample of size 1. The resulting model has a category of individuals who are likely to be fairly similar, with parameters for the category being "average" or "typical" values for members of the category. It is always possible to reduce model error by making a model more complex, but parameter error usually goes up because you have to somehow estimate more parameters from the same amount of data. We discuss this tradeoff more fully in Chapter 9.

More generally, individuals can be classified by their *stage* in the life cycle. Sometimes there really are discrete life stages, such as caterpillar-cocoon-butterfly (or more generally larva-pupa-adult in insects). But sometimes it is just a group of individuals defined by some measurable feature, such as length or weight, that is the best available attribute for predicting their fate over the next period of time.

The most common attributes for defining categories are measures of individual size. These have long been popular in the forestry literature, because size is generally much better than age for predicting tree growth and mortality. Now size is used also for animal populations, with recent examples including sea turtles, desert tortoise, geese, corals, copepods, and fish (Caswell 2001). Size categories are usually defined so that between one census and the next individuals can grow or shrink by at most one category, and all newborns are in the smallest category, but this is not always the case. For example, Valverde and Silvertown (1998) used size-classified matrix models for the woodland herbaceous plant *Primula vulgaris* in which individuals could grow by two categories, in order to study how *Primula* population growth was affected by the degree of forest canopy closure. For one of their study sites (Woburn Wood), the projection matrix for 1993 to 1994 was

$$\begin{bmatrix} 0 & 0 & 0.03 & 0.10 & 0.18 \\ 0.25 & 0.35 & 0.12 & 0.02 & 0 \\ 0.04 & 0.45 & 0.65 & 0.33 & 0.19 \\ 0 & 0 & 0.16 & 0.58 & 0.38 \\ 0 & 0 & 0 & 0.05 & 0.38 \end{bmatrix}$$

with the categories being defined by plant area (see the right-hand side of Table 2.1). As with our hypothetical model [2.19] we "read" this matrix by recognizing that each column specifies the contribution of one category to next year's population.

A useful way to graphically represent a stage-classified matrix model is the *life cycle graph* in which each "node" represents a stage, and arrows show possible changes in stage for individuals between one time interval and the next. Figures 2.1 and 2.2 show two examples. By convention, staying put in a stage is drawn

Category	Carapace length (mm)	Category	Plant area (cm^2)
Yearling		Seedling	0.5–5
Juvenile 1	<60	Juvenile	5.1–35
Juvenile 2	60–99	Adult 1*	35.1–200
Immature 1	100–139	Adult 2*	200.1–600
Immature 2	140–179	Adult 3*	> 600
Subadult*	180–207		
Adult 1*	208–239		
Adult 2*	>240		

Table 2.1 Two examples of stage classifications based in part on individual size. Asterisks indicate reproductive categories. The two left columns give the categories used by Doak et al. (1994) for desert tortoise in the western Mojave desert, which were the same as those used by the Bureau of Land Management in the population monitoring program that provided the data for the model. The two right columns give the categories defined by Valverde and Silvertown for the forest herb *Primula vulgaris*.

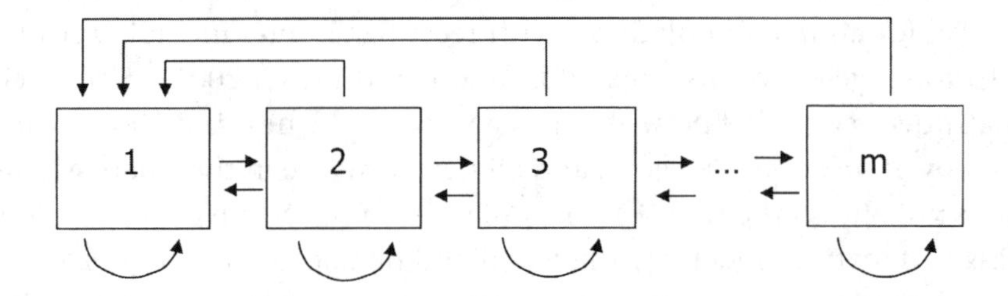

Figure 2.1 The standard size-class model. Size categories are broad enough that individuals can't change by more than one category between population censuses, and all newborn individuals are in the smallest size class. These all look the same apart from the number of "stages".

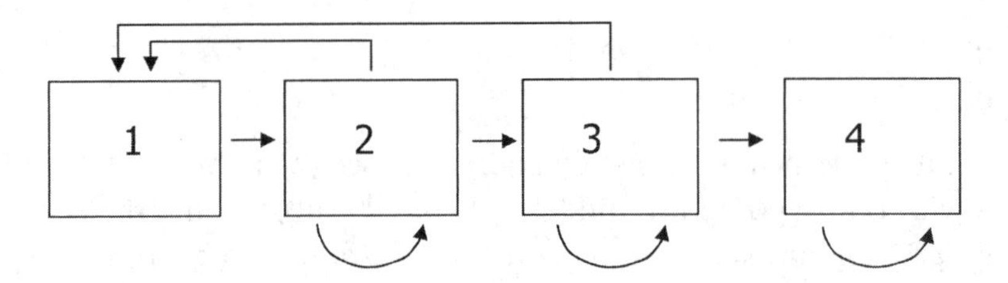

Figure 2.2 Stage-structured model for killer whales (from Brault and Caswell 1993). The stages recognized were 1 = yearling, 2 = juvenile, 3 = mature female, and 4 = postreproductive female.

as an arrow, but deaths are omitted on the assumption that no life stage is invulnerable. Also note that there is no distinction in the diagram between survival and fecundity. Since a basic premise of the model is that an individual's stage classification provides complete information about its future prospects, it does not matter (in the model) if a small individual is a newborn or an older individual who shrank back down to newborn size. The life cycle diagram represents your idea of a good way of classifying individuals. If there are discrete stages, it is probably a good idea to use those in the model. Otherwise, experience suggests that the most important issue is selecting which trait to use as the basis for classifying individuals (e.g., age versus size). The trait used for classifying individuals is sometimes called the *individual state variable* or the *i-state variable.*

Given the right data, alternative choices of *i*-state variable can be compared objectively. For example, if you know an individual's size, can you predict her fecundity more accurately if you also know her age? Caswell (2001, section 3.3) presents several examples of this kind of comparison. However, the classification is often dictated by circumstances. For example, Doak et al. (1994) based their model on data that had already been collected by the Bureau of Land Management. They had no choice but to use size as their *i*-state variable, with the categories used in the BLM surveys (Table 2.1). Valverde and Silvertown (1998) based their classification on knowledge of the species' natural history, with class boundaries chosen so that each category had sufficient sample size for estimating matrix entries.

Having chosen a stage classification, the model is completed by specifying the projection matrix entries a_{ij},

$$a_{ij} = \text{number of type-}i\text{ individuals at time } t+1\text{, per type-}j\text{ individual at time } t.$$

Contributions from j to i may be survival, fecundity, or a combination of these. As in the age-structured model, we assume (for now) that the a_{ij} are constant. The fundamental balance law is then

$$\begin{aligned} n_i(t+1) &= a_{i1}n_1(t) + a_{i2}n_2(t) + \cdots \\ &= \sum_{j=1}^{m} a_{ij}n_j(t). \end{aligned} \quad [2.20]$$

The *projection matrix* $\mathbf{A}$ is defined to be the $n \times n$ matrix with entries a_{ij}, and the model then becomes

$$\mathbf{n}(t+1) = \mathbf{A}\mathbf{n}(t). \quad [2.21]$$

Estimating the value of matrix entries is a subject in itself. Entire careers (and entire books, e.g., Williams et al. 2002) are devoted to methods for analyzing census data on populations in order to estimate demographic rates. Morris and Doak (2002, Chapter 6) give some guidelines on how to conduct field studies to estimate demographic rates. The ideal situation is if individuals can be given a unique tag or mark (or come with unique markings), and you can come back later to see what happened to them. Plants sit still and wait to be counted, but with animals it is often hard to distinguish between death and emigration out of the study area.

Similarly for fecundities, the best situation is if you can identify and count the offspring of each parent. This is often possible with large animals or animals that live in family groups. For plants, a common approach is to count seeds while they are still on the parent plant. Then, assuming that (once released) a seed is a seed is a seed, you can estimate

$$F_i = \text{(average number of seeds produced by a class-}i\text{ plant)}$$
$$\times \text{(fraction of all seeds that survive to be seedlings at the next census).}$$

The same idea has also been used for estimating fecundity in birds: do a census of offspring in their natal nest, and decrement those counts by the overall fraction of nestlings that survive to the next census.

There are also ways to indirectly estimate parameters from population count data. Indirect methods begin by assuming that the model is valid, and then asking: What must the survival (or growth or fecundity) parameters be, in order to generate the population that I observed? This is more difficult than direct estimates and less secure, because of the *a priori* assumption that the model is valid. Indirect methods for structured population models are reviewed by Wood (1997) and Caswell (2001, section 6.2).

Exercise 2.8. State in words the meaning of the second and fourth columns of the projection matrix given above for *Primula vulgaris* in Woburn Wood.

Exercise 2.9. Draw the life cycle graph for *Primula vulgaris* in Woburn Wood.

2.4 Matrices and Matrix Operations

Our goal now is to derive general properties of matrix models that allow us to make connections between the matrix entries and the long-term fate of a population governed by [2.21]. For example, in conservation planning it is important to know which matrix entries have the greatest impact on whether the population is growing or shrinking, so that those can be targets for remediation efforts (e.g., striving to increase the survival during particularly important stages of the life cycle).

Rewriting the balance equations [2.20] in matrix form [2.21] is more than a convenience, because the algebra of matrices (called *linear algebra*) has a lot to tell us about the equations. This section reviews some concepts and results from linear algebra that will give us much insight into the balance equations for populations. Moreover, these results will be employed in Chapter 3 in an entirely different setting, to model the gating of membrane channels in neurons.

2.4.1 Review of Matrix Operations

Addition and subtraction of matrices are done element by element, and are therefore only defined for matrices of the same size:

$$\mathbf{C} = \mathbf{A} \pm \mathbf{B} \text{ has entries } c_{ij} = a_{ij} \pm b_{ij}. \tag{2.22}$$

Multiplication of a matrix by a scalar (real number) is also element by element:

$$\mathbf{B} = c\mathbf{A} \text{ has entries } b_{ij} = ca_{ij}. \tag{2.23}$$

Examples are

$$\begin{bmatrix} 1 & 2 & 3 \\ 4 & 5 & 6 \end{bmatrix} + \begin{bmatrix} 2 & 6 & 10 \\ 4 & 8 & 12 \end{bmatrix} = \begin{bmatrix} 1+2 & 2+6 & 3+10 \\ 4+4 & 5+8 & 6+12 \end{bmatrix} = \begin{bmatrix} 3 & 8 & 13 \\ 8 & 13 & 18 \end{bmatrix}$$
$$2\begin{bmatrix} 1 & 2 \\ 3 & 4 \end{bmatrix} = \begin{bmatrix} 2 & 4 \\ 6 & 8 \end{bmatrix}. \tag{2.24}$$

Matrix multiplication is more complicated. The product $\mathbf{C} = \mathbf{AB}$ is defined if the number of columns of $\mathbf{A}$ is equal to the number of rows of $\mathbf{B}$. If $\mathbf{A} = (a_{ij})$ has size $m \times n$ and $\mathbf{B} = (b_{ij})$ has size $n \times r$, then $\mathbf{C} = \mathbf{A} \cdot \mathbf{B}$ has size $m \times r$ and

$$c_{ik} = \sum_{j=1}^{n} a_{ij} b_{jk}. \tag{2.25}$$

Note that if $\mathbf{B}$ has only one column, this reduces to the definition of matrix-vector multiplication [2.14]. Thus, another definition of matrix multiplication is the following:

$$k\text{th column of } \mathbf{AB} = \mathbf{A} \times (k\text{th column of } \mathbf{B}). \tag{2.26}$$

Our attitude is that matrix multiplication is usually best done on the computer. It is important to understand the conceptual definition [2.26] and the algebraic formula [2.25], but when working with actual numbers, it is easier and less error-prone to use a computer language that includes matrices and matrix multiplication.

Matrix operations share many properties with the familiar arithmetic of real numbers. For example,

- Matrix addition is associative $[\mathbf{A} + (\mathbf{B} + \mathbf{C}) = (\mathbf{A} + \mathbf{B}) + \mathbf{C}]$ and commutative $[\mathbf{A} + \mathbf{B} = \mathbf{B} + \mathbf{A}]$.

- Matrix multiplication is associative [$\mathbf{A}(\mathbf{BC}) = (\mathbf{AB})\mathbf{C}$] and distributive over addition [$\mathbf{A}(\mathbf{B}+\mathbf{C}) = \mathbf{AB}+\mathbf{AC}$, $(\mathbf{A}+\mathbf{B})\mathbf{C} = \mathbf{AC}+\mathbf{BC}$].

However, *matrix multiplication is not commutative*—typically $\mathbf{AB} \neq \mathbf{BA}$. Indeed, unless **A** and **B** are square matrices of the same size, either one of the products **AB** and **BA** will be undefined, or the two products will be matrices of different sizes. But even in the case of square matrices commutativity typically does not hold. Here is a simple example:

$$\begin{bmatrix} 1 & 0 \\ 0 & -1 \end{bmatrix} \begin{bmatrix} 0 & 1 \\ 1 & 0 \end{bmatrix} = \begin{bmatrix} 0 & 1 \\ -1 & 0 \end{bmatrix}$$

but

$$\begin{bmatrix} 0 & 1 \\ 1 & 0 \end{bmatrix} \begin{bmatrix} 1 & 0 \\ 0 & -1 \end{bmatrix} = \begin{bmatrix} 0 & -1 \\ 1 & 0 \end{bmatrix}.$$

However, scalar multiplication is commutative in the sense that $\mathbf{A}(c\mathbf{B}) = c(\mathbf{AB})$.

2.4.2 Solution of the Matrix Model

Having defined matrix multiplication, we can now easily write down the solution to the matrix model [2.21], in the same way that we solved the unstructured model [2.2]. Starting from some initial population vector $\mathbf{n}(0)$ we get

$$\begin{aligned} \mathbf{n}(1) &= \mathbf{An}(0) \\ \mathbf{n}(2) &= \mathbf{An}(1) = \mathbf{A}(\mathbf{An}(0)) = \mathbf{A}^2\mathbf{n}(0) \\ \mathbf{n}(3) &= \mathbf{An}(2) = \mathbf{A}(\mathbf{A}^2\mathbf{n}(0)) = \mathbf{A}^3\mathbf{n}(0) \\ &\vdots \end{aligned} \qquad [2.27]$$

leading to the general solution

$$n(t) = \mathbf{A}^t\mathbf{n}(0), \qquad [2.28]$$

where $\mathbf{A}^t$ denotes the product of **A** with itself t times. We can form these products because **A** is a square matrix, and the order of operations in computing the products does not matter because matrix multiplication is associative.

2.5 Eigenvalues and a Second Solution of the Model

The most basic question we can ask about a population is whether it will grow or become extinct in the long run. The solution of the matrix model shows that the answer depends on the behavior of $\mathbf{A}^t$, the powers of the projection matrix as t increases. We can determine the properties of $\mathbf{A}^t$ through the eigenvalues and eigenvectors of the matrix **A**.

A number (possibly complex) λ is an *eigenvalue* of $\mathbf{A}$ if there is a nonzero vector $\mathbf{w}$ such that $\mathbf{A}\mathbf{w} = \lambda\mathbf{w}$, and $\mathbf{w}$ is called the corresponding eigenvector. Eigenvectors are defined only up to scaling factors: if $\mathbf{w}$ is an eigenvector for λ then so is $c\mathbf{w}$ for any number $c \neq 0$. An $n \times n$ matrix $\mathbf{A}$ must have at least one eigenvalue-eigenvector pair, and it can have up to n (see this chapter's Appendix for an explanation of why this is true). The typical situation is to have n distinct eigenvalues each with a corresponding eigenvector—this is typical in the sense that if matrix entries are chosen at random according to some smooth probability distribution, the probability of the resulting matrix having n distinct eigenvalues is 1.

There is a useful formula for the eigenvalues of a 2×2 matrix $\mathbf{A}$. If $T = a_{11} + a_{22}$ is the *trace* (sum of diagonal elements) and $\Delta = a_{11}a_{22} - a_{12}a_{21}$ is the *determinant* then the eigenvalues are

$$\lambda_{1,2} = \frac{1}{2}\left(T \pm \sqrt{T^2 - 4\Delta}\right). \quad [2.29]$$

Back to equation [2.21]. Assuming there are n distinct eigenvalues, the corresponding eigenvectors $\mathbf{w}_i$ are *linearly independent*, which means that for any $\mathbf{n}(0)$ it is possible to find constants c_i such that

$$\mathbf{n}(0) = c_1\mathbf{w}_1 + c_2\mathbf{w}_2 + \cdots + c_n\mathbf{w}_n = \sum_i^n c_i\mathbf{w}_i. \quad [2.30]$$

Then

$$\begin{aligned} \mathbf{n}(1) = \mathbf{A}n(0) &= \mathbf{A}(c_1\mathbf{w}_1 + c_2\mathbf{w}_2 + \cdots + c_n\mathbf{w}_n) \\ &= c_1\mathbf{A}\mathbf{w}_1 + c_2\mathbf{A}\mathbf{w}_2 + \cdots + c_nA\mathbf{w}_n \\ &= c_1\lambda_1\mathbf{w}_1 + c_2\lambda_2\mathbf{w}_2 + \cdots + c_n\lambda_n\mathbf{w}_n. \end{aligned} \quad [2.31]$$

Comparing [2.30] with [2.31] we see that going forward one step in time corresponds to multiplying all the coefficients c_i by the corresponding eigenvalue λ_i. We can go from $t = 1$ to $t = 2$ in the same way, getting

$$\mathbf{n}(2) = c_1\lambda_1^2\mathbf{w}_1 + c_2\lambda_2^2\mathbf{w}_2 + \cdots + c_n\lambda_n^2\mathbf{w}_n \quad [2.32]$$

and so forth. Thus the solution of the matrix model is

$$\mathbf{n}(t) = \sum_i^n c_i\lambda_i^t\mathbf{w}_i. \quad [2.33]$$

An eigenvalue λ_1 is called *dominant* if $|\lambda_i| < |\lambda_1|$ for all other eigenvalues of A. If so, it follows from [2.33] that the long-run behavior of the population is determined by the dominant eigenvalue and its eigenvector:

$$\mathbf{n}(t) \sim c_1\lambda_1^t\mathbf{w}_1. \quad [2.34]$$

The meaning of $\sim$ in equation [2.34] is that as $t \to \infty$ the relative error goes to zero.

Equation [2.34] tells us two things about the population. First, in the long run the total population size grows exponentially at rate λ_1, just as in the unstructured model [2.2]. Second, the population vector becomes becomes proportional to $\mathbf{w}_1$; in particular, the relative numbers in each stage become constant. For that reason, $\mathbf{w}_1$ is called the *stable stage distribution*.

The Perron-Frobenius theorem from linear algebra provides an easy-to-check condition which guarantees existence of a dominant eigenvalue. A matrix $\mathbf{A}$ is called *power-positive* if there is an integer $m > 0$ such that all entries of the matrix $\mathbf{A}^m$ are strictly positive. The most important result is the following:

> If a non-negative, square matrix $\mathbf{A}$ is power-positive, then $\mathbf{A}$ has a unique dominant eigenvalue λ which is real and positive, and the eigenvector $\mathbf{w}$ corresponding to λ has all positive entries.

This criterion for existence of a dominant eigenvalue is especially useful because power-positivity depends only on which elements in the matrix are positive, *not on their numerical values*. It is also useful that there is a simple test to determine if a non-negative matrix is power-positive (Horn and Johnson 1985, p. 520), which is easy to implement on the computer:

> If $\mathbf{A}$ is a non-negative square matrix with n rows and columns, then $\mathbf{A}$ is power-positive if and only if all entries of $\mathbf{A}^{n^2-2n+2}$ are positive.

Because eigenvectors are defined only up to multiplication by a constant, the statement that the dominant eigenvector has all positive entries really means that all entries in any dominant eigenvector have the same sign. A software package may give you an eigenvector (call it $\mathbf{w}^*$) with all *negative* entries, in which case $\mathbf{w} = -\mathbf{w}^*$ is the strictly positive eigenvector guaranteed by Perron-Frobenius[1].

Exercise 2.10. Write a script to verify that the following projection matrix is power-positive:

$$\begin{bmatrix} 0 & 0 & 0 & 4 \\ .8 & .1 & 0 & 0 \\ 0 & .8 & .6 & .3 \\ 0 & .1 & .3 & .6 \end{bmatrix}.$$

Exercise 2.11. Find a 4×4 Leslie matrix $\mathbf{L}_1$ that is power-positive, and a second Leslie matrix $\mathbf{L}_2$ that is not power-positive. In the latter case, verify your conclusion by writing a script that computes and prints the smallest value in the matrix $\mathbf{L}_2^j$ for each $j = 1, 2, \ldots, n^2 - 2n + 2$. What happens to the age structure, starting from a single newborn, in your non-power-positive example? [Note: a Leslie matrix is a

[1]Power-positivity does not guarantee that there are n distinct eigenvalues, but so long as there is a dominant eigenvalue, equation [2.34] still holds—this is not hard to show using the Jordan Canonical Form for matrices.

matrix of the form [2.17], in which all of the p's are positive, and at least one of the f's must be positive.]

Exercise 2.12. According to Lande (1988) females of the Northern Spotted Owl begin breeding at age $a = 3$, and are estimated to have an average of 0.24 female offspring per year until they die ($f_a = 0.24$ for $a \geq 3$). The survival probability from birth to age 3 is estimated to be 0.0722, and the annual survival probability of adults (age 3 or older) is 0.942 (these values refer to the notational conventions that we used in the age-structured model, so that a newborn individual is 0). This owl has been controversial, because of the conflict between the need to preserve old-growth forests as habitat for spotted owl, and the interest of logging companies in harvesting those forests.

(a) We told you that $l_3 = p_0 p_1 p_2 = 0.0722$ but not the values of the individual p's. That is because any choice of p's with this product will result in the same population growth rate. Why is that true? (Note: the answer to this question should be verbal; no formulas are needed).

(b) Construct a projection matrix for the population based on the estimates above.

(c) Compute the owl's long-term growth rate λ from the projection matrix. Does it appear that the population is safe, or in danger of extinction?

Exercise 2.13. Killer whales (*Orcinus orca*) are long-lived marine mammals that live in stable social groups called "pods." Their stable social structure and the fact that individual whales can be photo-identified makes them especially well suited to scientific study. Demographic data on killer whale populations in the coastal waters of British Columbia and Washington state have been collected since 1973. Brault and Caswell (1993) used the 1973–1987 data and a stage-structured matrix model to investigate several demographic questions concerning the whales. They model the females with a mixed age-stage classification: yearlings, juveniles (past the first year, but not mature), mature, and postreproductive. The life cycle graph is shown in Figure 2.2 and the projection matrix $\mathbf{A}$ is given below:

$$\mathbf{A} = \begin{bmatrix} 0 & 0.0043 & 0.1132 & 0 \\ 0.9775 & 0.9111 & 0 & 0 \\ 0 & 0.0736 & 0.9534 & 0 \\ 0 & 0 & 0.0452 & 0.9804 \end{bmatrix}. \qquad [2.35]$$

Write a script file that

(a) computes the dominant eigenvalue λ and stable stage distribution $\mathbf{w}$ for the whale population;

(b) projects the population dynamics for the next 50 years assuming that the current population vector is $x_0 = (10, 60, 110, 70)$;

(c) plots on three separate graphs the projected changes over time in
- $N(t)$ = total population size in year t,
- the annual population growth rate $\lambda(t) = N(t+1)/N(t)$,
- the proportion of individuals in each stage.

Does the population structure become stable? How does it change over time? How quickly does the annual growth rate $\lambda(t)$ converge to the dominant eigenvalue λ?

Exercise 2.14. Rerun your script for killer whale population dynamics with the following initial population vectors:$x_0 = (250, 0, 0, 0), (0, 250, 0, 0), (0, 0, 250, 0)$, and $(0, 0, 0, 250)$. Compare and contrast the four population projections—for example, (a) consider the stage distribution and its stability; (b) which stage seems to be the most important in terms of the future growth of the population?

Exercise 2.15. Consider a possible harvest from the killer whale population, consisting of individuals from a single stage, for example, all juveniles or all reproductive adults. Suppose that the initial population structure is the stable distribution $\mathbf{w}$ with a total of 250 individuals. What is the maximum number of juveniles that can be taken each year such that the population is not driven to extinction? What is the maximum number of reproductive adults? Note: Assume that harvest will take place after the breeding season, so that the model becomes $\mathbf{x}(t+1) = \mathbf{A}\mathbf{x}(t) - h$ where $h = (h_1; h_2; h_3; h_4)$ is a vector of the number of individuals harvested from each stage each year, h_i. *Assume that h is constant: same harvest each year.*

2.5.1 Left Eigenvectors

There is a definition of left eigenvalues and eigenvectors analogous to that for right eigenvalues and eigenvectors: if $\mathbf{v}\mathbf{A} = \lambda\mathbf{v}$ (where $\mathbf{v}$ is a nonzero row vector of length n) then $\mathbf{v}$ is a left eigenvector and λ is the corresponding eigenvalues. There are three key properties:

- The left eigenvalues of a matrix $\mathbf{A}$ are the same as the right eigenvalues.
- The left eigenvectors of $\mathbf{A}$ are the right eigenvectors of its transpose $\mathbf{A}^T$. $\mathbf{A}^T$ is the matrix whose (i,j)th element is a_{ji}. That is, the rows of $\mathbf{A}$ become the columns of $\mathbf{A}^T$, for example,

$$\begin{bmatrix} 1 & 2 & 3 \\ 4 & 5 & 6 \\ 7 & 8 & 9 \end{bmatrix}^T = \begin{bmatrix} 1 & 4 & 7 \\ 2 & 5 & 8 \\ 3 & 6 & 9 \end{bmatrix}.$$

- If A is power-positive so is $\mathbf{A}^T$, and hence the dominant left eigenvalue has all positive entries.

Left eigenvectors are important for the long-term dynamics of matrix models. First, they determine eigenvalue sensitivity: the extent to which each matrix entry affects the dominant eigenvalue. Let **v** and **w** be the left and right eigenvectors corresponding to the dominant eigenvalue λ of a power-positive projection matrix **A**. Then

$$\frac{\partial \lambda}{\partial a_{ij}} = \frac{v_i w_j}{\mathbf{v} \cdot \mathbf{w}} \qquad [2.36]$$

where $\mathbf{v} \cdot \mathbf{w}$ is the dot product defined above [equation (2.15)].

Second, the dominant left eigenvector (when it exists) has a biological interpretation as the "reproductive value" of different stages, a concept due to R. A. Fisher. Think of different stages as alternate "investments" in long-term population growth. If you could put one dollar into any one of these investments (~ one individual in any of the stages) what is their relative payoff in the long run (relative size of the resulting population in the distant future)? The answer is that the "payoff" from a stage-j individual is proportional to v_j (see the Appendix of this chapter). In age-structured models it is conventional to scale **v** so that $v_0 = 1$, that is, so that its entries are reproductive values relative to that of a newborn.

Exercise 2.16. Compute the reproductive value vector **v** for the killer whale model [2.35], scaled so that $v_1 = 1$. How does this relate to your conclusions from simulating the population starting from different initial populations?

Exercise 2.17. Construct the projection matrix **A**, and then find $\lambda, \mathbf{v}$, and **w** for an age-structured model with the following survival and fecundity parameters. Age-classes 0–5 are genuine age classes with survival probabilities $[p_0, p_1, \ldots, p_5] = [0.3, 0.4, 0.5, 0.6, 0.6, 0.7]$. Note that $p_j = a_{j+1,j}$, the chance of surviving from age j to age $j+1$, for these ages. Age-class 6 are adults (age 6 or older), with survival 0.9 and fecundity 12.

2.6 Some Applications of Matrix Models

Table 2.2 summarizes the main theoretical results for the case of a power-positive projection matrix **A**. From here on $\lambda, \mathbf{v}, \mathbf{w}$ without subscripts will refer to the dominant eigenvalue (formerly λ_1) and corresponding left and right eigenvectors, whose existence is guaranteed by the Perron-Frobenius theorem. We now present two applications of these results to biological questions that have been addressed using matrix population models.

2.6.1 Why Do We Age?

Evolutionary biologists distinguish between *proximate* and *ultimate* explanations for phenomena. A proximate explanation tells us *how* the phenomenon occurs—

Model	$\mathbf{n}(t+1) = \mathbf{A}\mathbf{n}(t), \mathbf{n}(0) = \mathbf{n}_0$
Solution	$\mathbf{n}(t) = \mathbf{A}^t\mathbf{n}_0$
Eigenmode expansion	$\mathbf{n}(t) = \sum c_i\lambda_i^t\mathbf{w}_i$
Long-term exponential growth	$\sum_i n_i(t) \sim c\lambda^t$ as $t \to \infty$
Stable stage distribution	$\mathbf{n}(t) \sim c\lambda^t\mathbf{w}$
Stage-specific reproductive value	Proportional to $\mathbf{v}$
Eigenvalue sensitivity formula	$\dfrac{\partial\lambda}{\partial a_{ij}} = \dfrac{v_i w_j}{\mathbf{v}\cdot\mathbf{w}}$

Table 2.2 Main properties of a matrix model with power-positive projection matrix $\mathbf{A}$

for example, the physical and biochemical processes involved in meiotic cell division as part of sexual reproduction. An ultimate explanation attempts to say *why* the phenomenon occurs—for example, why some species have evolved to have sexual reproduction while others have not.

Models are important for developing ultimate explanations, because they let us consider the consequences of the alternatives that are not seen in nature. As R. A. Fisher observed, if we want to understand why humans have two sexes rather than three or more, we must "work out the detailed consequences experienced by organisms having three or more sexes" (Fisher 1930). To understand the "why" of traits molded by evolution, we first need to compare things as they are with the other ways things might have been. Only then can we start to hypothesize *why* evolution produced one outcome rather than the other.

For modeling evolution of the life cycle, the growth rate λ can be identified with Darwinian *fitness*: the contribution of offspring to future generations. On the reasonable assumption that matrix entries are determined by the organism's genotype, and if multiple genotypes are present within a population, then population genetic models predict (with some caveats) that the genotype with the largest λ for its matrix becomes fixed in the population (Charlesworth 1994). The main caveat is a standard one in population genetics theory: if the most fit genotype is a heterozygote, then a stable polymorphism is maintained.

An ultimate explanation for sexual reproduction still eludes us: theories abound, and new ones are proposed as quickly as old ones are rejected. But for aging (technically called senescence), there is a widely accepted theory based on the eigenvalue sensitivity [equation [2.36]]. This explanation is derived by modeling a life cycle without aging—the alternative that is not seen in nature—and then asking whether a little bit of aging would lead to increased Darwinian fitness.

Life without aging means that females start reproducing at some age m (for "maturity"), and thereafter have constant fecundity $f_j = f$ and survival $p_j = p < 1$

for all ages $j \geq m$. We have assumed that $p < 1$ to represent an age-independent rate of deaths unrelated to aging.

The eigenvalue sensitivity formula lets us compute the relative eigenvalue sensitivities at different ages for this life cycle without any hard calculations, so long as $\lambda = 1$. Populations cannot grow or decline without limit, so λ must be near 1. The reproductive value of adults ($v_i, i \geq m$) is independent of age because all adults have exactly the same future prospects and therefore make the same long-term contribution to future generations. On the other hand, the stable age distribution w_j goes down with age. With $\lambda = 1$ the number of m year olds is constant, so we can compute $n_{m+k}(t) = n_m(t-k)p^k = n_m(t)p^k$. That is, in order to be age $(m+k)$ now, you must have been m years old k years ago, and you must have survived for the k years between then and now. Therefore

$$w_j \propto p^{j-m} \text{ for } j \geq m.$$

Consequently, the relative sensitivity of λ to changes in either the fecundity $a_{1,j}$ or survival $a_{j+1,j}$ of age-j females, is proportional to p^{j-m}. In both cases, as j changes the relevant w_j is proportional to p^{j-m} while the reproductive value v_j stays the same. This has two consequences:

1. The strength of selection against deleterious mutations acting late in life is weaker than selection against deleterious mutations acting early in life.
2. Mutations that increase survival or fecundity early in life, at the expense of an equal decrease later in life, will be favored by natural selection.

These are known, respectively, as the *mutation accumulation* and *antagonistic pleiotropy* theories of aging. In addition there is a particular form of antagonistic pleiotropy, the *disposable soma* hypothesis, which posits that the connection between early and late vigor is mediated by investment in maintenance and repair mechanisms at the cellular and molecular levels, such as DNA repair and antioxidant systems.

Distinguishing between these theories is difficult because they agree on the fundamental prediction: If the level of unavoidable extrinsic mortality is high then the organism is predicted to be short lived even in a protected environment, while low levels of unavoidable extrinsic mortality should lead to potentially long-lived organisms. Experiments—mainly on *Drosophila*—have uniformly supported this prediction (Kirkwood and Austad 2000). In addition, there is some direct support for each mechanism.

Antagonistic pleiotropy.

There is abundant evidence for antagonistic tradeoffs (e.g., Roff 2001, Chapter 3). The ideal organism would mature instantly, live forever, breed often, and have many offspring each time. In reality we never see this because improvements on one front are paid for on another. For example, early maturation typically entails smaller adult body size and hence lower fecundity. Conversely, experimental selection against early fecundity in *Drosophila* led to increased fecundity later in life. Similar results have been obtained on other insects, birds, and mice (Roff 2001).

Disposable soma.
A unique prediction of this theory, also supported by numerous studies, is that intrinsically long-lived organisms should have higher levels of cellular-level maintenance and repair processes. For example, DNA repair capacity correlates with lifespan in mammals, as does the level of poly(ADP-ribose) polymerase, an enzyme that is important in maintaining DNA integrity. Intrinsic longevity also correlates with levels of defense against thermal extremes and chemical toxins (Kirkwood and Austad 2000).

Mutation accumulation.
A unique prediction of this theory is that genetic variability should increase with age. There is some evidence in *Drosophila* for genetic variance in male mating success and mortality rate, but experiments on other traits have found no evidence of mutation buildup with age.

So it seems likely that all three hypothesis play some role in actual patterns of aging. As usual in biology, it's not *quite* that simple. An essential assumption of the theory is that age-independent mortality is unavoidable. If some mortality risks can be reduced by retaining youthful vigor—for example, predator avoidance—then populations exposed to higher mortality might have reduced senescence in traits that reduce the avoidable mortality (Abrams 1993). This idea has recently been invoked to explain a "mosaic" pattern of senescence observed in guppies, in which some traits exhibit more rapid aging in populations exposed to high predation, while other traits do not (Reznick et al. 2004).

The more general message of this section is that structured population models provide a framework for understanding the life cycles of organisms as adaptations for maximizing fitness subject to tradeoffs and constraints. This topic, called *life history theory*, has been an active research area since the 1960s. Stearns (2000) gives a good short overview, and Roff (2001) is a recent comprehensive text.

2.6.2 Elasticity Analysis and Conservation Biology

The dominant eigenvalue-eigenvector pair summarize what will happen to the population if nothing changes. A value of $\lambda > 1$ implies a growing population, and $\lambda < 1$ means that the population is predicted to decline to extinction.

In the latter case, the practical issue is, what can we do to improve things? One approach to that question was based on using the eigenvalue sensitivity formula to identify matrix entries with the biggest effect on λ. Fairly soon, the objection was raised that survival and fecundity entries are on intrinsically different scales: a survival must lie between 0 and 1 while fecundities can be enormous (balanced by high mortality between birth and maturation). As a result, survival rates often have higher sensitivity than fecundity: changing newborn survival from 0.1 to 0.4 will probably have a large impact, but changing adult fecundity from 1000.1 to 1000.4 will not do much at all. As this example indicates, a better measure is the proportional sensitivity or *elasticity*, defined as

$$
\begin{aligned}
e_{ij} &= \frac{\text{fractional change in } \lambda}{\text{fractional change in } a_{ij}} = \frac{\partial\lambda/\lambda}{\partial a_{ij}/a_{ij}} \\
&= \frac{a_{ij}}{\lambda}\frac{\partial\lambda}{\partial a_{ij}}.
\end{aligned}
\qquad [2.37]
$$

The value of e_{ij} says nothing about which matrix entries actually could be changed, or by how much, but it does identify potential targets of opportunity. So in many applications of matrix population models, the main goal of building the model is to compute the elasticities. Note also that [2.37] applies only to small changes in matrix entries, and effects of large changes have to be computed directly, by modifying the matrix and recomputing λ.

Desert tortoise.

A structured population model for the desert tortoise *Gopherus agassizii* (Doak et al. 1994) illustrates how an imperfect model can still be valuable because its relative predictions are robust in the face of uncertainty about parameter values. The desert tortoise was listed as endangered in 1989 and a draft recovery plan was issued in 1993. Particularly severe declines were occurring in the western Mojave desert. Direct human impacts on the tortoise include

- habitat degradation by off-road vehicles
- habitat loss to urban or agricultural uses
- deliberate hunting (up to 14% of mortality in some areas)
- getting run over by cars or off-road vehicles

There are also indirect impacts, including

- habitat degradation by sheep or cattle grazing
- predation by ravens (which are associated with human presence and attack yearlings and juveniles)
- an upper respiratory tract infection that may have been introduced by release of pet tortoises into the wild

Doak et al. (1994) had two goals. The first was to assess the potential threat to the tortoise posed by the U.S. Army's proposed expansion of Fort Irwin. The second was to compare two management scenarios being considered or implemented: reducing human disturbance and removing ravens. Human disturbance mainly affects larger individuals, while raven predation is limited to smaller ones.

The model was based mainly on government reports and previously unanalyzed mark-recapture data at eight Bureau of Land Management permanent study plots in the western Mojave. Individuals were classified based on size and life stage (Table 2.1). The data included multiple (site × year) combinations for which stage-specific growth or survival rates could be estimated (6–18 combinations for the different stages and rates). However, data on fecundity were limited. For the Mojave there were no direct observations of individual fecundity. Instead, the modelers divided yearling counts by the number of females censused at the time the yearlings would have been born. Because yearlings are much harder to find than adults, this value was regarded as an underestimate and Doak et al. (1994) applied an arbitrary tenfold factor to compensate for undercounting of yearlings. In addition, they considered fecundity estimates based on direct observations of egg production and two estimates of survival to hatching at a different site in the eastern Mojave, where tortoise populations were not in decline. This gave a total of four fecundity estimates for the breeding classes. The overall matrix model is

$$\mathbf{A} = \begin{bmatrix} 0 & 0 & 0 & 0 & 0 & f_6 & f_7 & f_8 \\ 0.716 & 0.567 & 0 & 0 & 0 & 0 & 0 & 0 \\ 0 & 0.149 & 0.567 & 0 & 0 & 0 & 0 & 0 \\ 0 & 0 & 0.149 & 0.604 & 0 & 0 & 0 & 0 \\ 0 & 0 & 0 & 0.235 & 0.560 & 0 & 0 & 0 \\ 0 & 0 & 0 & 0 & 0.225 & 0.678 & 0 & 0 \\ 0 & 0 & 0 & 0 & 0 & 0.249 & 0.851 & 0 \\ 0 & 0 & 0 & 0 & 0 & 0 & 0.016 & 0.860 \end{bmatrix}$$

$$\begin{aligned} (f_6, f_7, f_8) = \; & [0.042, 0.069, 0.069] \\ & \text{or } [0.42, 0.69, 0.69] \\ & \text{or } [1.30, 1.98, 2.57] \\ & \text{or } [2.22, 3.38, 4.38]. \end{aligned} \qquad [2.38]$$

Even this enormous range of possible fecundity estimates is not necessarily catastrophic, because the predictions that you care about may not be affected. Beginning modelers often doubt such claims, but sometimes you get lucky and the parameters that you know the least about turn out to be the least important. In this case, Figure 2.3 shows that the eigenvalue elasticities are consistently highest for survival of larger individuals. The management implication is to forget about ravens, and concentrate on reducing the impacts of humans on larger individuals.

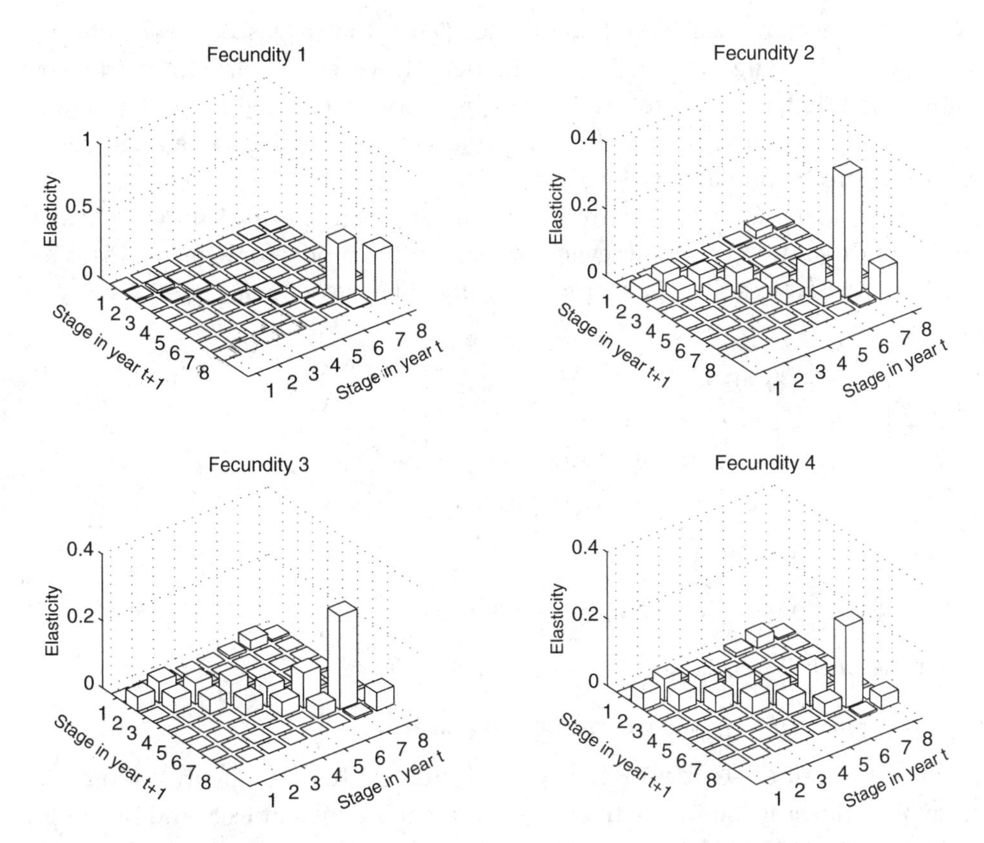

Figure 2.3 Eigenvalue elasticities for the Doak et al. (1994) stage-structured model for desert tortoise. The stages are 1 = yearling, 2, 3 = juveniles, 4, 5 = immature, 6 = subadult, 7 = smaller adult, 8 = larger adult.

The recovery plan for desert tortoise, summarized by Berry (1997), accepted this conclusion and proposed to create fourteen reserves where the tortoise would be protected from detrimental human activities, including cattle grazing. Some of the proposed reserve areas have been established, while others (at this writing) are in dispute because court-ordered grazing restrictions have not been implemented.

Loggerhead sea turtles.

These studies were also intended to evaluate two different management strategies, in this case for loggerhead sea turtles *Caretta caretta* in the southeastern United States (Heppell et al. 1998). Loggerheads are listed as threatened under the U.S. Endangered Species Act. Conservation efforts for marine turtles had focused on reducing egg mortality on human-impacted beaches, but after twenty or thirty years of effort the numbers of nesting turtles were not showing any increases. In addition, incidental trapping and drowning of sea turtles in commercial fishing gear, especially shrimp trawlers, led the National Marine Fisheries Service

(NMFS) to develop a turtle excluder device (TED) that released 97% of trapped turtles while keeping most shrimp in the net. However, the shrimping industry complained that TEDs led to loss of valuable harvest, damage to their gear, and crew injuries, whereas nest protection projects were yielding large increases in hatchling production at very low cost.

Crouse et al. (1987) used very rough estimates of age-specific survival, growth, and fecundity rates to derive a seven-class structured model for loggerheads. This was later revised to a five-class structured model (Crowder et al. 1994).

$$\mathbf{A} = \begin{bmatrix} 0 & 0 & 0 & 4.665 & 61.896 \\ 0.675 & 0.703 & 0 & 0 & 0 \\ 0 & 0.047 & 0.657 & 0 & 0 \\ 0 & 0 & 0.019 & 0.682 & 0 \\ 0 & 0 & 0 & 0.061 & 0.8091 \end{bmatrix}$$

$\sigma_i =$ survival probability

$\gamma_i =$ growth probability, conditional on survival

On diagonal: $P_i = \sigma_i(1 - \gamma_i)$ survive and remain in stage

Subdiagonal: $G_i = \sigma_i \gamma_i$ survive and grow to next stage.

The eigenvalue elasticities for the five-stage model, shown in Figure 2.4, provide an explanation for the ineffectiveness of strategies aimed at eggs and nestlings, and suggest that TEDs will be far more effective. Note that an increase in a stage-specific survival (σ_i) will increase both P_i and G_i values by the same proportional amount. If TEDs were to increase the annual survival of all individuals in stages 2–5, then the proportional sensitivity would be given by the sum of the diagonal and subdiagonal bar heights, which is 0.88. That is, a 10% increase in annual survival applied to all these stages would lead to an $0.88 \times 10\% = 8.8\%$ increase in λ, which would be sufficient to bring the model's λ above 1. In contrast, even if the stage-1 survival were increased to 100%, λ would still be less than 1. So nest protection is helpful but not sufficient to reverse the population decline. The same was found to be true for "head-starting," measures aimed at increasing the survival of hatchlings (Heppell et al. 1996).

Based in large part on the original analysis, the National Academy of Sciences recommended requiring TEDs, and the National Marine Fisheries Service expanded seasonal TED requirements to all southeastern shrimp trawls starting in December 1994. By 1998, loggerhead populations were found to be stable or increasing on most monitored nesting beaches (Heppell et al. 1998). Because this was an "uncontrolled experiment" we cannot have full scientific certainty, but it is highly suggestive that a change in management plans prescribed on the basis of population models allowed the population to quickly rebound.

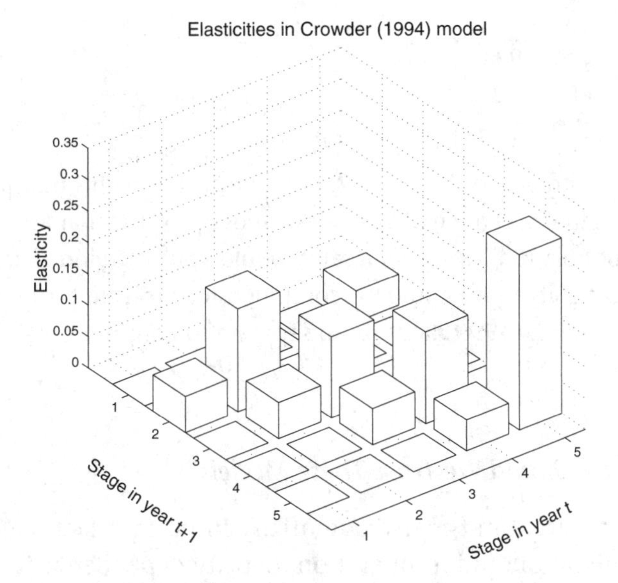

Figure 2.4 Eigenvalue elasticities for the Crowder et al. (1994) stage-structured model for loggerhead sea turtles. Stages [stage durations] in the model are 1 = egg and hatchling [1 yr], 2 = small juvenile [7 yr], 3 = large juvenile [8 yr], 4 = subadult [6 yr], 5 = adult [indefinite].

Biological control and pest management.
Matrix models can also be used to compare alternative options for controlling an undesired species, but for some reason such applications have been rare until very recently. Rockwell et al. (1997) developed a matrix model for the lesser snow goose, which has overgrazed and damaged areas of salt marsh in Canada large enough to be clearly visible in satellite images. As one author of that study puts it, "If you can see it from space, it's a real problem" (Evan Cooch, personal communication). Rockwell et al. (1997) found that the elasticity of adult survival was 87%, which implies that the only means for controlling the geese is to reduce adult survival. As a result of this and other analyses, the U.S. Fish and Wildlife Service relaxed its restrictions on goose hunting in order to increase the harvest of migrating adults. Similarly Shea and Kelly (1998), McEvoy and Coombs (1999), and Parker (2000) describe matrix models for the control of invasive and other undesirable plant populations.

Exercise 2.18. Johnson and Braun (1999) constructed a matrix model for an exploited population of sage grouse; in this and subsequent exercises we consider a simplified version of their model. The mean matrix estimated from 23 years of field survey data is

$$\mathbf{A}_0 = \begin{bmatrix} 0.18 & 0.60 & 0.60 \\ 0.33 & 0 & 0 \\ 0 & 0.73 & 0.73 \end{bmatrix} \qquad [2.39]$$

with the stages being age 0–1, age 1–2, and age > 2. For this matrix, compute the dominant eigenvalue, and the matrix of elasticities e_{ij}. As the manager responsible for survival of this population, which vital rates would you be trying to improve? [Note: the yearling and adult stages could be combined for projecting population growth, but Johnson and Braun (1999) kept them separate to consider possible age-selective harvesting policies.]

2.6.3 How Much Should We Trust These Models?

Simple matrix models omit many potentially important factors for population persistence: emigration and immigration, density dependence in fecundity and survival, environmental variability, and effects of finite population size (demographic stochasticity). The use of these models to guide conservation policy, despite their limitations, brings to the fore the issue of their reliability.

Our examples focused on predicting the relative effectiveness of possible actions, but models are also asked to make absolute predictions. For example, Kareiva et al. (2000) used a matrix model to predict whether dam removals (a proposed but highly controversial action) would be sufficient to reverse the precipitous decline in salmon stocks in the Columbia River basin. Quantitative predictions of absolute extinction risk are also among the listing criteria for the IUCN Red List of Threatened Species (IUCN 2001; also at http://www.redlist.org), which plays a major role in guiding conservation efforts worldwide. Appreciable risk of extinction is also the main criterion for listing under the U.S. Endangered Species Act (USFWS 1988). Consequently, one of the main uses for models in conservation is to make quantitative predictions about extinction risk (Morris and Doak 2002).

We still have little evidence as to the accuracy of relative predictions, because we usually only know what happened under the one policy that was actually implemented. Comparisons of absolute predictions with actual outcomes are also rare. The most comprehensive study is by Brook et al. (2000); they compared observed and predicted population growth rates and risk of extinction (more precisely "quasiextinction," meaning decline below some threshold density at which extinction is considered to be inevitable). They compiled population studies for twenty-one animal species, of sufficient duration that they could use the first half of each populations' data to parameterize a simple structured population model. They then used the model to predict population changes over the second half of the time period covered by the data, and compared these with the actual data.

Brook et al. (2000) found remarkably close agreement between the observed and predicted total number of extinctions, and no tendency to systematically under- or overestimate the final population size. Holmes and Fagan (2002) performed a similar test, with similarly good results, for U.S. Pacific Northwest salmon stocks.

These tests indicate that good modelers, with access to a few decades of good data, can make reliable predictions about groups of related species or populations, such as the total number of bird or amphibian species expected to go extinct in the next fifty years. Unfortunately the tests do not tell us about the reliability of predictions for each individual species (Ellner et al. 2002). Theoretical analysis (Fieberg and Ellner 2000) and simulation studies (Ludwig 1999; Ellner et al. 2002) indicate that species-by-species predictions of long-term extinction risk will not be very accurate, given the amounts of data generally available.

On the other hand, the Doak et al. (1994) analysis illustrates that the important predictions may still be robust, especially comparisons of options within a single population. A recent simulation study suggests that relative predictions will often be reliable even when absolute predictions are not (McCarthy et al. 2003). The only way to find out in any particular case is to build the model, and then quantify how much uncertainty in the relevant predictions is produced by your uncertainty about parameter values and other aspects of the model. Some general computational methods for quantifying prediction uncertainty are reviewed by Ellner and Fieberg (2003), and illustrated on models for salmon stocks in the Northwest United States. Decisions have to be made, and an imperfect model based on limited data is better than none at all, so long as you examine model predictions across the range of plausible parameter values and model assumptions (Morris and Doak 2002; Reed et al. 2002).

2.7 Generalizing the Matrix Model

Before moving on to other models we briefly mention three important ways in which matrix models can be made more realistic. The first two involve dropping the assumption that matrix entries are constant.

2.7.1 Stochastic Matrix Models

In natural animal and plant populations, fecundity has been observed to vary enormously between years (by factors of up to 333 in plants, 38 in terrestrial vertebrates, and 2200 in birds; Hairston et al. 1996). When sufficient data are available, random variability in transition rates can be incorporated into a matrix model. There are two main ways of doing so.

1. A nonparametric "bootstrap" approach can be used if a population has been studied for a series of years, resulting in a series of different estimates of the

matrix, $\mathbf{A}_1, \mathbf{A}_2, \ldots, \mathbf{A}_k$. The model is then

$$\mathbf{n}(t+1) = \mathbf{A}(t)\mathbf{n}(t)$$

where $\mathbf{A}(t)$ is drawn at random from $\mathbf{A}_1, \mathbf{A}_2, \ldots, \mathbf{A}_k$. Some of the variability among the estimated $\mathbf{A}$'s will typically be sampling error rather than real variation in vital rates, so this approach will tend to overestimate variability.

2. The "parametric" approach is to fit statistical distributions to the observed patterns of variation and covariation in matrix entries, and simulate the model by drawing an $\mathbf{A}$ matrix for each year from the fitted distributions. Morris and Doak (2002) review methods for estimating stochastic matrix models from empirical data.

The theory of stochastic matrix models requires mathematics beyond that presented in this chapter. We refer interested readers to Caswell (2001) for a readable and practically oriented summary, and to Tuljapurkar (1990) for a comprehensive review of the theory. One very important result is that there is typically still a long-term growth rate, analogous to the dominant eigenvector λ. Knowing that this rate exists, it is then possible to compute its numerical value by simulation, and also elasticities and the like.

Exercise 2.19. This exercise continues our study of the (simplified) Johnson-Braun (1999) sage grouse model. Johnson and Braun also estimated the variability over time in matrix entries; the matrix S below gives the standard deviations:

$$S = \begin{bmatrix} 0.04 & 0.13 & 0.13 \\ 0.09 & 0 & 0 \\ 0 & 0.15 & 0.15 \end{bmatrix}.$$

Write a script to simulate the sage grouse model with random variability in the vital rates, using the standard deviations above and assuming a normal distribution for each entry, and initial population vector $\mathbf{n}(0) = (430, 140, 430)$. Since each simulation run will have a different outcome, have your program do 1000 simulations, record (in a vector) the minimum population size (total number of individuals in all size classes) over the course of 100 years in each run, and plot a histogram of the minimum population sizes in the 1000 runs.

Exercise 2.20. When you computed elasticities for the deterministic version of the sage grouse model, you should have found that the highest elasticity is for adult (age >2) survival. Therefore, a 20% increase in the mean adult survival has a larger beneficial impact than a 20% increase in mean adult fecundity, based on the mean matrix $\mathbf{A}_0$. Use simulations to check if this is also true in the stochastic model of the previous exercise, by comparing (in some informative way) the results from 1000 simulations each of the two scenarios (1, 20% higher mean adult fecundity; 2, 20% higher mean adult survival).

2.7.2 Density-Dependent Matrix Models

As a second extension, we can allow density-dependent limits to population growth. Here the changes in vital rates are assumed to occur due to endogenous feedbacks, such as decreased survival when there is stiff competition for resources, or difficulty in finding mates when the population is too sparse. In contrast to random variation, these feedbacks can give rise to completely new kinds of dynamic behavior. We will soon explore these behaviors in a different context—differential equation models—so for the moment we give only one illustrative example. Through a series of elegant experiments a simple three-stage model has been shown to give remarkably accurate predictions for laboratory populations of flour beetles *Tribolium castaneum* (Cushing et al. 2002; Dennis et al. 1997, 2001; Henson et al. 2001). The model is

$$\begin{aligned} L(t+1) &= bA(t)\exp(-c_{el}L(t) - c_{ea}A(t)) \\ P(t+1) &= (1-\mu_l)L(t) \\ A(t+1) &= P(t)\exp(-c_{pa}A(t)) + (1-\mu_a)A(t). \end{aligned} \tag{2.40}$$

Here L, P, and A are the numbers in the larval, pupal, and adult stages of the beetle life cycle. A time-step of the model corresponds to two weeks of real time, which is the approximate duration of the larval and pupal stages under laboratory conditions. Population growth is limited by cannibalism. Foraging adults consume eggs and pupae (which are immobile), and foraging larvae consume eggs. The model parameters are the birth rate b in the absence of cannibalism, intrinsic larval mortality μ_l and adult mortality μ_a, and cannibalism rate parameters c_{el}, c_{al}, c_{pa}.

Figure 2.5 shows some of the possible dynamical behaviors. We plot the number of larvae, which shows the patterns most clearly; the total population size has the same qualitative dynamics in each case. In panel (A), the population quickly reaches a steady state. Matrix entries are then constant (because L, P, and A are all constant), and such that the dominant eigenvalue is $\lambda = 1$ exactly: there is neither growth nor decline. The population self-regulates to a state where each individual (on average) replaces itself exactly. In panel (B) the population oscillates: it grows to such a high density that almost all new eggs are cannibalized, and consequently then drops to a lower density at which higher egg survival allows the population to rebound. Panels (C) and (D) show more complicated patterns of overgrowth, crash, and recovery, with periodic oscillations in (C) and aperiodic *chaotic* oscillations in (D). Note that there is nothing random in this model: the erratic behavior in (D) is entirely due to the nonlinear feedback of population density on population growth rate mediated by cannibalism.

The LPA model has been extensively validated by experiments in which parameter values of the laboratory population are manipulated (e.g., removing or

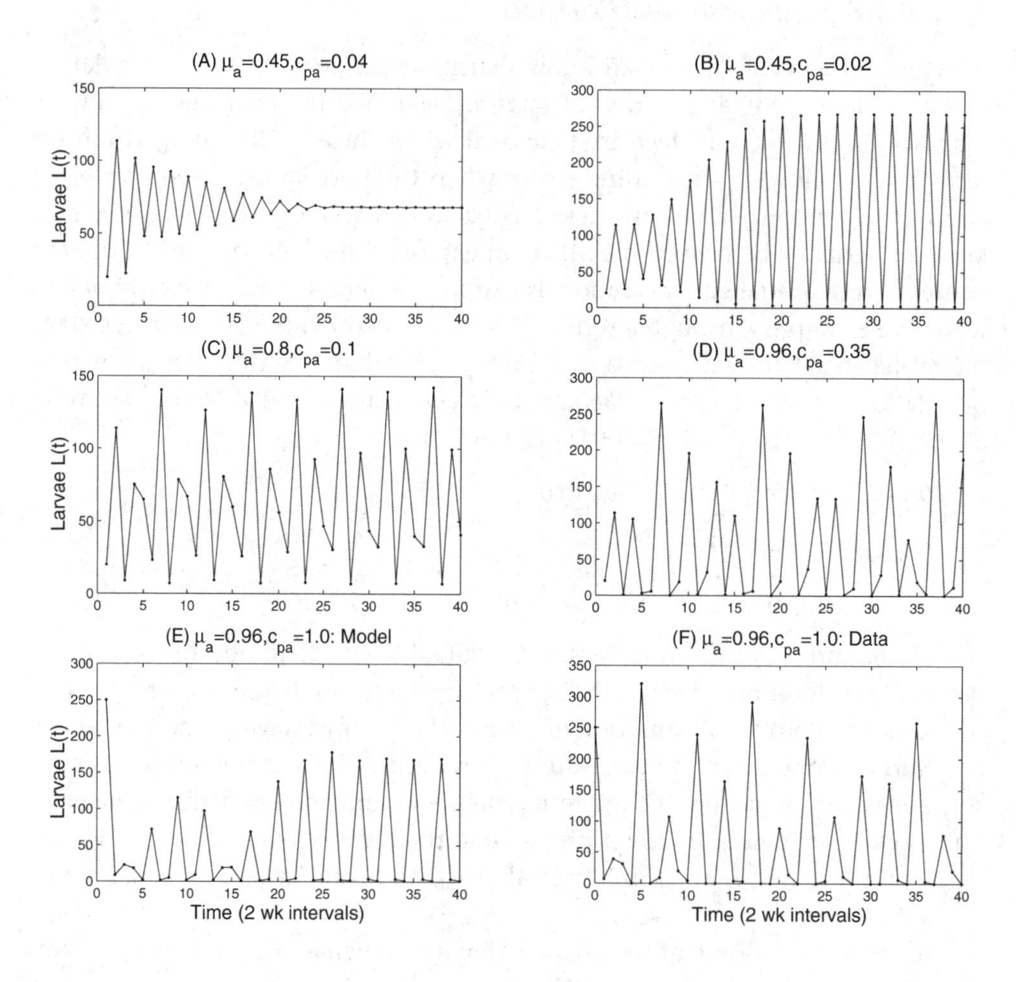

Figure 2.5 The LPA model. Panels (A)–(D) show simulations started with 20 individuals in each life stage and running for 40 two-week time steps of the model. Panels (E) and (F) compare model simulations with experimental results for high values of μ_a and c_{pa}; model simulations used the same initial conditions as the experiments, $[L(0), P(0), A(0)] = [250, 5, 100]$.

adding adults to alter the effective values of μ_a and c_{pa}), and experimental data are compared with model predictions. Panels (E) and (F) show one example out of many (Cushing et al. 2002). Setting high values of μ_a and so c_{pa} and using other parameter values estimated from experimental data, cycles of period 3 with an *on/off/off* sequence of egg laying are predicted in model simulations (panel E) and observed in experiments (panel F). The *on*'s are variable in the data, rather than constant as predicted by the model. To capture this variability we would need a more complex model that simulated variability in egg-laying rates, and the actual process of cannibalism upon random encounters of beetles burrowing through flour. But clearly the simple model [2.40] has captured the essentials of the experimental dynamics. In other species, however, a stage-structured matrix model was inadequate because there was too much within-stage variation among individuals (Benton et al. 2004), so more complex models (such as those described in the next section) would be needed.

The exercises below are an introduction to how nonlinearity can affect population dynamics. First, consider a population model for discrete, nonoverlapping generations defined by

$$x(t+1) = bx(t)\exp(-cx(t)) = f(x(t)), \quad b, c, > 0. \qquad [2.41]$$

Models of this kind have been used for insects like gypsy moths that have one generation each year, with adults that do not survive from one year to the next, and eggs that overwinter. The parameter b is the "intrinsic" birth rate that holds when the population size is small, and the term $\exp(-cx(t))$ represents density-dependent decreases in the birth rate. A *fixed point* x for the model is a value of x for which $f(x) = x$. If $x(t)$ is a fixed point, then the population size remains constant: $x(t+1) = x(t)$.

Exercise 2.21. (a) For which values of the parameters b and c does the model have a positive fixed point?

(b) Can the model have more than one positive fixed point? [There is a fixed point for this model where $x = bx\exp(-cx)$; this occurs if $x = 0$ or if ...].

A fixed point x_e is *stable* if nearby values of the population evolve to the fixed point: if $x = x(0)$ is close to x_e, then the sequence $x(t)$ has x_e as a limit. When a fixed point is unstable, the behavior of the population model is more complicated.

Exercise 2.22. Write a script file to simulate model [2.41]. For parameter values $c = 0.01$ and $b = 4, 8, 12, 16, 20, 24$, and initial population size $x(1) = 1$, have the script compute and graph $x(t)$ for $1 \leq t \leq 100$. Have all six graphs appear in a single window, and put the value of b in the title of each graph.

Exercise 2.23. Write a script file to simulate the LPA model and replicate the results in Figure 2.5.

2.7.3 Continuous Size Distributions

In many applications of matrix models individuals are categorized based on a continuously varying attribute such as body size, rather than by discrete life stages. These modeler-defined "stages" are an artifice imposed to allow the convenience of using a matrix model. Our final case study in this chapter is a cautionary tale about the limits of this approach and a possible solution.

Northern Monkshood *Aconitum noveboracense* is an herbaceous perennial plant listed as threatened under the U.S. Endangered Species Act (Dixon and Cook

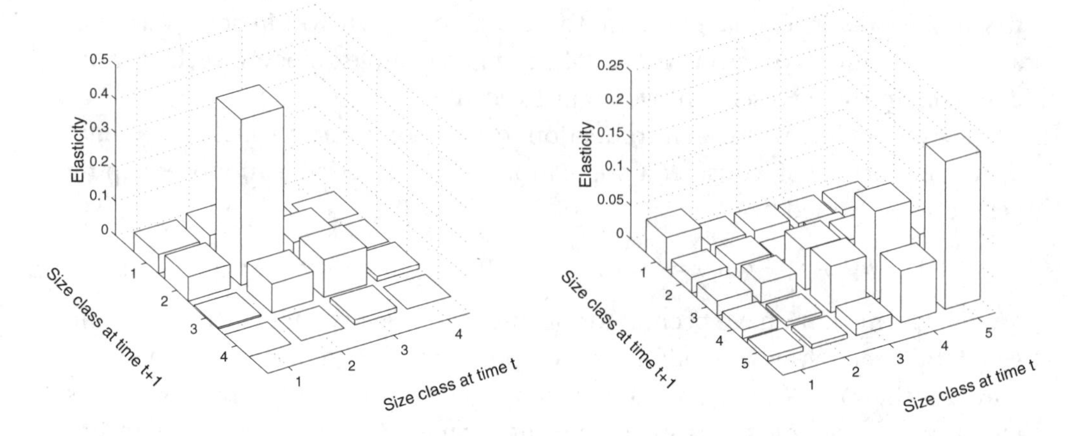

Figure 2.6 Left panel: Eigenvalue elasticities for a matrix model of northern monkshood using the size-based classification of Dixon and Cook: seedling, juvenile, small (< 2 mm), medium (2–4 mm), and large (>4 mm). The seedling stage is omitted because the subset of populations used to parameterize the model did not have any reproduction by seed during the study period. Right panel: Eigenvalue elasticities when class boundaries are set based on stem diameter, so that each class contained the same number of censused individuals.

1990). Transition rates were estimated by repeated census of marked individuals in a series of populations in the Catskill mountains, with stem diameter and number of leaves recorded as measures of size. Dixon and Cook (1990) and Dixon et al. (1997) analyzed the data using matrix models with a small number of size classes. Figure 2.6 shows elasticity analysis for two matrix models differing only in their choice of class boundaries. The left panel shows the elasticities using the "stages" selected by Dixon and Cook (1990), with parameters estimated from three years of data at one of their sites (Easterling et al. 2000). The right panel shows elasticities based on exactly the same data, but with boundaries set so that there were an equal number of observations for each stage class.

The elasticities and their implications for managing the population appear to be very different. This discrepancy occurs because the size categories used to build the population model are also used for sensitivity analysis. These objectives may conflict. To predict future population trends, size categories should contain individuals who are similar in survival and fecundity under current conditions. To predict the effect of management actions, size categories should contain individuals who are similar in their response to possible actions. In principle we could achieve both goals by using a large number of small categories containing very similar individuals. But then we would have very little data on each category and therefore poor parameter estimates—another example of the tradeoff between model error and parameter error.

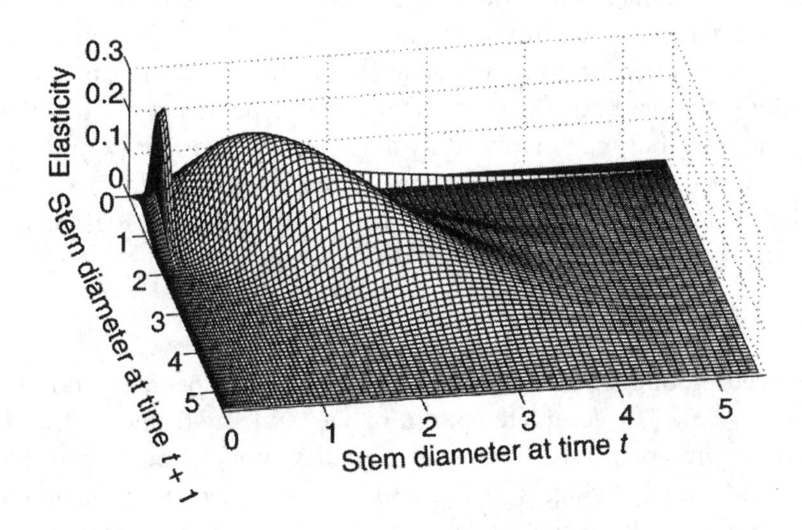

Figure 2.7 Elasticity surface for the integral projection model for northern monkshood.

To avoid these problems, Easterling et al. (2000) proposed that the matrix model should be replaced by an integral projection model (IPM) in which size is a continuous variable, and the population state is described by a continuous size distribution $n(y,t)$ such that $\int_a^b n(y,t)dy$ is the number of individuals whose size is between a and b at time t. Instead of a projection matrix $\mathbf{A}$ the IPM has a "projection kernel" function K, defined so that

$$n(y,t+1) = \int_s^S K(y,x)n(x,t)dx \qquad [2.42]$$

where s and S are the minimum and maximum possible sizes of individuals. This integral model does the same thing as the matrix model [equation [2.20]], computing $n(y,t+1)$ as the total contribution of size-y individuals "now" from individuals of any size x "last year," through either fecundity or individual survival and growth.

Easterling et al. (2000) describe how a projection kernel can be estimated from the same data on monkshood that were used to estimate a projection matrix. Eigenvalue sensitivity and elasticity can then be calculated using a formula similar to [2.36] (Easterling et al. 2000). For monkshood (Figure 2.7) the elasticity surface has regions of high elasticity that could be the focus of management actions. The narrow, high ridge occurs at the typical size of newborn plants and thus corresponds to survival from age 1 to age 2. The broader diagonal mound corresponds to survival of mid-size individuals, with stem diameter up to 2 or 2.5 mm.

Mark Rees and collaborators (Rees and Rose 2002; Rees et al. 2004; Childs et al. 2003, 2004) have recently extended the integral model to include size- and age-dependent demography, as well as stochastic variation. A benefit of the integral model in these cases is that age and size dependence are described by smooth functions that may involve only a few parameters. For example, in the Childs et al. (2004) model for the monocarpic thistle *Carlina vulgaris*, data on the probability of flowering as a function of age a and size x (the log-transformed length of the longest leaf) were fitted by the function

$$p_f(a, x) = \frac{\exp(\beta_0 + \beta_a a + \beta_x x)}{1 + \exp(\beta_0 + \beta_a a + \beta_x x)}$$

while the seed production of flowering plants was size dependent and fitted by $s(x) = \exp(A + Bx)$. The complete pattern of age- and size-dependent fecundity is specified by five parameters, whereas a matrix model would require cross-classifying individuals by age and size and a separate fecundity parameter for each age-size class. Using the model to predict how natural selection acts on the value of the flowering function parameters, it was possible to show that the observed strategy, where flowering depends on both age and size, is an adaptation to random variation across years in mortality and growth—models ignoring random variation gave incorrect predictions of the distribution of plant size at flowering, whereas a model incorporating observed levels of variation gave very accurate predictions (Childs et al. 2004).

2.8 Summary and Conclusions

We have covered a lot of ground in this chapter, for two reasons. First, to understand matrix models for structured populations we had to review some matrix algebra. The payoff for this mathematical investment has only begun. In the next chapter we will see that the eigenvalues and eigenvectors that characterize the long-term behavior of a matrix population model also summarize important predictions of Markov chain models for ion channels. In later chapters they will be essential for understanding differential equation models of gene regulation and infectious disease dynamics—the universality of mathematics can sometimes be astonishing. Second, we have tried to illustrate how these general properties are useful in both basic and applied settings, and how simple matrix models are being used as the starting point for more general models of structured populations. Structured populations are everywhere, even inside you. The neurons in your brain, the T-cells in your immune system, the mercury atoms that you've absorbed from tuna fish sandwiches—what characteristics distinguish one T-cell (or neuron or $\cdots$) from another? What processes cause those characteristics to change? How much of the resulting dynamics of T-cell diversification (or neuron aging or $\cdots$) could you summarize in a matrix model? How rapidly does mercury move among different tissues in your body, and how much of it is lost each year—it's not *exactly* like asking how many sea turtles will still be alive and how big they will be, but it's not entirely different. Once you get started, it becomes natural to think of just about *everything* as a structured population of its constituent components, and then to start wondering what you can learn by modeling from that perspective.

Exercise 2.24. Give an example of a structured population that you have seen in the last month, ideally one very different from the ones presented in this chapter. Would a simple matrix model be appropriate? Why or why not?

2.9 Appendix

Here we fill in some mathematical details for readers who have had a course in linear algebra.

2.9.1 Existence and Number of Eigenvalues

λ is an eigenvalue of $\mathbf{A}$ if and only if it satisfies the *characteristic equation* $\det(\mathbf{A} - \lambda\mathbf{I}) = 0$, since this is equivalent to $(\mathbf{A} - \lambda\mathbf{I})\mathbf{w} = 0$ having a nonzero solution $\mathbf{w}$. Here $\mathbf{I}$ is the *identity matrix* having 1's on its diagonal (running from top left to bottom right) and all other elements zero. It is so named because $\mathbf{AI} = \mathbf{A}$ and $\mathbf{IB} = \mathbf{B}$ for any matrices $\mathbf{A}, \mathbf{B}$. The characteristic equation is a polynomial of degree n, so it has at least 1 and at most n distinct solutions in the set of complex numbers.

2.9.2 Reproductive Value

Here we show that reproductive values v_i from the left eigenvector of the projection matrix give the long-term relative sizes of populations descended from a single founding individual in stage i. First, observe that $\mathbf{v}$ defines an exponentially growing weighted sum of the population vector $\mathbf{n}(t)$. That is,

$$V(t+1) \equiv \mathbf{v} \cdot \mathbf{n}(t+1) = \mathbf{v} \cdot \mathbf{A}\mathbf{n}(t) = \mathbf{v} \cdot \lambda\mathbf{n}(t) = \lambda\mathbf{v} \cdot \mathbf{n}(t) = \lambda V(t).$$

So if $V_j(t)$ is the value of $V(t)$ when the population is started from a single stage-j individual at time 0, we have

$$V_j(t) = \lambda^t v_j.$$

But we also have

$$V_j(t) = \mathbf{v} \cdot \mathbf{n}(t) \sim \mathbf{v} \cdot C_j\lambda^t\mathbf{w} = C_j\lambda^t\mathbf{v} \cdot \mathbf{w}$$

where the constant C_j depends on the stage j of the founding individual. Equating the two expressions above for $V_j(t)$ we see that C_j is proportional to v_j, as claimed.

Another way of expressing this result is that for large t,

$$\mathbf{A}^t \sim C\lambda^t(\mathbf{w} * \mathbf{v}) \qquad [2.43]$$

regarding $\mathbf{w}$ as a column vector (size $n \times 1$) and $\mathbf{v}$ as a row vector (size $1 \times n$)) so that $\mathbf{w} * \mathbf{v}$ is a size-n square matrix. From the definition of matrix multiplication we see that $\mathbf{w} * \mathbf{v}$ is the matrix whose jth column is $\mathbf{w} \times v_j$. If we start with a single single stage-j individual, the population at time t is the jth column of $\mathbf{A}^t$. Since the population structure converges to $\mathbf{w}$ for any initial population, and the total number of individuals is proportional to $v_j\lambda^t$, the conclusion is that A^t is given by (2.43).

2.10 References

Abrams, P. A. 1993. Does increased mortality favor the evolution of more rapid senescence? Evolution 47: 877–887.

Anderson, R. M. and R. M. May. 1992. *Infectious Disease of Humans: Dynamics and Control.* Oxford University Press, Oxford.

Benton, T. G., T. C. Cameron, and A. Grant. 2004. Population responses to perturbations: Predictions and responses from laboratory mite populations. Journal of Animal Ecology 73: 983–995.

Berry, K. H. 1997. The desert tortoise recovery plan: An ambitious effort to conserve biodiversity in the Mojave and Colorado Deserts of the United States. Pages 430–440 in *Proceedings: Conservation, Restoration, and Management of Tortoises and Turtles: An International Conference.* New York Turtle and Tortoise Society.

Brault, S., and H. Caswell. 1993. Pod-specific demography of killer whales (*Orcinus orca*). Ecology 74: 1444–1454.

Brook, B. W., J. J. O'Grady, A. P. Chapman, M. A. Burgman, R. Akcakaya, and R. Frankham. 2000. Predictive accuracy of population viability analysis in conservation biology. Nature 404: 385–387.

Caswell, H. 2001. *Matrix Population Models: Construction, Analysis, and Interpretation*, 2nd edition. Sinauer Associates, Sunderland, MA.

Charlesworth, B. C. 1994. *Evolution in Age-Structured Populations*, 2nd edition. Cambridge University Press, Cambridge U.K. (1st edition 1980).

Childs, D. Z., M. Rees, K. E. Rose, P. J. Grubb, and S. P. Ellner. 2003. Evolution of complex flowering strategies: An age- and size-structured integral projection model. Proceedings of the Royal Society B 270: 1829–1839.

Childs, D. Z., M. Rees, K. E. Rose, P. J. Grubb, and S. P. Ellner. 2004. Evolution of size dependent flowering in a variable environment: Construction and analysis of a stochastic integral projection model. Proceedings of the Royal Society B 271: 425–434.

Crouse, D. T., L. B. Crowder, and H. Caswell. 1987. A stage-based population model for loggerhead sea turtles and implications for conservation. Ecology 68: 1412–1423.

Crowder, L. B., D. T. Crouse, S. S. Heppell, and T. H. Martin. 1994. Predicting the impact of turtle excluder devices on loggerhead sea turtle populations. Ecological Applications 4: 437–445.

Cushing, J. M., R. F. Costantino, B. Dennis, R. A. Desharnais, and S. M. Henson. 2002. *Chaos in Ecology: Experimental Nonlinear Dynamics.* Academic Press, San Diego.

Dennis, B., R. A. Desharnais, J. M. Cushing, and R. F. Costantino. 1997. Transitions in population dynamics: equilibria to periodic cycles to aperiodic cycles. Journal of Animal Ecology 66: 704–729.

Dennis, B., R. A. Desharnais, J. M. Cushing, S. M. Henson, and R. F. Costantino. 2001. Estimating chaos and complex dynamics in an insect population. Ecological Monographs 71: 277–303.

Dixon, P. M., and R. E. Cook. 1990. Life History and Demography of Northern Monkshood (*Aconitum noveboracense*) in New York State. Report, Cornell Plantations, Ithaca, NY.

Dixon, P. M., N. Friday, P. Ang, S. Heppell, and M. Kshatriya. 1997. Sensitivity analysis of structured-population models for management and conservation. Pages 471–513 in S. Tuljapurkar and H. Caswell (eds.), *Structured Population Models in Marine, Terrestrial and Freshwater Systems.* Chapman and Hall, London.

Doak, D., P. Kareiva, and B. Klepetka. 1994. Modeling population viability for the desert tortoise in the Western Mojave desert. Ecological Applications 4, 446–460.

Easterling, M. R., S. P. Ellner, and P. Dixon. 2000. Size-specific sensitivity: Applying a new structured population model. Ecology 81: 694–708.

Ellner, S. P., J. Fieberg, D. Ludwig, and C. Wilcox. 2002. Precision of population viability analysis. Conservation Biology 16: 258–261.

Ellner, S. P., and J. Fieberg. 2003. Using PVA for management despite uncertainty: Effects of habitat, hatcheries, and harvest on salmon. Ecology 84: 1359–1369.

Fieberg, J., and S. P. Ellner. 2000. When is it meaningful to estimate an extinction probability? Ecology 81: 2040–2047.

Fisher, R. A. 1930. *The Genetical Theory of Natural Selection.* Oxford University Press, Oxford. Second ed. Dover, New York, 1958.

Hairston, N. G., S. Ellner, and C. M. Kearns. 1996. Overlapping generations: the storage effect and the maintenance of biotic diversity. Pages 109–145 in O. E. Rhodes, R. K. Chesser, and M. H. Smith (eds), *Population Dynamics in Ecological Space and Time.* University of Chicago Press, Chicago.

Henson, S. M., R. F. Costantino, J. M. Cushing, R. A. Desharnais, B. Dennis, and A. A. King, 2001. Lattice effects observed in chaotic dynamics of experimental populations. Science 294: 602–605.

Heppell, S. S., L. B. Crowder, and D. T. Crouse. 1996. Model to evaluate headstarting as a management tool for long-lived turtles. Ecological Applications 6: 556–565.

Heppell, S. S., D. R. Crouse, and L. B. Crowder. 1998. Using matrix models to focus research and management efforts in conservation. Pages 148–168 in S. Ferson and M. Burgman (eds.), *Quantitative Methods for Conservation Biology.* Springer, New York.

Holmes, E. E., and W. F. Fagan. 2002. Validating population viability analysis for corrupted data sets. Ecology 83: 2379–2386.

Horn, R. A. and C. R. Johnson. 1985. *Matrix Analysis.* Cambridge University Press, Cambridge U.K.

IUCN 2001. IUCN Red List Categories and Criteria: Version 3.1. Report, IUCN Species Survival Commission. IUCN, Gland, Switzerland and Cambridge, U.K. Online at http://www.iucn.org/themes/ssc/realists/RLcats2001booklet.html

Johnson, K. H., and C. E. Braun. 1999. Viability and conservation of an exploited sage grouse population. Conservation Biology 13: 77–84.

Kareiva, P., M. Marvier, and M. McClure. 2000. Recovery and management options for spring/summer Chinook salmon in the Columbia River basin. Science 290: 977–979.

Kirkwood, T.B.L., and S. N. Austad. 2000. Why do we age? Nature 408: 233–238.

Lande, R. 1988. Demographic models of the Northern Spotted Owl (*strix occidentalis caurina*). Oecologia 7S: 601–607.

Ludwig, D. 1999. Is it meaningful to estimate a probability of extinction? Ecology 80: 298–310.

McCarthy, M. A, S. J. Andelman, and H. P. Possingham. 2003. Reliability of relative predictions in population viability analysis. Conservation Biology 17: 982–989.

McEvoy, P. B. and E. M. Coombs. 1999. Biological control of plant invaders: Regional patterns, field experiments, and structured population models. Ecological Applications 9: 387–401.

Morris, W. F. and D. F. Doak. 2002. *Quantitative Conservation Biology: Theory and Practice of Population Viability Analysis.* Sinauer, Sunderland, MA.

Parker, I. M. 2000. Invasion dynamics of *Cytisus scoparius*: A matrix model approach. Ecological Applications 10: 726–743.

Reed, J. M., L. S. Mills, J. B. Dunning, Jr., E. S. Menges, K. S. McKelvey, R. Frye, S. R. Beissinger, M.-C. Anstett, and P. Miller. 2002. Emerging issues in population viability analysis. Conservation Biology 16: 7–19.

Rees, M., and K. E. Rose. 2002. Evolution of flowering strategies in *Oenothera glazioviana*: An integral projection model approach. Proceedings of the Royal Society B 269: 1509–1515.

Rees, M., D. Z. Childs, K. E. Rose, and P. J. Grubb. 2004. Evolution of size dependent flowering in a variable environment: Partitioning the effects of fluctuating selection. Proceedings of the Royal Society B 271: 471–475.

Reznick, D. N. M. J. Bryant, D. Roff, C. K. Ghalambor, and D. E. Ghalambor. 2004. Effect of extrinsic mortality on the evolution of senescence in guppies. Nature 431: 1095–1099.

Rockwell, R. F., E. G. Cooch, and S. Brault. 1997. Dynamics of the mid-continent population of lesser snow geese: Projected impacts of reductions in survival and fertility on population growth rates. Pages 73–100 in B. Batt (ed.), Arctic Goose Habitat Working Group Report. U.S. Fish and Wildlife Service, Washington, DC, and Canadian Wildlife Service, Ottowa, ON.

Roff, D. A. 2001. *Life History Evolution*. Sinauer Associates, Sunderland, MA.

Shea, K., and D. Kelly. 1998. Estimating biocontrol agent impact with matrix models: *Carduus nutans* in New Zealand. Ecological Applications 8: 824–832.

Stearns, S. C. 2000. Life history theory: Successes, limitations, and prospects. Naturwissenschaften 87: 476–486.

Tuljapurkar, S. 1990. *Population Dynamics in Variable Environments*. Springer-Verlag, New York.

USFWS. 1988. Endangered Species Act of 1973 as amended through the 100th Congress. U.S. Department of Interior, Fish and Wildlife Service, Washington, DC.

Valverde, T. and J. Silvertown. 1998. Variation in the demography of a woodland understorey herb (*Primula vulgaris*) along the forest regeneration cycle: Projection matrix analysis. Journal of Ecology 86: 545–562.

Williams, B. K., J. D. Nichols, and M. J. Conroy. 2002. *Analysis and Management of Animal Populations*. Academic Press, New York.

Wood, S. N. 1997. Inverse problems and structured population dynamics. Pages 555–586 in S. Tuljapurkar and H. Caswell (eds.), *Structured Population Models in Marine, Terrestrial and Freshwater Systems*. Chapman and Hall, New York.

3 Membrane Channels and Action Potentials

This chapter continues our study of matrix models, but applies them to study molecular structures called membrane channels. The counterparts of different age classes of a population are different conformations of a channel. The transitions between age-classes for an individual and those between conformations for a channel are very different. Individuals never become younger, returning to a former age-class, while channels may revisit each conformation many times. Nonetheless, the theory of power-positive matrices is just the right tool for predicting the behavior of a large collection of channels, based on the behavior of individual channels.

Membrane channels are pores in the cell membrane that selectively allow ions to flow through the membrane. These ionic flows constitute electrical currents that are biologically important. This is especially apparent in electrical excitable tissue like skeletal muscles, the heart, and the nervous system. All of these tissues have *action potentials*, impulses of electrical current that affect ionic flows through membrane channels. Action potentials are a primary means of signaling within the nervous system. They are triggered by inputs to sensory neurons, communicate information among neurons throughout the nervous systems, and transmit motor commands to the musculoskeletal system. Action potentials coordinate the heartbeat, and malfunction of cardiac electrical activity is quickly fatal if proper function is not restored.

Dynamic models built upon extensive knowledge about the nervous system and other organs help us to understand action potentials and the electrical excitability of these complex systems. This chapter provides a brief overview of the electrical excitability of membranes at molecular and cellular levels. We use the mathematics of matrix algebra introduced in Chapter 2 to analyze models of how individual membrane channels function. There is also new mathematics here, namely, aspects of probability. The specific times that channels change their conformational states are unpredictable, so we introduce probabilistic models for these switching times. As in tossing a large number of coins, we can predict the average behavior of a large population of channels even though we cannot predict the behavior of individual channels. The measurements that can be made give us only partial information about the state of individual channels. Models and their analysis provide a means for "filling the gaps" and deducing information about things that cannot be observed directly.

We consider models at three different levels: a single ion channel, a population of channels of a single type, and multiple populations of several different channel types in a cell, whose interactions lead to the generation of action potentials. In each case, we review the biology underlying the models. The next level up—the patterns that can arise in a population of interconnected cells—is touched on in Chapter 7.

3.1 Membrane Currents

The membranes surrounding cells are for the most part impermeable, but they incorporate a variety of molecular complexes designed to transfer material across the membrane. The membrane allows cells to maintain different concentrations of ions inside the cells than outside. The imbalance of electric charge across the cell membrane creates an electrical potential difference. There are molecular complexes that use inputs of energy to maintain the concentration and electrical potential differences across the membrane. Membrane *channels* are different structures through which ions cross the membrane. Channels are pores that can open to allow ionic flows driven by diffusive and electrical forces (see Figure 3.1). You can imagine a channel as an electrical valve that allows ions to flow through when it is open and blocks ionic flows when it is closed. The electrical excitability of membranes relies on *gating*, the opening and closing of different types of channels. Channels have remarkable properties, notably:

- Many channels are selective, allowing significant flows only of particular ionic species. Channels that are specifically selective for sodium (Na^+), potassium (K^+), and calcium (Ca^{2+}) ions are ubiquitous in the nervous system and important in the generation of action potentials.
- Channels have only a few distinct conformational states. Unlike a valve in a pipe that regulates the flow rate to any specified value up to a maximum, a channel only allows ions to flow at a rate determined by its conformational state and the driving forces. As we describe below, each channel state has a fixed conductance. The switching time between states is so rapid that we regard it as instantaneous. At fixed membrane potential, the current observed to flow has a small number of values, and the channel jumps abruptly from one to another.

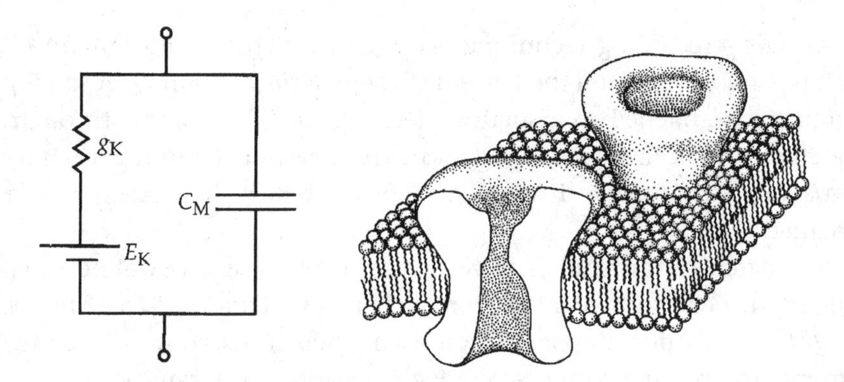

Figure 3.1 Artist's representation of a membrane-spanning channel with a diagram representing an equivalent electrical circuit for the channel (from Hille 2001, p. 17).

- Switching of channels among states appears to happen randomly. However, there are different probabilities for switching between each pair of states. These switching probabilities typically depend upon membrane potential (*voltage dependence*) or upon the binding of neurotransmitters to the membrane (*ligand dependence*).

Learning, memory, dexterity, and sensory perception are all mediated by the electrical signals within the nervous system. The "architecture" of this system is enormously complex, both at the cellular level and as a network of cells. For example, the human brain is estimated to have approximately 10^{12} neurons. The geometry of most individual neurons is elaborately branched, with long processes called *dendrites* and *axons*. Pairs of neurons make contact at specialized structures called *synapses*, which can be either chemical or electrical. At chemical synapses, action potentials traveling along an axon of the presynaptic cell trigger the release of neurotransmitter into the intercellular space of the synapse. The neurotransmitter, in turn, binds to receptor sites on dendrites of the postsynaptic cell, stimulating a postsynaptic current.

Channels play a central role in the propagation of action potentials within neurons, along axons, and in synaptic transmission between nerve cells. Techniques for identifying the genes that code for channels and comparative studies of channels from different organisms are rapidly leading to catalogs of hundreds of different types of channels. Alternative splicing of these genes and modification of channels by events such as phosphorylation contribute additional diversity. Nonetheless, the basic structure and function of channels has been conserved throughout evolution. One place to start in understanding how the nervous system works is by considering an individual ion channel.

3.1.1 Channel Gating and Conformational States

The *patch clamp* recording technique invented by Neher and Sakmann (1976) makes it possible to measure the current through a *single* channel. A small piece of membrane is attached to a suction electrode and the current through this tiny patch of membrane is measured. If the patch is small enough, then it may contain a single channel and the measurements reflect the current flowing in that channel.

The first patch clamp studies were of the nicotinic acetylcholine receptors (nAChR) in the neuromuscular junction (Neher and Sakmann 1976). The recordings (Figure 3.2) exhibit the properties of channels listed above—in particular, the current switches between discrete levels at apparently random times. The nAChR receptor has two possible conductance levels, zero when it is closed and approximately 30 pS when open. From these recordings we obtain sequences of successive *dwell times*, the amounts of time that the channel remains open or closed before switching. These appear to be irregular, so we will turn to mathematical models that assume that the *probability* that the channel will switch states in a small interval depends on only its current state, and not its past history.

Biophysical models for channel gating are based upon the physical shape, or *conformation*, of the channels. Channel proteins can be modeled as a collection of balls and springs representing individual atoms and the forces between them. The springs correspond to covalent bonds, the strongest forces in the molecule. There are additional forces that depend upon the environment of the molecule and the three-dimensional shape of the molecule. Thermal fluctuations are always present, driving small energy changes through collisions and causing the molecule to vibrate. There is a *free-energy* function assigned to each spatial configuration of the atoms within the channel, and the physical position of the molecule is normally at states of smaller energy. The molecule spends relatively long periods of time near local minima of the free energy, undergoing only small vibrations. However, sometimes the thermal fluctuations are sufficiently large to cause the molecule to move from one local minimum to another.

The behavior of this system can be interpreted in terms of an "energy landscape." Corresponding to any set of positions for the atoms is a value of the free energy, which we regard as defining a height at each point of this space. The resulting landscape has hills and valleys (see Figure 3.3). Imagine the state of the molecule as a particle vibrating on this landscape like a ball oscillating on a frictionless surface under the influence of gravity and being given small random kicks. The amplitude and frequency of vibrations of the particle around a minimum are determined by the steepness of the sides of the valley in the energy landscape. Occasionally, the molecule receives a kick that is hard enough to make it cross the ridge between one valley and another. The height of the ridge determines how frequently this occurs. When the molecule crosses to a new valley,

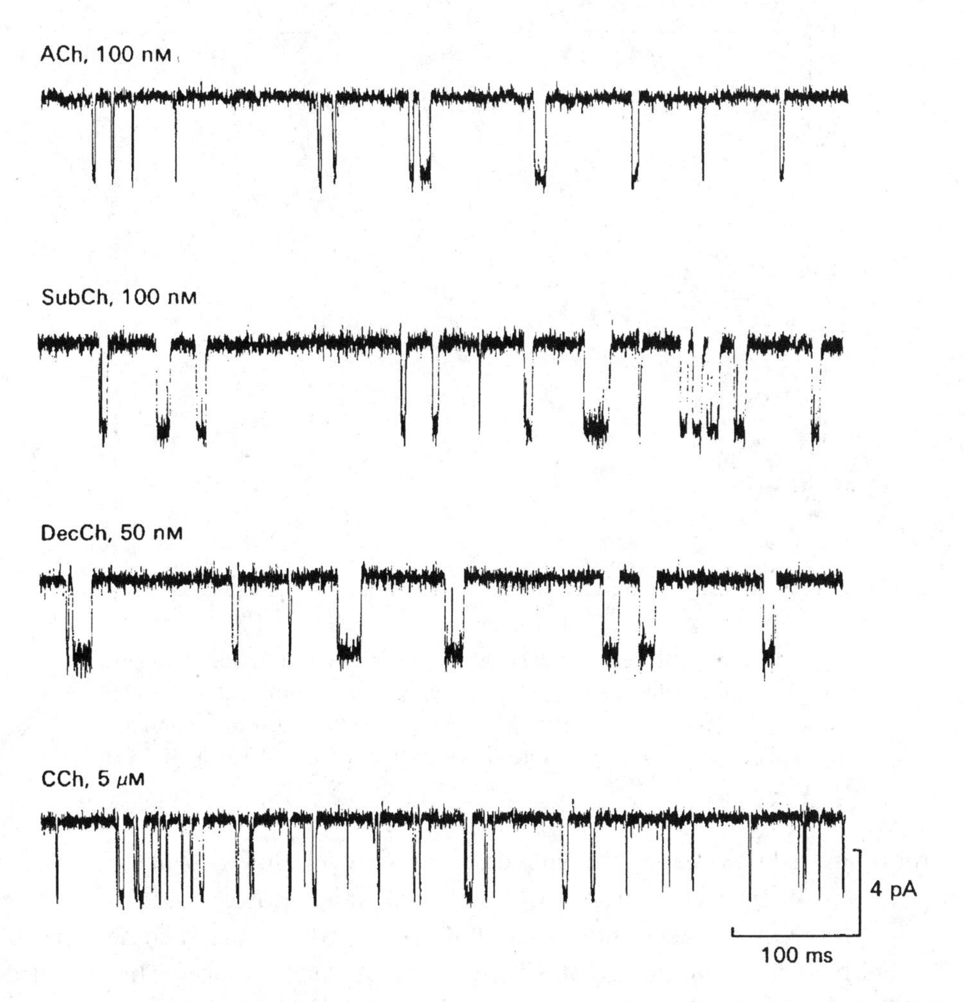

Figure 3.2 Data from a single channel recording of the nicotinic acetylcholine receptor (nAChR) channel in the neuromuscular junction, in the presence of four different agonists that can bind to the receptor. Note that the open channel current is the same for each agonist, but the dwell times are different (from Colquhoun and Sakmann 1985).

it will settle at the new local minimum until it receives a kick that knocks it out of the valley surrounding that minimum. For any pair of local minima, there is a transition probability that the molecule will cross from the valley surrounding the first minimum to the valley surrounding the second. This transition probability depends *only* on the pair of states; in particular, it does not depend upon the past history of the particle. This leads to a model for the switching dynamics of the channel as a Markov chain, the type of mathematical model we study in the next section.

One important use of these models is to infer information about the conformational states of channels from measurements of membrane current. If each

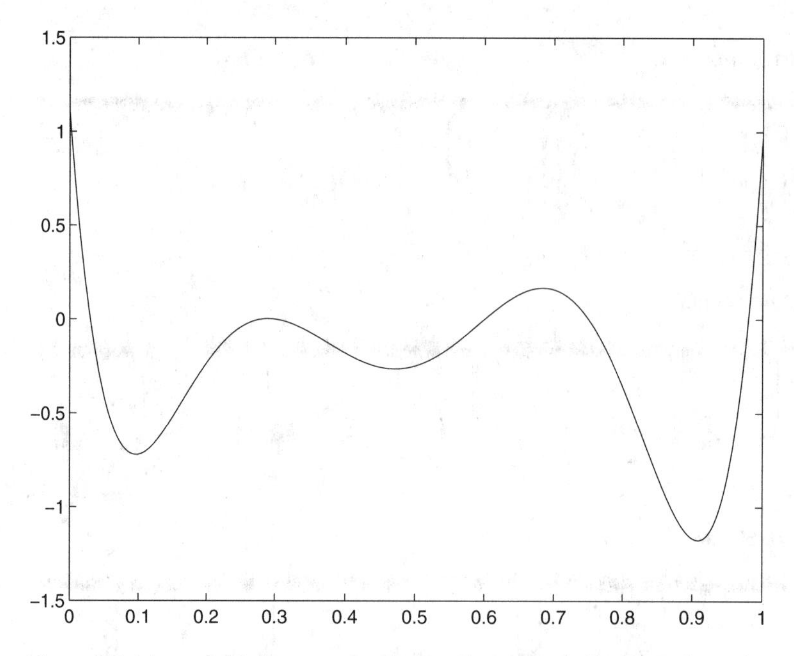

Figure 3.3 A hypothetical energy landscape. Imagine a ball rolling along this curve under the influence of gravity and collisions with other, much smaller particles. When the ball sits at a local minimum, it stays near this minimum until it receives a "kick" hard enough to knock it over one of the adjacent local maxima.

conformational state has a different conductance level, then the membrane current tells us directly about the conformational state. However, this is seldom true. Most channels have more than a single closed state, and some types of channel have multiple open states with the same conductance. The *selectivity filter* that determines the ion selectivity and permeability may be a small portion of the channel. Conformational changes that have little effect on this part of the channel are likely to have little effect on conductance.

To simplify matters, we shall only consider situations in which the channel conductance has just two values, one for open states and one for closed states. When we have only partial information about the molecular dynamics, we are left with the fundamental question as to whether we can model the dwell times of the channel from the information we do have. This is where mathematical analysis enters the picture.

Let us be careful here to specify in mathematical terms the problem that we want to solve. The channel that we want to model has conformational states $C_1, C_2, \ldots, C_k, O_1, \ldots, O_l$. The C_j are closed states and the O_j are open states. We do not know the numbers k and l of closed and open states: they are part of what we are trying to determine. The observations we make yield a *time series* of the membrane current at discrete times $t_0, t_1, \ldots, t_N$. These measurements tell us whether the channel is in a closed state or an open state at each time. If the temporal resolution $\tau = t_{i+1} - t_i$ is fine enough, then the probability of two or more transitions occurring in a single time step is very small. Assuming that this is the case, our data tell us when the channel makes transitions from an open state to a closed state, or vice versa. What information about the channel can we extract from these transition times? Can we determine values of k, l, and transition probabilities between states? How large are the expected uncertainties in our estimates associated with the limited amount of data that we have? What kinds of experiments can we do that will help us reconstruct more information about the switching process? These are the questions we shall examine more closely.

Exercise 3.1. Discuss how you expect the transition rates between two states to depend upon the height of the energy barrier between them. How do you think the *distribution* of amplitudes and frequency of the kicks affects the transition rates? Design a computer experiment to test your ideas.

3.2 Markov Chains

This section discusses the probability theory we use in models for membrane channels. Only in the next section do we return to the biology of channels.

Individual membrane channels are modeled as objects that have several conformational states, divided into two classes: open and closed. In open states, the channel allows a current flow of ions, while in closed states, it does not. If a channel has more than one open state, we assume that the current flow in each of the different open states is the same. We observe switching times between open and closed states of a channel, and then attempt to produce stochastic[1] models that reproduce the observed statistics of switching. Using simple examples of probability, we illustrate the principles and methods that lead to successful stochastic models of switching in some cases.

The stochastic models we discuss are *Markov chains*. Markov chain models are framed in terms of probabilities. We assume that a physical system has a finite number of states $S_1, S_2, \ldots, S_n$, and that the system switches from one state to another at random times. The fundamental Markov assumption is that the switching probabilities depend only upon the current state of the system. Past history has no influence in determining the switching probability. Markov mod-

[1] A *stochastic* model is one that contains random elements. We do not give a formal definition of randomness, but intuitively we think of a quantity as random if repeated measurements of that quantity give different results. We characterize a random quantity by the distribution or histogram of results we obtain when we make a large number of repeated measurements. We assume that this distribution is hardly changed if we make a second set of repeated measurements.

els can be formulated with either continuous time or discrete time. We shall emphasize discrete-time models in this chapter.

3.2.1 Coin Tossing

The simplest example of a Markov chain is a thought experiment of probability theory: flipping coins repeatedly. We think of a "fair" coin as having equal probability of landing heads or tails, but what does this mean since each toss must produce either heads or tails? The mathematical description of probability is built upon the idea of repetition of events. If we flip a coin once, there is no way to test the probabilities of the two different outcomes, heads and tails. From an experimental perspective, we view the probabilities as statements about the frequencies of these outcomes if we repeat an experiment many times. If we flip a coin N times, with N a large number, then we expect that a probability of 0.5 for each outcome will be reflected in roughly the same number of outcomes that are heads and outcomes that are tails. However, the exact numbers will fluctuate if we repeat the experiment of flipping a coin N times. Can we test whether observations are consistent with the hypothesis that a coin has equal probabilities of 0.5 for heads and tails by flipping it many times and observing the relative proportion of heads and tails? If the results for the series of experiments are highly improbable, then we reject them as being plausible. What we need here is a mathematical analysis of the probability *distribution* for sequences of N coin flips. This distribution is derived from the probabilities for an individual flip. In the mathematical idealization of the experiment of flipping a fair coin N times, all sequences of heads and tails of length N are equally likely outcomes and all are equally probable. By counting the number of sequences of length N with each number of heads and tails, we can assign probabilities to different outcomes that can be used in testing statistical hypotheses. Let us examine the case $N = 4$ to develop an intuition for the general case.

The possible outcomes of tossing the coin four times would be

hhhh	*hthh*	*thhh*	*tthh*
hhht	*htht*	*thht*	*ttht*
hhth	*htth*	*thth*	*ttth*
hhtt	*httt*	*thtt*	*tttt*

where h represents the outcome of a head and t represents the outcome of a tail. Note that there are 2^N possible such sequences and so if the coin is fair, each of these sequences has equal probability, namely, $1/2^4 = 1/16$. Looking at the list, we count that there is 1 outcome with 0 heads, 4 outcomes with 1 head, 6 outcomes with 2 heads, 4 outcomes with 3 heads, and 1 outcome with 4 heads. Even though each (ordered) sequence of tosses is equally likely, it is six times more likely to observe 2 heads and 2 tails than to observe 4 heads. So we hypothesize that the probabilities of $0, 1, 2, 3, 4$ heads are $1/16, 4/16, 6/16, 4/16, 1/16$, respectively. Even though each sequence of tosses is equally likely, we expect to observe sequences with 2 heads and 2 tails about six times as often as we see sequences with 4 heads. If we perform a large number of sequences of four tosses and make a histogram for the number of trials with $0, 1, 2, 3, 4$ heads, we expect that the bins of the histogram will have relative proportions close to $1/16, 4/16, 6/16, 4/16, 1/16$.

The numbers $1, 4, 6, 4, 1$ are the *binomial coefficients* that appear in expanding the polynomial

$$(H + T)^4 = \sum_{k=0}^{4} c_k H^k T^{4-k}$$

for $k = 0, 1, 2, 3, 4$. For the general case of sequences of N coin tosses, we use the expansion

$$(H + T)^N = \sum_{k=0}^{N} c_k H^k T^{N-k}$$

where the c_k are the binomial coefficients

$$c_k = \frac{N!}{k!(N-k)!}.$$

The distribution in the outcomes of flipping a *fair* coin N times is given by the *binomial distribution* defined by

$$b_N(k) = \frac{N!}{2^N k!(N-k)!}.$$

This is justified by the following argument. There are 2^N possible outcomes for sequences of N coin tosses. If the coin is fair, then each of these sequences has equal probability, namely, $1/2^N$. The number of sequences with k heads and $N - k$ tails is given by the binomial coefficient c_k, so the probability of obtaining exactly k heads is b_k.

What happens if a coin is not fair? We want to compute the probabilities of different outcomes in a sequence of tosses of an unfair coin. Let H and $T = 1 - H$ denote the probabilities of tossing a head or a tail, respectively. In the case of a fair coin, we have $H = T = \frac{1}{2}$. The probability that a particular sequence of heads and tails occurs is given by the product of the corresponding probabilities. This is the assumption that the successive tosses are *independent* of each other. Thus the probability of the sequence *hhth* is $HHTH$. Placing subscripts on the H and T to denote the probabilities of the outcome of each toss, we write the formula

$$1 = (H_1 + T_1)(H_2 + T_2) \cdots (H_N + T_N)$$

which expresses that either heads or tails is sure to be the outcome on each toss. Expanding the right-hand side of this formula, each of the 2^N sequences of heads and tails is represented by a single term that is the product of the probabilities of each outcome in that sequence of tosses. To determine the probability of all sequences with specified numbers of heads and tails, we drop the subscripts in the polynomial to get the formula

$$1 = (H + T)^N = \sum_{k=0}^{N} c_k H^k T^{N-k}$$

where the c_k are the binomial coefficients above. The number of sequences with exactly k heads is not affected by the probabilities of heads or tails, but the probability that a particular sequence of heads and tails with exactly k heads occurs is $H^k T^{N-k}$. Therefore, the probability of obtaining exactly k heads in a sequence of N independent tosses is $c_k H^k T^{N-k}$. To know what to expect when N is large, we compute and evaluate histograms of $c_k H^k T^{N-k}$.

It is convenient to plot these histograms with the horizontal axis scaled to be $[0, 1]$: at k/N we plot the probability of having k heads in N tosses. Numerical computations with different values of H and N large suggest that the histograms are always approximated by a "bell shaped" curve with its peak at $k/N = H$ (see Figure 3.4). This observation is confirmed by the *DeMoivre-Laplace limit theorem*, a special case of the central limit theorem of probability. This theorem states the following.

Consider a sequence of N independent coin tosses, each with probability H of obtaining heads. The probability of obtaining k heads with

$$NH + z_l\sqrt{NH(1-H)} < k < NH + z_u\sqrt{NH(1-H)}$$

approaches

$$\frac{1}{\sqrt{2\pi}} \int_{z_l}^{z_u} \exp\left(-\frac{1}{2}x^2\right) dx$$

as $N \to \infty$.

The integrand in this formula is called the *normal distribution*. Other names for this important function are the Gaussian distribution and a bell-shaped curve. The DeMoivre-Laplace limit theorem tells us that in our coin-tossing experiment, as the length N of the sequence increases, the proportion k/N of heads is more and more likely to be close to H. The expected deviation of k/N from H is comparable to $1/\sqrt{N}$.

Diaconis, Holmes, and Montgomery (2004) reexamined the tossing of real coins and came to surprising conclusions. They built a coin-flipping machine and conducted experiments by repeatedly flipping a coin that starts heads up. The proportion of coin tosses in which the coin landed heads up was larger than 1/2 by an amount that was statistically significant. In other words, if one assumes that the probability of landing heads was 1/2 and that the tosses were independent of each other, then the DeMoivre-Laplace theorem implies that the results they observed were very unlikely to have occurred. They went further with a

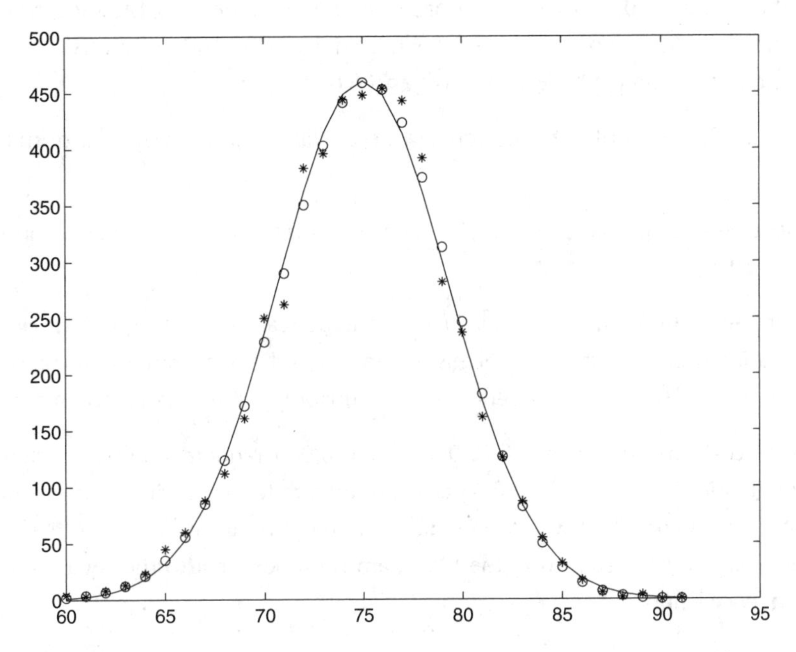

Figure 3.4 A histogram from a computer experiment: 5000 repetitions of tossing an unfair coin 100 times. The probability of heads is 0.75 and the probability of tails is 0.25. The asterisks show the number of experiments that produced k heads as a function of k, the circles are the values of the binomial distribution, and the solid line is the approximation to the binomial distribution from the bell shaped Gaussian function. Note that the minimal number of heads obtained in our trials was 60 and the maximum was 91. There are small deviations from the predicted distribution that change when we rerun the experiment with different sequences of the random numbers used to generate the experimental outcome.

careful examination of the physics of coin tossing. They found that reasonable hypotheses lead to the conclusion that it is more probable that a coin lands with the same side up that was face up at the beginning of the toss. Here we rely on computer experiments, using algorithms that generate *pseudorandom numbers*—sequences of numbers that look as if they are independent choices from a uniform probability distribution on the interval $[0, 1]$ (Knuth 1981). There are different approaches to the generation of pseudorandom numbers, and some widely used generators produce poor results.

Exercise 3.2. Consider a sequence of 100 tosses of a fair coin. Use the DeMoivre-Laplace limit theorem to check the following statements. The odds (probability) of obtaining fewer than 20 heads is less than one in a billion. The odds of obtaining fewer than 30 heads is about one in 25,000, and the odds of obtaining fewer than 40 heads is less than three in 100. The odds of obtaining between 40 and 60 heads is more than 94 in 100, *but* there is more than a 5% chance of obtaining fewer than 41 or more than 59 heads. If one tosses a coin 10,000 times, there is more than a 95% chance of obtaining between 4900 and 5100 heads.

Exercise 3.3. Experiment with sequences of coin flips produced by a random number generator:

(a) Generate a sequence `r` of 1000 random numbers uniformly distributed in the unit interval $[0, 1]$.

(b) Compute a histogram for the values with ten equal bins of length 0.1. How much variation is there in values of the histogram? Does the histogram make you suspicious that the numbers are not independent and uniformly distributed random numbers?

(c) Now compute sequences of 10,000 and 100,000 random numbers uniformly distributed in the unit interval $[0, 1]$ and a histogram for each with ten equal bins. Are your results consistent with the prediction of the central limit theorem that the range of variation between bins in the histogram is proportional to the square root of the sequence length?

Exercise 3.4.

(a) Convert the sequence of 1000 random numbers `r` from the previous exercise into a sequence of outcomes of coin tosses in which the probability of heads is 0.6 and the probability of tails is 0.4.

(b) Recall that this coin-tossing experiment can be modeled by the binomial distribution: the probability of k heads in the sequence is given by

$$c_k(0.6)^k(0.4)^{1000-k}$$

where $c_k = 1000!/k!(1000 - k)!$. Calculate the the probability of k heads for values of k between 500 and 700 in a sequence of 1000 independent tosses. Plot your results with k on the x-axis and the probability of k heads on the y-axis. Comment on the shape of the plot.

(c) Now test the binomial distribution by doing 1000 repetitions of the sequence of 1000 coin tosses and plot a histogram of the number of heads obtained in each repetition. Compare the results with the predictions from the binomial distribution.

(d) Repeat this experiment with 10,000 repetitions of 100 coin tosses. Comment on the differences you observe between this histogram and the histogram for 1000 repetitions of tosses of 1000 coins.

3.2.2 Markov Chains

A small generalization of the coin-flipping experiment produces a two-state Markov chain. In this generalization, one has two coins, each with its own heads versus tails probabilities. We will call one coin a penny and the other a quarter. The two possible states are that we are flipping either a penny or a quarter at each turn. We switch states according to the following:

- After a heads, we flip the penny
- After a tails, we flip the quarter

Given states S_1 and S_2, which may be the same, we call the *transition probability* from S_1 to S_2 the probability that we are in state S_1 at time t and in state S_2 at time $t+1$. In the example, there are four transition probabilities associated with the current state and the outcome of the flip we do now. Since the coin we toss depends on the previous outcome, the transition probabilities are the probabilities of heads previous and heads now, heads previous and tails now, tails previous and heads now, and tails previous and tails now. These can be written in a 2×2 matrix called the *transition* matrix of the Markov chain.

Let us pick specific probabilities for the two coins to make our discussion more concrete. Assume that the probability of getting heads when we toss the penny is 0.4 and so the probability of getting tails is $1 - 0.4 = 0.6$, and that the probability of getting heads when we toss the quarter is 0.7 and so the probability of getting tails is $1 - 0.7 = 0.3$. Then the transition matrix for this system is

$$\begin{array}{cc} \textit{penny} & \textit{quarter} \end{array}$$
$$\begin{bmatrix} 0.4 & 0.7 \\ 0.6 & 0.3 \end{bmatrix} \quad \begin{array}{c} \textit{penny} \\ \textit{quarter}. \end{array}$$

This means that if we are currently flipping the penny, then the probability that next time we will flip a penny is 0.4 and the probability that next time we will flip a quarter is 0.6. If we are currently flipping a quarter, the probability that next time we will flip a penny is 0.7 and the probability that next time we will flip a quarter is 0.3.

The behavior of the Markov chain can be different from flipping a single coin. For example, if the probability of flipping heads for the penny and the probability of flipping tails for the quarter are both less than 0.5 as in the example above, then we expect a larger number of alternating strings of penny:tails, quarter:heads than of successive quarter:heads or successive penny:tails. Repeated runs of heads and tails are less frequent than when flipping a single coin. On the other hand, if the probability of flipping tails for the penny is very low and the probability of flipping heads for the quarter is very low, then we expect a large number of repeated runs of both heads and tails. When the probabilities of heads and tails are independent of the current state, that does not happen. In that case, only one outcome can have high probability since the sum of the probabilities is 1. Runs of the outcome with low probability are very rare. The lengths of runs are directly relevant to our observations of membrane channels: if one thinks of the states of the Markov chain as being open and closed states of a membrane channel, the lengths of "runs" of the Markov chain correspond to the dwell times of the channel in open or closed states.

A Markov chain model can be represented by its transition matrix: the $n \times n$ matrix A is defined by setting the entry A_{ij} to be the probability that state S_j switches to state S_i. The sum of the entries in each column of A adds to 1 since state S_j surely switches to some state S_i at each time-step. This way of looking at the Markov chain allows us to relate its properties to those of matrix population models. We compare the states of the Markov chain with stages of a population, and the transition matrix of the Markov chain with the transition matrix of the population. We imagine a large population of individuals distributed among the states of the Markov chain. Each individual independently makes a transition to a new state using the probabilities in the transition matrix. The sample paths of the Markov chain are paths taken by individuals, and the transition matrix coefficient A_{ij} approximates the proportion of individuals in state S_j at time t who move to state S_i at time $t+1$.

The matrix makes it easy to compute the transition probabilities for paths of length longer than 2. Consider the transition from state S_k to S_j to S_i in two steps. The Markov assumption implies that the probability for the two transitions to take place at successive time-steps is the product $A_{ij}A_{jk}$. Also, the probability that state S_k will make a transition to state S_i in two steps is

$$\sum_{j=1}^{n} A_{ij}A_{jk} = (A^2)_{ik},$$

the (i,k) component of the matrix product of A with itself. Extending this argument, the probability that a sequence of length l beginning at state S_j ends at state S_i is $(A^l)_{ij}$, the (i,j) component of the matrix product A^l of A with itself l times.

The matrix A has non-negative entries. If it is power-positive, then we can use the Perron-Frobenius theorem to tell us about the evolution of our population: it will have a dominant eigenvalue and the eigenvector of the dominant eigenvalue is the only eigenvector with all positive entries. Since the sum of each column of A is 1, the row vector $(1, 1, \ldots, 1)$ is a left eigenvector with eigenvalue 1. Intuitively, growth rate 1 makes sense since our population is being redistributed among the states with no births or deaths in the population. Thus, the total population size remains constant. In formulas, if $(w_1, w_2, \ldots, w_j)$ is the population vector at time t, then the population vector at time $t+1$ is

$$\left(\sum_j A_{1j}w_j, \sum_j A_{2j}w_j, \ldots, \sum_j A_{nj}w_j\right)^t$$

and the total population is

$$\sum_i \sum_j A_{ij}w_j = \sum_j \sum_i A_{ij}w_j = \sum_j w_j,$$

unchanged from time t. The Perron-Frobenius theorem implies that the population distribution will converge to the stable stage distribution, the right eigenvector of 1 for the transition matrix A. In the setting of membrane channels, there is an actual population of individual channels and the stable stage distribution gives the expected number of channels in each state under steady-state conditions. Moreover, the conductance of the membrane is the conductance of a single channel times the number of open channels.

Single-channel recordings of membrane currents show us transitions between open and closed states. Therefore, we are particularly interested in the *residence* or dwell times of a Markov chain, the times that a sample path remains in a particular state. In the channel, these transitions are the result of stochastic fluctuations and we assume that the time between measurements is short enough that the probability of multiple transitions between successive measurements is negligible. When the time-step is small enough, during most time-steps a channel remains in the same state. This implies that the transition matrix has diagonal entries A_{jj} close to 1 and the remaining entries are small.

We seek to determine the distribution of residence times for each state. If a channel is in state S_j at time t_0, the probability that it remains in state S_j at time $t_0 + 1$ is A_{jj}. The probability that it remains in the state S_j at times $t_0 + 1$ and $t_0 + 2$ is A_{jj}^2. Continuing, the probability that the channel remains in state S_j for times $t_0 < t \leq t_0 + k$ is A_{jj}^k. Similarly, the probability that the system switches from state S_j to some other state at each time is $1 - A_{jj}$. The channel has a residence time of k in state S_j beginning at time t_0 if it is in state S_j at times $t_0 < t < t_0 + k$ *and* it switches to another state in the step from time $t_0 + k - 1$ to $t_0 + k$. The probability associated with this sequence of states is $A_{jj}^{k-1}(1 - A_{jj})$. We conclude that the residence times have an exponential distribution: the probability of residence time k in state S_j decreases exponentially with k. If the time between measurements is δ and $A_{jj} = \exp(-\lambda_j \delta)$, then the probability that a channel in state S_j makes its first transition to another state between times $k\delta$ and $(k+1)\delta$ is $(1 - A_{jj})\exp(-k\lambda_j\delta)$.[2]

These results yield predictive information that we can utilize in analyzing the data from single-channel patch clamp measurements of membrane current. In the situation where we measure just two current levels (from open and closed states of the channel), our analysis of residence times predicts that histograms of the residence times will be sums of exponentials, with one exponential term for each channel state. Given time series data from a single-channel recording, we construct a histogram for the residence times of open states and the residence times of closed states. If there is a single open state, then the residence time histogram for the open state can be fitted by a single exponential function. If we

[2]We can let δ tend to zero in this formula while holding $k\delta$ constant and obtain the result that the probability density of residence times in state S_j is an exponential function proportional to $\exp(-t\lambda_j)$. The distributions of dwell times for each state in a continuous-time Markov chain are exponential functions.

need to sum more than one exponential to fit the histogram, then there must be more than one open state. The same conclusions apply to closed states. The number of exponential functions we need to fit the residence time histogram of the closed states gives a lower bound on the number of closed states of the channel.

Exercise 3.5. The histogram of expected residence times for each state in a Markov chain is exponential, with different exponentials for different states. To observe this in the simplest case, use a random number generator to produce data for a two-state Markov chain. Next, determine the *transitions* that occur in these data as the state changes, and the residence times between these. Finally, compute histograms of the residence times.

3.2.3 The Neuromuscular Junction

Using the theory that we have developed for Markov chain models, we now want to use single-channel patch clamp recordings to infer properties of the molecular mechanisms underlying the gating (opening and closing) of nAChR channels at the neuromuscular junction. Since the only step in this multistep process that is observed directly in these recordings is the final opening of the channel, analysis of the dwell times is required to deduce information about the steps taking place while the channel is closed.

The simplest possible model is that receptors with bound acetylcholine agonist are open, while unbound receptors are closed. The total membrane current would then be perfectly correlated with the amount of bound acetylcholine. This model is *much* too simple to explain two important observations. First, at low agonist concentrations, the current from a population of channels is seen to be roughly proportional to the square of agonist concentration rather than linearly proportional (Katz and Thesleff 1957). This *cooperativity* suggests that enhancement of the opening probability of a channel depends on a *pair* of agonist molecules binding to a channel, rather than a single bound molecule. Second, the dwell time distributions have "bursts" typified by several successive very short closings interspersed with long ones (Colquhoun and Sakmann 1985). This distribution of dwell times is not well approximated by a single exponential, implying that a Markov model for a single channel must have two or more closed states.

Colquhoun and Sakmann (1985) studied the dwell time distributions of open and closed states for the nAChR channels and how these vary with different experimental conditions. They varied membrane potential, agonist concentration, and the agonist itself. Acetylcholine is not the only molecule that binds to the receptors and acts as an agonist for the receptors, so use of different agonists is a way of exploring the dynamics of binding. Figure 3.2 shows recordings of single-channel recordings from their paper. They found in these studies that a

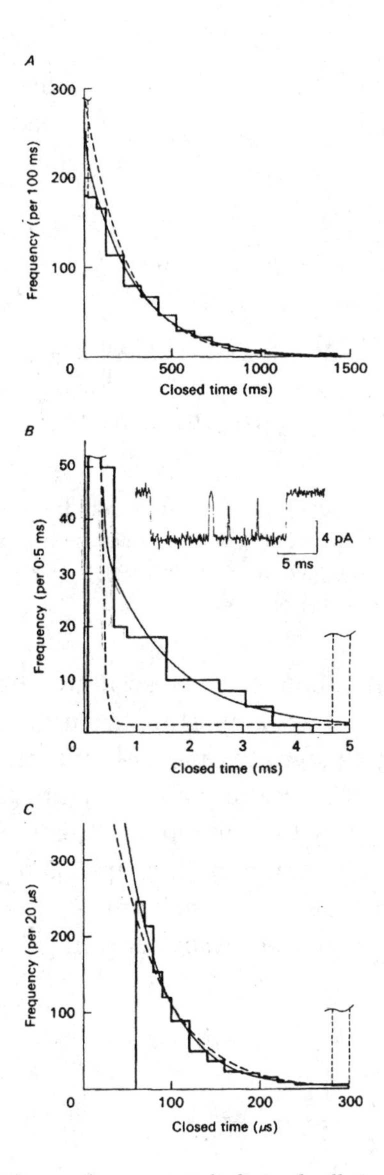

Figure 3.5 Sums of exponentials fit to dwell time distributions of open states in an nAChR channel. The three histograms show distributions of dwell times from three ranges of times: slow (top), medium (middle), and fast (bottom). The dashed curves come from a fit to a sum of two exponentials, the solid curves from a fit to a sum of three exponentials (from Colquhoun and Sakmann 1985).

sum of at least three exponentials was required to fit the dwell time distributions for closed states and a sum of at least two exponentials was required to fit the dwell time distributions for open states (see Figure 3.5). They postulated that the closed states are receptors with 0, 1, and 2 agonists bound, and that the open

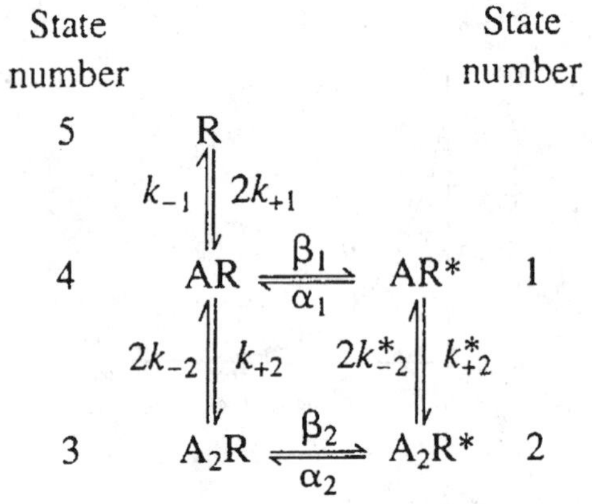

State number			State number
5	R		
	k_{-1} ⇅ $2k_{+1}$		
4	AR $\underset{\alpha_1}{\overset{\beta_1}{\rightleftharpoons}}$	AR*	1
	$2k_{-2}$ ⇅ k_{+2}	$2k^*_{-2}$ ⇅ k^*_{+2}	
3	A_2R $\underset{\alpha_2}{\overset{\beta_2}{\rightleftharpoons}}$	A_2R^*	2

Figure 3.6 A diagram of the states and transitions in the Markov chain model used to model the nAChR channel (from Colquhoun and Sakmann 1985). The states labeled with an $*$ are the open states.

states have 1 or 2 agonists bound. Of special note here is that the transitions between open and closed states are distinct from the binding events of the agonist. Agonist binding changes the probability that a channel will open, but the binding itself does not open the channel. Figure 3.6 shows a diagram of the five-state model proposed by Colquhoun and Sakmann. The transition matrix for a Markov model corresponding to this graphical representation of the model has the approximate form $I + \delta A$ for small time steps δ, where A is the following matrix expressing the *rates* of transitions between pairs of states.[3]

$$\begin{bmatrix} -(\alpha_1 + k^*_{+2}x_A) & 2k_{-2} & 0 & \beta_1 & 0 \\ k^*_{+2}x_A & -(\alpha_2 + 2k^*_{-2}) & \beta_2 & 0 & 0 \\ 0 & \alpha_2 & -(\beta_2 + 2k_{-2}) & k_{+2}x_A & 0 \\ \alpha_1 & 0 & 2k_{-2} & -(\beta_1 + k_{+2}x_A + k_{-1}) & 2k_{+1}x_A \\ 0 & 0 & 0 & k_{-1} & -2k_{+1}x_A \end{bmatrix}$$

with x_A the agonist concentration. The diagonal term A_{jj} of A gives the rate of decrease of channels in state j which remain in state j. Since the total number of channels does not change, the magnitude of this rate is the sum of the rates with which channels in state j make a transition to some other state. Colquhoun and Sakmann (1985) discuss parameter fits to this model from the dwell time distributions for different agonists. They find a good fit to the dwell times for acetylcholine by ignoring the state with a singly bound opening (AR^* in the

[3]The exact form of the transition matrix is $\exp(\delta A)$, expressed using the matrix exponential function. One way of defining the matrix exponential is as a power series: $\exp(\delta A) = \sum \delta^k A^k / k!$

diagram) and the remaining parameters set to $k_{-1} = k_{-2} = 8150\ \mathrm{s}^{-1}$, $k_{+1} = k_{+2} = 10^8\ \mathrm{s}^{-1}(\mathrm{M})^{-1}$, $\alpha = 714$, and $\beta = 30,600$.

Exercise 3.6. This exercise examines Markov chains with three states: one "open" state O and two "closed" states C_1 and C_2. We assume that measurements do not distinguish between the states C_1 and C_2, but we want to demonstrate that there are more than two states.

(a) Generate a set of 1,000,000 samples from the Markov chain with transition matrix

$$\begin{array}{ccc} C_1 & C_2 & O \end{array}$$
$$\begin{bmatrix} 0.98 & 0.1 & 0 \\ 0.02 & 0.7 & 0.05 \\ 0 & 0.2 & 0.95 \end{bmatrix} \begin{array}{c} C_1 \\ C_2. \\ O \end{array}$$

You can see from this matrix that the probability 0.7 of staying in state C_2 is much smaller than the probability 0.98 of staying in state C_1 or the probability 0.95 of remaining in state O.

(b) Compute the eigenvalues and eigenvectors of the matrix A. Compute the total time that your sample data in the vector `states` spends in each state and compare the results with predictions coming from the dominant right eigenvector of A.

(c) Produce a new vector `rstates` by "reducing" the data in the vector `states` so that states 1 and 2 are indistinguishable. The states of `rstates` will be called "closed" and "open."

(d) Plot histograms of the residence times of the open and closed states in `rstates`.

(e) Comment on the shapes of the distributions for open and closed states. Show that the residence time distribution of the closed states is not fitted well by an exponential distribution. Using your knowledge of the transition matrix A, make a prediction about what the residence time distributions of the open states should be.

The nAChR channel illustrates the complexity of gating of membrane channels. Measurements of unitary currents through individual channels are conceptually simple, but the behavior of the channel is not. Abstract Markov chain models help answer questions about a complex multistep process without incorporating detail about the physics of binding or gating. Our discussion has brought forward basic principles about how these stochastic models are fit to the data, but we have just touched the surface of these principles are applied. Here are three mathematical questions arising from the limitations of the experimental techniques. They give pointers to how the analysis becomes more complicated.

1. The instrumental noise in patch clamp current measurements may be comparable to the current flowing through an open channel. How do we distinguish the times when the channel is truly open and truly closed?
2. If a channel opens and closes between two successive measurements, we do not observe these transitions. What effect do such short visits to a closed or open state have on our estimates?
3. We assume that there is a single channel in the patch of membrane in our apparatus. How do we tell? If we see the combined current from two open channels simultaneously, this is clear evidence for multiple channels in the patch. However, if openings are short and infrequent, we may have to watch for a long time before two channels are open simultaneously. How long is long enough to make us confident that we have a single channel?

These questions are discussed by Colquhoun and Hawkes (1995).

3.3 Voltage-Gated Channels

We now move up a level—from models of a single channel to models for a population of channels that are *voltage gated*, meaning that their transition probabilities between conformational states are a function of the potential difference across the membrane. The statistical summation of all ion currents from the channels in a patch of membrane behaves in a much more predictable fashion than the instantaneous on/off current through a single channel. The central limit theorem of probability theory predicts that the relative magnitude of the fluctuations in the current will be comparable to the square root of the number of channels contributing to the current. Since the total current is proportional to the number of channels, the *relative* magnitude of the fluctuations decreases as the population size grows.

What are the quantitative relations between the ionic current flowing through a population of channels, the fluctuations in this current, the unitary current through an individual channel, and the population size? This is the question we now explore using the probability theory in section 3.2. To begin, we make several assumptions about the channels and the membrane in which they are embedded:

- The voltage potential across the membrane is held constant (this experimental manipulation is called a *voltage clamp*, and the experimental methodology to achieve it was first developed by Hodgkin and Huxley (1952)).
- There are N channels in a membrane randomly switching between open and closed states.
- The current through each open channel is the same.
- The state of each channel is independent of the others.

These assumptions lead to similarities with the coin-tossing problem. Let the probability that a single channel is open be p and the probability that it is closed be $q = 1 - p$. At each time, we can represent the state of the system as a vector $(a_1, \ldots, a_N)$ where each a_i is a C for closed or O for open. There are 2^N different sequences representing the possible states of the system at each time. The assumption that the states of the individual channels are independent means that the probabilities of the population being in a given state is the product of the probabilities for the individual channels: if there are k open channels and $N - k$ closed channels, the probability of the state $(a_1, \ldots, a_N)$ is $p^k(1-p)^{(N-k)}$. Using the same combinatorics as in coin tossing, the probability that there are exactly k open channels at each time is

$$\frac{N!}{k!(N-k)!}p^k(1-p)^{(N-k)}.$$

Even though analysis of dwell times may require a complicated Markov model with multiple states of each channel, independence implies that the current flow in the population has a binomial distribution. If we sample the population at times that are long compared to the switching time of the individual channels, then the samples will be independent of each other. This provides helpful information that relates the population current, the individual channel current, and the fluctuations in the channel current. We derive below a formula for this relationship.

Let the current of a single open channel in our population be i. This depends on the membrane potential, but that is held constant in voltage clamp. If k channels in the population are open, then the population current is ki. The DeMoivre-Laplace Limit Theorem stated earlier tells us that as N grows large, the expected value of k is pN with fluctuations in k of magnitude proportional to $\sqrt{N}$. An alternative calculation works directly with the population current measurements. When we measure the current in a population of N channels repeatedly, $\bar{I} = pNi$ is expected to be the *mean* population current. Think of this current as the analog of tossing N coins simultaneously, with p the probability of heads for each. Measuring the population currents M times, obtaining values I_j, we expect the average population current

$$\bar{I} = \frac{1}{M}\sum_{j=1}^{M} I_j$$

to be closer to the mean than the typical I_j. Fluctuations in the measurements of the population current can be measured by the *variance* σ, defined by the formula

$$\sigma^2 = \frac{1}{M}\sum_{j=1}^{M}(I_j - \bar{I})^2.$$

Expanding the term for $\bar{I}$ in this formula and a bit of algebra give the alternative formula

$$\sigma^2 = \frac{1}{M}\sum_{j=1}^{M} I_j^2 - \bar{I}^2 \qquad [3.1]$$

for σ. These population level measurements of mean and variance can be related to those for a single channel by the following argument.

If we measure current from a single channel M times, then we expect that the channel will be open approximately $k = pM$ times, so the mean current from individual channels will be $pi = (1/M)ki$. If k of the individual channels are open, the variance of the single-channel measurements is

$$\frac{1}{M}ki^2 - (pi)^2 = pi^2 - p^2i^2 = p(1-p)i^2.$$

Now, if each of our population of N channels behaves in this manner, the mean current will be $I = Npi$ and the variance will be obtained from adding the variances of the individual channels in the equation [3.1] for σ^2:

$$\sigma^2 = \frac{1}{M}kNi^2 - N(pi)^2 = \frac{1}{M}pMNi^2 - \frac{1}{N}(Npi)^2 = iI - \frac{1}{N}I^2.$$

We eliminate N from this last formula using the expression $N = I/(pi)$ to obtain $\sigma^2 = iI - piI$ and then solve for i:

$$i = \frac{\sigma^2}{I(1-p)}.$$

This is the result that we are after. It relates the single-channel current to the population current and its variance. The term $(1-p)$ in this expression is not measured, but it may be possible to make it close to 1 by choosing a voltage clamp potential with most of the channels in the population closed at any given time.

Fluctuation analysis based upon this argument preceded the development of single-channel recordings based upon patch clamp. Thus, one can view the formula for the single-channel current as the basis for *predictions* of the unitary conductance of single channels. This analysis was carried out by Anderson and Stevens (1973) on the nAChR channel, giving an estimate of 30 pS for the unitary conductance. Subsequent direct measurements with patch clamp gave similar values, in accord with the theoretical predictions. Thus, the theory of coin tossing applied to the fluctuations in total current from a population of membrane channels gave successful predictions about the number of individual channels and their conductances. This was science at its best: data plus models plus theory led to predictions that were confirmed by new experiments that directly tested the predictions.

3.4 Membranes as Electrical Circuits

We now go up one more level, to consider the interactions among several different types of voltage-gated channels in a single neuron, and the patterns of neural activity that can result, such as action potentials. The interaction between voltage-gated channels is indirect, mediated by changes in the electrical potential (voltage difference) across the membrane. As ions flow through channels this changes the electrical potential across the membrane, which in turn changes the transition probabilities for all voltage-gated channels. In the last section, we saw that the relative magnitude of the fluctuations in a population of identical channels tends to 0 as the size of the population increases. In other words, large populations of thoroughly random individual channels behave like deterministic entities, so we model them that way. Models at this level are based on regarding the membrane and its channels as an electrical circuit—such as that depicted in Figure 3.1—so it is helpful to begin by reviewing a few concepts about circuits.

Electric charge is a fundamental property of matter: electrons and protons carry charges of equal magnitude, but opposite sign. Electrons have a negative charge, protons a positive charge. Charge adds: the charge of a composite particle is the sum of the charges of its components. If its charges do not sum to zero, a particle is an *ion*.

Charged particles establish an *electric field* that exerts forces on other charged particles. There is an electromotive force of attraction between particles of opposite electric charge and a force of repulsion between particles of the same charge. The electric field has a *potential*. This is a scalar function of position whose *gradient*[4] is the electric field. The movement of charges creates an electrical *current*. *Resistivity* is a material property that characterizes the amount of current that flows along an electrical field of unit strength. Resistance is the corresponding bulk property of a substance. The resistance of a slab of material is proportional to its resistivity and thickness, and inversely proportional to its surface area. *Conductance* is the reciprocal of resistance, denoted by g_K in Figure 3.1.

Channels in a membrane behave like resistors. When they are open, channels permit a flow of ions across the membrane at a rate proportional to $v - v_r$ where v is the membrane potential and v_r is the *reversal potential* discussed below. The membrane potential relative to v_r makes the membrane act like a battery, providing an electrodiffusive force to drive ions through the channel. This potential is labeled E_K in Figure 3.1. Even though membranes are impermeable, an electric potential across them will draw ions of opposite polarity to opposite sides of the membrane. The accumulation of ions at the membrane surface makes it an electrical capacitor (with capacitance C_M in Figure 3.1). The movement of these electrical charges toward or away from the membrane generates a current, the *capacitive current* CdV/dt, despite the fact that no charge flows *across* the membrane. Charge movement is induced by changes in the membrane potential,

[4]The gradient of a function is a vector field that is aligned with the direction in which the function changes fastest. The magnitude of the gradient is the rate of change of the function along a line in the gradient direction.

proportional to the change in the membrane potential. This leads to *Kirchhoff's current law*, the fundamental equation describing the current that flows between the two sides of the membrane:

$$C\frac{dv}{dt} = \sum_j -g_j(v - v_j). \qquad [3.2]$$

In [3.2], C is the membrane capacitance and each term $g_j(v - v_j)$ represents the current flowing through all the channels of type j. Here v_j is the reversal potential of the channels of type j and g_j is their conductance. For voltage-gated channels, g_j itself changes in a way that is determined by differential equations. This aspect of channel behavior will be investigated later in Section 3.4.2.

Exercise 3.7. Using electrodes inserted into a cell, it is possible to "inject" current into a cell, so that [3.2] becomes

$$C\frac{dv}{dt} = I_i + \sum_j -g_j(v - v_j).$$

A "leak" current is one with constant conductance g_L. Show that changes in the reversal potential v_L of the leak current and changes in I_i, the injected current, have similar effects on the membrane current. Discuss the quantitative relationship between these.

3.4.1 Reversal Potential

The ionic flows through an open channel are affected by diffusive forces as well as the electrical potential across the membrane. Membrane *pumps* and *exchangers* can establish concentration differences of ions inside and outside a neuron. They rely upon active shape changes of the transporter molecules and inputs of energy rather than diffusive flow of ions through the pores of membranes. Pumps and exchangers operate more slowly as ion transporters than channels, but they can maintain large concentration differences between the interior and exterior of a cell. In particular, the concentration of K^+ is higher inside the cell than outside, while the concentrations of Na^+ and Ca^{2+} ions are higher outside.

Diffusive forces result from the thermal motions of molecules. In a fluid, particles are subject to random collisions that propel them along erratic paths called *Brownian motion*. The resulting *diffusion* acts to homogenize concentrations within the fluid. Near a channel, the intracellular and extracellular spaces are maintained at different concentrations, so the concentration within the channel cannot be constant: there must be a concentration gradient. At equilibrium and in the absence of other forces, the concentration of an ion in a tube connected to two reservoirs at differing concentrations will vary linearly, and there will be a constant flux of ions flowing from the region of high concentration to the region of low concentration.

The electrical force across a membrane is proportional to its potential. This electrical force may be directed in the same direction as the diffusive forces or in the opposite direction. The diffusive and electrical forces combine to determine the motion of ions in an open channel. As the membrane potential varies, there will be a specific potential, the *reversal potential*, at which the two forces balance and there is no net flux of ions. At higher potentials, the net flow of ions is from the interior of the neuron to the exterior. At lower potentials, the ionic flux is from the exterior to the interior. The *Nernst equation* gives a formula for the reversal potential through a channel selective for a single species of ions: the reversal potential is proportional to $\ln(c_1/c_2)$ where c_1 is the external concentration and c_2 is the internal concentration.

Even though the ionic flow through channels is faster than the transport of ions by the pumps and exchangers, it is small enough that it has a minimal effect on the relative internal and external concentrations of K^+ and Na^+ ions. Ca^{2+} ions play essential roles as signaling molecules, and large increases in local concentrations of Ca^{2+} ions due to the influx through calcium channels have important effects on electrical excitability in some cases. Accordingly, internal calcium ion concentration is used as a variable in many models of membrane excitability, but the concentrations of sodium and potassium are constant in virtually all models. The dynamics of calcium concentration within cells is an active area of intense research, aided by the ability to visualize calcium concentrations with special dyes.

Exercise 3.8. The full expression for the reversal potential in the Nernst equation is

$$v_j = \frac{RT}{zF} \ln\left(\frac{c_1}{c_2}\right)$$

where R is the gas constant 1.98 cal/K-mol, F is the Faraday constant 96, 840 C/mol, T is the temperature in K, and z is the valence of the ion. The value of RT/F is approximately 26.9 mV at $T = 37°$ C $= 310$ K. Typical values of c_1 and c_2 for sodium ions in a mammalian cell are 145 and 15 mM. Compute the reversal potential for sodium in these cells.

3.4.2 Action Potentials

We now return to considering the total current flowing through all the channels in a membrane. Typical neurons and other excitable cells have several different types of voltage-gated channels whose currents influence one another. In our representation of the membrane as a circuit, each type of channel is represented by a resistor. Current through each type of channel contributes to changing the membrane potential, and the conductance of the channel is voltage dependent. Quantitative models are needed to predict the effects of this cycle of interactions in which current affects potential which affects current. As the number of channels of each type in a membrane increases, the *relative* magnitude of the fluctuations in the population current decrease. In many systems, the relative fluctuations are small enough that we turn to *deterministic* models that represent the total current of populations of voltage gated channels explicitly as a function of membrane potential. This leads to a class of models first proposed by Hodgkin and Huxley (1952) to model action potentials.

The primary object of study in the Hodgkin-Huxley model is the time-dependent membrane potential difference $v(t)$. We assume a "space clamped" experimental setup in which the potential difference is uniform across the membrane, so spatial variables are not included in the model. The potential difference results from ionic charges on the opposite sides of the membrane. The membrane itself acts like a capacitor with capacity C. The specific capacitance (capacitance per unit area) is usually given a nominal value of $1\ \mu F/cm^2$. Membrane channels act like voltage-dependent resistors, and support current flowing through the membrane.

The first ingredient in the Hodgkin-Huxley model is simply Kirchhoff's law, [3.2], which says that Cdv/dt is given by the sum of currents flowing through the membrane channels. However, to complete the model we also have to specify how the changes in potential affect the conductances g for each type of channel. The remainder of this section describes the formulation of models using the approach established by Hodgkin and Huxley. The mathematical and computational methods used to derive predictions from these models are discussed in Chapter 5.

Hodgkin and Huxley observed two distinct types of voltage-dependent membrane currents: one permeable to potassium ions and one permeable to sodium ions. There is a third type of channel, called the "leak," with conductance that is not voltage dependent. Channels had not been discovered when they did their research. One of their contributions was the development of "voltage clamp" protocols to measure the separate conductances and their voltage dependence.

To fit their data for the sodium and potassium currents, they introduced the concept of *gating variables*, time-dependent factors that modulated the conductance of the currents. Today, these gating variables are associated with the probabilities of channel opening and closing.

In the Hodgkin-Huxley model, there are three gating variables. The gating variables m and h represent *activation* and *inactivation* processes of the sodium channels while the gating variable n represents activation of the potassium channels. The gating variables themselves obey differential equations that Hodgkin and Huxley based on experimental data. Figure 3.7 shows a set of current traces from voltage clamp measurements of the squid axon.

To understand the kinetics of channel gating, a few more facts and a bit more terminology are needed. As in all electrical circuits, membrane potential has an arbitrary baseline. The convention in neuroscience is to set the potential outside

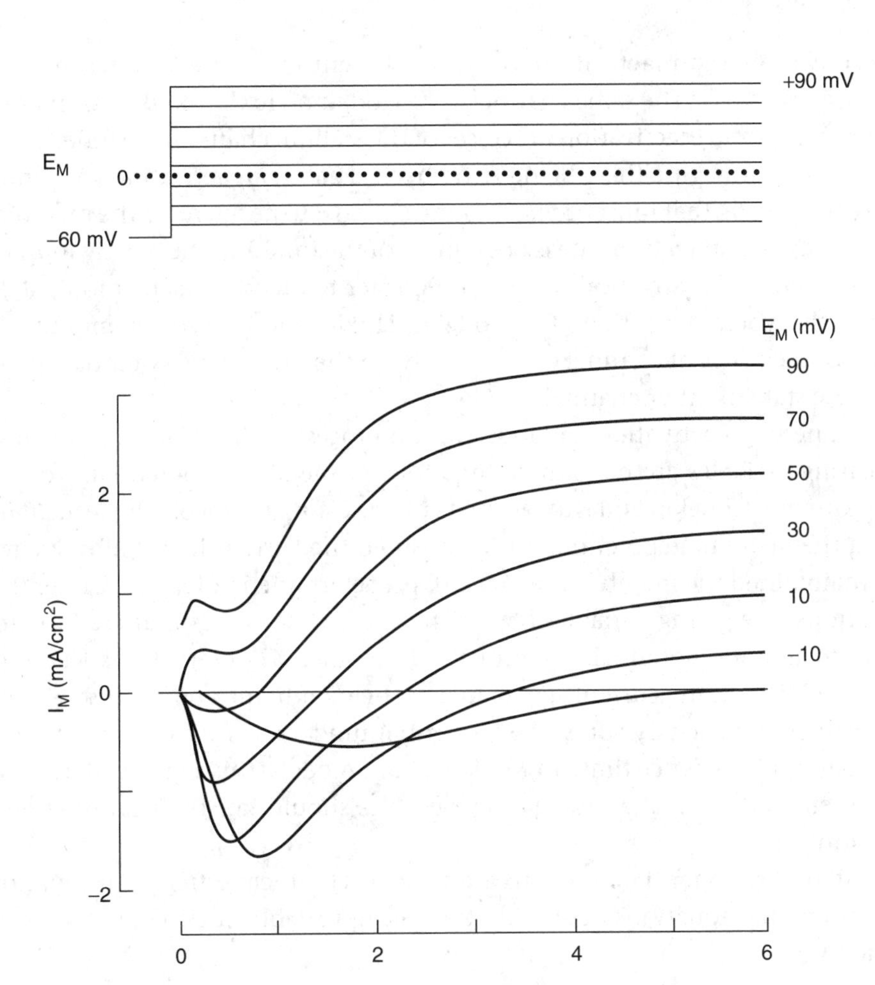

Figure 3.7 Data from a series of voltage clamp measurements on a squid giant axon. The membrane is held at a potential of −60 mV and then stepped to higher potentials. The traces show the current that flows in response to the voltage steps (from Hille 2001, p. 38).

the membrane to be 0 and measure potential as inside potential relative to the outside. The *resting potential* of most neurons, as well as the squid giant axon, is negative and in the range of −50 to −80 mV. If the membrane potential is made more negative, the membrane is said to be *hyperpolarized.* If the potential is made more positive, the membrane is *depolarized.* Processes that tend to increase the probability of a channel opening with depolarization are called activation processes. Processes that tend to decrease the probability of a channel opening with depolarization are called inactivation processes. The reversal of of an activation process with hyperpolarization is called *deactivation* and the reversal of an inactivation process with depolarization is called *deinactivation.*

Deactivation and inactivation are very different from one another, but both processes can act on the same channel. The Hodgkin-Huxley model assumes that the activation and inactivation processes of the sodium channels are independent of each other. Imagine there being two physical gates, an activation gate and an inactivation gate, that must be open for the channel to be open. As the membrane depolarizes, the activation gate is open more of the time and the inactivation gate is closed more. Deactivation reflects the greater tendency of activation gates to close with hyperpolarization. The Hodgkin-Huxley model also assumes that the gating of each type of channel depends only on the membrane potential and not upon the state of other channels.

The kinetics of activation and inactivation processes are not instantaneous, so the gating variables are not simple functions of membrane potential. To study these kinetics, Hodgkin and Huxley used voltage clamp protocols. Instead of measuring the membrane potential, they measured the current in a feedback circuit that maintained the membrane at a fixed, predetermined potential. Under these conditions, the gating variables approach steady-state values that are functions of the membrane potential. One of the fundamental observations leading to the models is that the currents generated as the membrane approaches its steady state can be modeled by sums of exponential functions. In a similar fashion to the analysis of residence times in Markov chain models, this suggests that, at any fixed membrane potential, the gating variables should satisfy linear differential equations.

Recall that exponential functions are solutions of linear differential equations. The differential equations describing the gating variables are frequently written in the form

$$\tau \dot{x} = -(x - x_\infty) \tag{3.3}$$

where τ and x_∞ are functions of the membrane potential. The solution of this equation is

$$x(t) = x_\infty + (x(0) - x_\infty)\exp(-t/\tau). \tag{3.4}$$

This formula expresses quantitatively the exponential relaxation of the variable x to its steady state (or "infinity") value x_∞ with the time constant τ. The Hodgkin-Huxley model has equations of this form for each of the gating variables m, n, h. For each gating variable, the voltage-dependent steady-state values and time constants are measured with voltage clamp experiments. This is easier for the potassium currents that have only one type of gate than for the sodium currents that have two types of gates. To measure the values for the sodium currents, the contributions of the activation and inactivation gating to the measured currents in voltage clamp must be separated.

To estimate the parameters in a model for the currents, the contributions from the different kinds of channels need to be separated. Ideally, the membrane is treated so that only one type of channel is active at a time, and the changes observed in the membrane come from changes in a single gating variable. We then take observed values for the gating variables $x(t)$ from the voltage clamp data and fit exponential functions to estimate x_∞ and τ. The actual data come from voltage clamp protocols in which the membrane is initially held at one potential, and then abruptly changed to another potential. Transient currents then flow as the gating variables tend to their new steady states as described by equation [3.4]. This experiment is repeated for a series of different final membrane potentials, with the exponential fits to each producing estimates for x_∞ and τ as functions of membrane potential. Figure 3.7 illustrates the results of this voltage clamp protocol when applied to the squid axon without any separation of currents.

In squid axon, there are two principal types of voltage-dependent channels, sodium and potassium. The conductance of a leak current is independent of voltage, so it can be determined in squid axon by measurements at hyperpolarized potentials when neither sodium or potassium currents are active. The separation of voltage dependent currents is now commonly done pharmacologically by treating the membrane with substances that block specific channels. The sodium channels of the squid axon are blocked by tetrodotoxin (TTX), a poison that is collected from pufferfish, and the potassium channels are blocked by tetraethylammonium (TEA) (Figure 3.8). Blockers were not available to Hodgkin and Huxley, so they use a different method to measure the separate potassium and sodium currents.

Hodgkin and Huxley proceeded to develop quantitative models for each gating variable based on their voltage clamp measurements. However, as an introduction to this type of model we will now consider a simpler model formulated by Morris and Lecar (1981) for excitability of barnacle giant muscle fiber. This system has two voltage-dependent conductances, like the squid giant axon, but their voltage dependence is simpler. Nonetheless, the system displays varied oscillatory behaviors that we examine in Chapter 5.

The inward current of barnacle muscle is carried by Ca^{2+} ions rather than Na^+ ions as in the squid axon. In both cases the outward currents flow through potassium channels. Neither the calcium nor potassium channels in the muscle barnacle show appreciable inactivation. Thus the Kirchhoff current equation for this system can be written in the form

$$C\dot{v} = i - mg_{Ca}(v - v_{Ca}) - wg_K(v - v_K) - g_L(v - v_L). \qquad [3.5]$$

Here C is the capacitance of the membrane, g_{Ca}, g_K, and g_L are the maximal conductances of the calcium, potassium, and leak conductances, v_{Ca}, v_K, and v_L are their reversal potentials, m is the activation variable of the calcium channels, w is the activation variable of the potassium variable, and i represents current injected via an intracellular electrode.

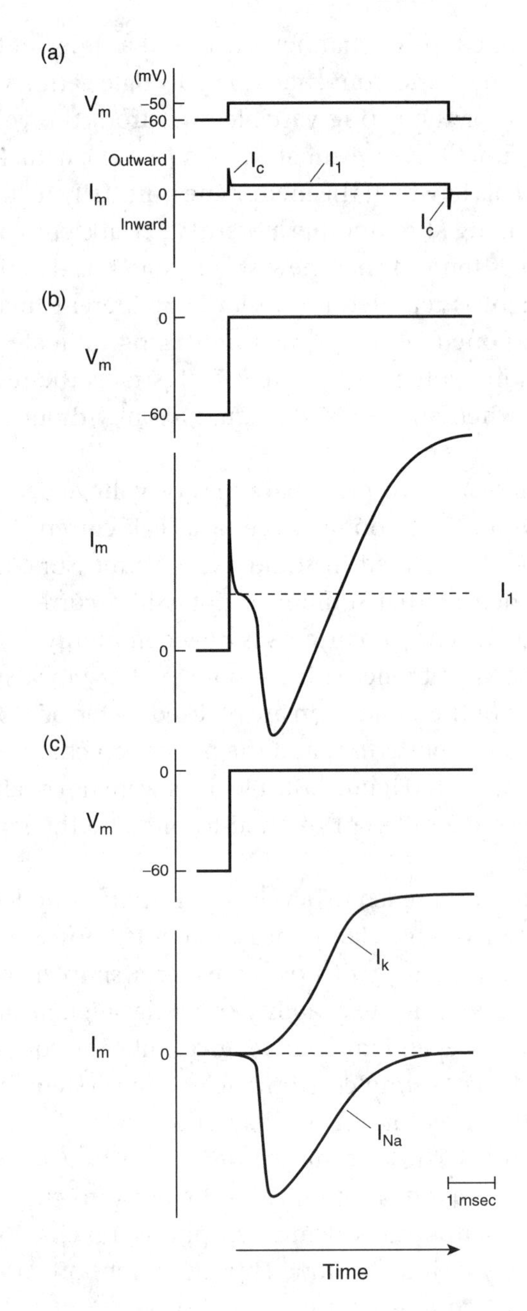

Figure 3.8 The separation of voltage clamp currents in the squid axon into sodium and potassium currents. Pharmacological blockers (TTX for sodium, TEA for potassium) are used to block channels and remove the contribution of the blocked channels from the voltage clamp current. The spikes at the beginning of the step in the second and fourth traces and at the end of the second trace is capacitative current due to the redistribution of ions on the two sides of the membrane when its potential changes (from Kandel et al. 1991, p. 107).

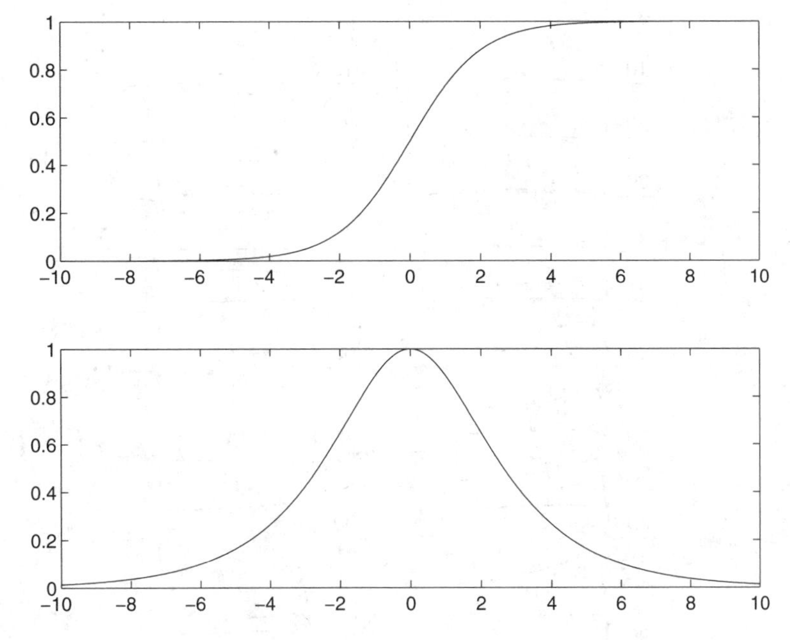

Figure 3.9 Plots of the functions $1/1 + \exp(-v)$ and $1/\cosh(-v/2)$.

Morris and Lecar used a theoretical argument of Lecar et al. (1975) to determine the functions $x_\infty(v)$ and $\tau(v)$ in the differential equations [3.3] for the gating variables. Lecar et al. (1975) argued that the functions $x_\infty(v)$ and $\tau(v)$ in these equations should have the forms

$$m_\infty(v) = 0.5\left(1 + \tanh\left(\frac{v - v_1}{v_2}\right)\right) = \frac{1}{1 + \exp(2(v_1 - v)/v_2)}$$
$$\tau(v) = \phi \frac{1}{\cosh((v - v_1)/2v_2)} \qquad [3.6]$$

for channels that satisfy four properties: (1) they have just one closed state and one open state, (2) the ratio of closed to open states is an exponential function of the difference between energy minima associated with the two states, (3) the difference between energy minima depends linearly upon the membrane potential relative to the potential at which half the channels are open, and (4) transition rates for opening and closing are reciprocals up to a constant factor. In these equations, v_1 is the steady-state potential at which half of the channels are open, v_2 is a factor that determines the steepness of the voltage dependence of m_∞, and ϕ determines the amplitude of the time constant. Figure 3.9 shows plots of these two functions for $v_1 = 0$ and $v_2 = -1$ and $\phi = 1$.

Morris and Lecar (1981) fitted voltage clamp data of Keynes et al. (1973), shown in Figure 3.10, to this model for the gating variable equations. Thus the full model

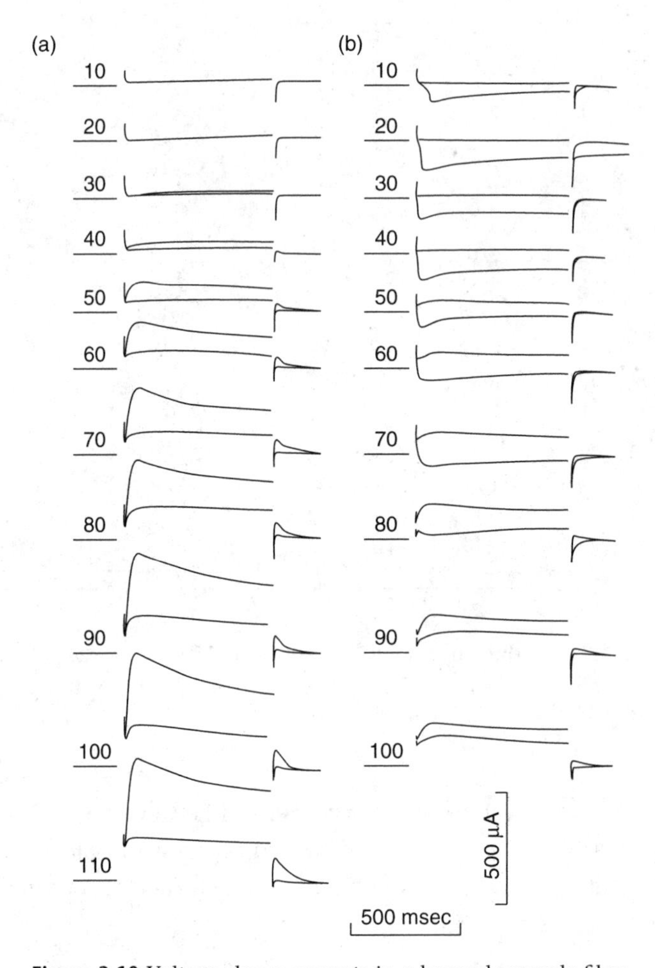

Figure 3.10 Voltage clamp currents in a barnacle muscle fiber. The left hand column shows currents elicited in a Ca^{2+} free bathing medium, thereby eliminating calcium currents. The right hand column shows currents elicited in a muscle fiber perfused with TEA to block potassium currents. Note that the currents that follow the depolarizing voltage step are shown as well as those during the step (from Keynes et al. 1973, Figure 8).

that they studied is the following system of three differential equations:

$$C\dot{v} = i - g_{Ca}m_\infty(v)(v - v_{Ca}) - g_K w(v - v_K) - g_L(v - v_L)$$

$$\tau_m(v)\dot{m} = \phi_m(m_\infty(v) - m) \qquad [3.7]$$

$$\tau_w(v)\dot{w} = \phi_w(w_\infty(v) - w),$$

g_L	g_{Ca}	g_K	v_L	v_{Ca}	v_K	v_1	v_2	v_3	v_4	ϕ_m	ϕ_w	C
2	4	8	−50	100	−70	0	15	10	20	1	10	20

Table 3.1 Parameter values of the Morris-Lecar model producing sustained oscillations (Morris and Lecar 1981, Figure 6)

where

$$\begin{aligned} m_\infty(v) &= 0.5\left(1+\tanh\left(\frac{v-v_1}{v_2}\right)\right) \\ \tau_m(v) &= \frac{1}{\cosh((v-v_1)/2v_2)} \\ w_\infty(v) &= 0.5\left(1+\tanh\left(\frac{v-v_3}{v_4}\right)\right) \\ \tau_w(v) &= \frac{1}{\cosh((v-v_3)/2v_4)}. \end{aligned} \qquad [3.8]$$

Typical parameter values for the model are shown in Table 3.1. Note that $\phi_m/\phi_w = 0.1$ and $\phi_m/C = 0.05$ are much smaller than 1 in this set of parameters. Consequently, the gating variable m tends to change more quickly than w or v. For this reason, a frequent approximation, similar to the analysis of enzyme kinetics in Chapter 1, is that $\tau_m = 0$ and m tracks its "infinity" value $m_\infty(v)$. This reduces the number of differential equations in the Morris-Lecar model from three to two. In Chapter 5, we shall use this reduced system of two differential equations as a case study while we study techniques for solving nonlinear systems of differential equations.

3.5 Summary

This chapter has applied the theory of power-positive matrices to models for membrane channels. Channels switch between conformational states at seemingly random times. This process is modeled as a Markov chain, and the steady-state distribution of a population of channels converges to the dominant eigenvector of its transition matrix as the population size grows. While the behavior of individual channels appears random, the fluctuations in the current from a population of channels are related to the population size by the central limit theorem of probability theory. The relative magnitude of the fluctuations decreases with population size, so large channel populations can be modeled as deterministic currents. Hodgkin and Huxley introduced a method to construct differential equations models for the membrane potential based on gating variables. The gating variables are now interpreted as giving steady-state probabilities for channels to be in an open state, and their rates of change reflect the convergence rate to the dominant eigenvector. The steady states and time constants of the gating variables are calculated from voltage clamp data.

The systems of differential equations derived through the Hodgkin-Huxley formalism are highly nonlinear and cannot be solved analytically. Instead, numerical algorithms are used to calculate approximate solutions. Systems of differential equations are one of the most common type of model for not only biological phenomena, but across the sciences and engineering. Chapter 4 examines additional examples of such models that have been used to "design" and construct gene regulatory networks that display particular kinds of dynamical behavior. We use these examples to introduce mathematical concepts that help us to understand the solutions to differential equations models. Chapter 5 reexamines the Morris-Lecar model while giving a more extensive introduction to numerical methods for solving differential equations and to the mathematical theory used to study their solutions.

3.6 Appendix: The Central Limit Theorem

The Central Limit Theorem is a far-reaching generalization of the DeMoivre-Laplace limit theorem for independent sequences of coin tosses. We will meet the Central Limit Theorem again in Chapter 7. The intuitive setting for the central limit theorem is an experiment that is repeated many times, each time producing a numerical result X. We assume that the results vary from trial to trial as we repeat the experiment, perhaps because of measurement error or because the outcome is unpredictable like the toss of a fair coin. However, we assume that the probability of each outcome is the same on each trial, and that the results of one trial do not affect those of subsequent trials. These assumptions are expressed by saying that the outcomes obtained on N different trials are *independent* and *identically distributed* random variables $X_i,\ i = 1, 2, \ldots, N$.

If the experiment has only a finite number of possible values $v_1, \ldots, v_n$, we can think of an X_i as yielding these outcomes with probabilities $p_1, \ldots, p_n$. Over many trials, we expect to get outcome v_1 in a fraction p_1 of the trials, v_2 in a fraction p_2 of the trials, and so on. The average value of the outcomes is then approximately

$$\frac{p_1 N v_1 + p_2 N v_2 + \cdots + p_n N v_n}{N} = \sum_{k=1}^{N} p_k v_k \qquad [3.9]$$

when N is large. This motivates the definition that the *expectation* or *expected value* of the random variable X is given by the formula

$$E(X) = \sum_{k=1}^{N} p_k v_k.$$

The expectation is also called the *mean*.

The Central Limit Theorem tells us about how well [3.9] approximates the average as the number of trials N becomes larger and larger. This naturally depends on how unpredictable each of the individual experiments is. The unpredictability of a single experiment can be measured by the *variance*

$$\mathrm{Var}(X) = E((X-\mu)^2) = \sum_{k=1}^{N} p_k(v_k-\mu)^2. \qquad [3.10]$$

A small variance means that X is usually close to its expected value; a high variance means that large departures from the expected value occur often. The symbols μ_X and σ_X^2 are often used to denote the mean and variance of a random variable X.

Let

$$S_N = X_1 + X_2 + \cdots + X_N \qquad [3.11]$$

be the total of all outcomes in N trials, so that S_N/N is the average outcome. Using properties of the mean and variance (which can be found in any introductory probability or statistics text), it can be shown that the rescaled sum

$$Z_N = \frac{S_N - N\mu_X}{\sqrt{N}\sigma_X} \qquad [3.12]$$

always has mean 0 and variance 1. The Central Limit Theorem states not only that this is true, but that the distribution of the scaled deviation Z_N approaches a normal (Gaussian) distribution with mean 0 and variance 1 as N becomes larger and larger. That is, the probability of Z_N taking a value in the interval $[a, b]$ converges to

$$\frac{1}{\sqrt{2\pi}} \int_a^b e^{-x^2/2} dx. \qquad [3.13]$$

This expresses the remarkable fact that if we average the results of performing the same experiment more and more times, not only should we expect the average value we obtain to converge to the mean value, but deviations from this mean value converge a normal distribution—regardless of the probability distribution of the outcomes on any one repetition of the experiment.

We can informally summarize the Central Limit Theorem by writing [3.12] in the form

$$S_N = N\mu_X + \sqrt{N}\sigma_X Z_N \qquad [3.14]$$

where Z_N is approximately a Gaussian random variable with mean 0 and variance 1. The right-hand side of [3.14] consists of two parts: the "deterministic" part $N\mu_X$ and the "random" part $\sqrt{N}\sigma_X Z_N$. As the number of trials increases, the deterministic part grows proportionally to N while the random part only grows proportionally to $\sqrt{N}$, so in a very large number of trials—such as the very large numbers of membrane channels in a neuron—the deterministic part will be dominant.

Versions of the Central Limit Theorem also apply to random variables with a continuous distribution such as the Gaussian, meaning that they can take any value in some segment (finite or infinite) of the real line; to random variables that are independent but not necessarily identical in distribution; and to many cases where the random variables are not completely independent.

3.7 References

Anderson, C., and C. Stevens. 1973. Voltage clamp analysis of acetylcholine produced end-plate current fluctuations at frog neuromuscular junction, Journal of Physiology. 235: 655–691.

Colquhoun, D., and A. Hawkes. 1995. The principles of the stochastic interpretation of ion-channel mechanisms. Pages 397–481 in *Single Channel Recording*, 2nd edition, B. Sakmann and E. Neher, eds., Plenum, New York.

Colquhoun, D., and B. Sakmann. 1985. Fast events in single-channel currents activated by acetylcholine and its analogues at the frog muscle end-plate, Journal of Physiology 369: 501–557.

Colquhoun, D., and B. Sakmann. 1998. From muscle endplate to brain synapses: A short history of synapses and agonist-activated ion channels. Neuron 20: 381–387.

Diaconis, P., S. Holmes, and R. Montgomery. 2004. Dynamical bias in the coin toss. Online at http://www-stat.stanford.edu/ cgates/PERSI/by_year.html

Hille, B. 2001. *Ion Channels of Excitable Membranes*, 3rd edition, Sinauer Associates, Sunderland, MA, Chapter 2 and pp. 172–185.

Hodgkin, A., and A. Huxley. 1952. A quantitative description of membrane current and its application to conduction and excitation in nerve. Journal of Physiology 117: 500–544.

Johnston, D., and S. Wu. 1995. *Foundations of Cellular Neurophysiology*, MIT Press, Cambridge, MA.

Kandel, E., J. Schwartz, and T. Jessell. 1991. *Principles of Neural Science*. Appleton and Lange.

Katz, B., and S. Thesleff. 1957. A study of the "desensitization" produced by acetylcholine at the motor end-plate. Journal of Physiology 138: 63–80.

Keynes, R., E. Rojas, R. Taylor, and J. Vergara. 1973. Calcium and potassium systems of a giant barnacle muscle fiber under membrane potential control. Journal of Physiology 229: 409–455.

Knuth, D. E. 1981. *The Art of Computer Programming*. Vol. 2, *Seminumerical Algorithms*. 2nd edition. Addison-Wesley, Reading, MA.

Lawler, G. 1995. *Introduction to Stochastic Processes* Chapman & Hall, London, pp. 53–61.

Lecar, H., G. Ehrenstein, and R. Latorre. 1975. Mechanism for channel gating in excitable bilayers. Annals of the New York Academy of Science 264: 304–313.

Morris, C., and H. Lecar. 1981. Voltage oscillations in the barnacle giant muscle fiber. Biophysical Journal, 35: 193–213.

Neher, E., and B. Sakmann. 1976. Single channel currents recorded from membrane of denervated frog muscle fibres. Nature 260: 799–802.

4 Cellular Dynamics: Pathways of Gene Expression

The dynamics of neurons and neural networks are sufficiently complicated that we turn to simple systems of gene regulation for our first studies of nonlinear differential equations. The interiors of cells are filled with complex structures that are constantly changing. Modern molecular biology has developed astounding technologies to determine the constituents of cells, including the ability to sequence the entire genomes of organisms. The "central dogma" of molecular biology describes how proteins are produced from genes. Gene expression has two main steps: (1) transcription of DNA produces mRNA with complementary nucleotide sequences and (2) translation produces proteins with amino acid sequences corresponding to sequential triplets of nucleotides in the mRNA. Proteins also play a key role in the regulation of gene expression by binding to DNA to either block or enhance the expression of particular genes. Thus, there are *feedback* loops in which protein induces or inhibits gene expression which produces protein. There are large numbers of proteins, and these interact with each other as well.

We model these feedback processes in the same way we model chemical reactions. The variables in the models are concentrations of mRNA and proteins. We assume that the rates of production of these molecules are functions of their concentrations. However, the complexity of the reaction networks is daunting. Few reaction rates have been measured, and the protein interactions are far more complicated than the elementary reactions described by the law of mass action. Without well-established laws for deriving reaction rates, we make simplified models that incorporate basic aspects of gene expression and regulation. The models are unlikely to be quantitatively accurate, but they can give insight into qualitative features of the dynamics of the regulatory networks.

We will study two simple examples of synthesized networks of gene regulation. These are systems that have been constructed with plasmids and inserted into bacteria as demonstrations that it may be possible to engineer gene regulatory networks in analogy with the way we design electrical circuits or industrial chemical reactors. There is much interest in developing these techniques of "synthetic biology" into a new domain of bioengineering (Ferber 2004). Differential equations models provide the starting point for this engineering. Beginning with components whose characteristics we understand, we want to build systems from these components that accomplish a desired task. Here we define the desired tasks by the system dynamics, in one case building a gene regulatory circuit that oscillates and in the other building a circuit that acts like a switch or memory element in a computer. Transcription and translation are simpler in bacteria than in eukaryote cells where much larger molecular complexes carry out transcription and translation. In this chapter, we consider gene expression in bacteria only.

The gene regulation models we study are nonlinear systems of differential equations. It is seldom possible to find explicit expressions for the solutions of such equations. In our discussion of enzyme kinetics in Chapter 1, we performed an analysis that led to approximate solutions. Here we confront the typical situation in which we rely upon numerical methods to solve *initial value problems* for the equations. The initial value problem specifies a starting state for the system at an initial time and then uses the differential equations to predict the state of the system at later times. The methods we use employ a time-stepping procedure. We do a computation with the equations that predicts the state of the system a short time after the initial time. Then we update the state to the predicted state and repeat the procedure, computing a new predicted state a short time later than the last update. Iterating the computation of new predicted states and updates many times, we arrive at a prediction of the state of the system at much later times.

Frequently, we are interested in the *asymptotic* behavior of the system—the state(s) of the system after very long times. Will the system approach an equilibrium where the state no longer changes, will it continue to change in regular periodic oscillations, or will it continue to change in more complicated ways? This chapter and the next introduce mathematical theory that helps us to answer these questions. The mathematics characterizes patterns of dynamical behavior that are found repeatedly in different systems and provides guidelines for the numerical investigation of specific systems. This chapter uses examples of synthetic gene regulation networks to introduce the phenomena that are studied more systematically from a mathematical perspective in Chapter 5.

4.1 Biological Background

Regulation of transcription by molecules that bind to DNA plays a critical role in the development of each cell and in determining how it responds to its environment. Transcription and translation can be compared with an assembly line: the steps on the assembly line are the individual reactions that happen during the entire process. Genes are "switched" on and off. However, unlike on an assembly line, there is no supervisor flipping the switch. Instead, there are complex networks of chemical reactions that underlie the regulation of gene expression. Transcription of a gene requires that the polymerase locate the beginning and the end of the gene. In addition to *coding sequences* that contain the templates for proteins, the DNA has regulatory sequences of nucleotides. Polymerases attach to *promoter* regions of DNA that lie near the beginning of a coding region for a gene (or an *operon*, a group of genes that are transcribed together) and detach at *terminator* regions. The rate at which gene transcription occurs is determined largely by the binding rate of polymerase to promoter. This varies from gene to gene, and it is also actively regulated by other proteins. *Repressor* proteins bind to the promoter of a gene, preventing transcription. *Activators* are proteins that increase transcription rates.

Pathways are reaction networks that connect gene products with activators and repressors. They can be enormously complex, involving hundreds of chemical species (Kohn 1999). Pathways can be viewed graphically as a depiction of the chemical species that participate in varied reactions. Loops within these networks indicate feedback, in which a sequence of reactions starting with a particular chemical species affects the rate of production of that species. The simplest loops are ones in which a gene codes for a repressor of that gene. Pathways of gene expression and regulation are central to many fundamental biological processes, including cell division, differentiation of cells during development and the generation of circadian rhythms.[1] Schematic pathway information is adequate for many purposes, such as identifying mutations that are likely to disrupt a pathway, or potential targets for drugs. However, we also need to know the rates of reactions if we are to predict quantitative properties of the system. For example, we might want to model the effect of a nonlethal mutation, a change in substrate on the doubling time of a bacterial population, or the free-running period of a circadian rhythm oscillator. Additionally, many gene regulatory processes support several different dynamical behaviors. In these cases dynamical models are needed in order to understand the processes.

To understand what is happening in the cell, we would like to measure fluctuating chemical concentrations and reaction rates. There have been breathtaking improvements in biotechnologies during the past fifty years, but we are still far from being able to observe the details needed to construct accurate dynamical models of these cellular processes. That makes it difficult to construct dynamic models that reproduce observed phenomena accurately. A few researchers have

---

[1]Interest in pathways is intense: the Science Magazine Signal Transduction Knowledge Environment provides an online interface (http://stke.sciencemag.org/cm/) to databases of information on the components of cellular signaling pathways and their relations to one another.

begun to circumvent these difficulties by turning the problem around: instead of developing models to fit experimental data on cells, they have used the new technologies to engineer biological systems that correspond to simple models. This approach is attracting attention in the scientific press as "synthetic biology" (Ferber 2004). In this chapter we shall look at two examples of synthesized networks that perform different functions. The first is a "clock" and the second is a "switch."

4.2 A Gene Network That Acts as a Clock

Elowitz and Leibler (2000) constructed an oscillatory network based upon three transcriptional repressors inserted into *E. coli* bacteria with a plasmid. They chose repressors *lacI*, *TetR*, and *cI*. The names and functions of the repressors are unimportant; what matters is that *lacI* inhibits transcription of the gene coding for *TetR*, *TetR* inhibits transcription of the gene coding for *cI*, and *cI* inhibits transcription of the gene coding for *lacI*. This pattern of inhibition describes a *negative feedback loop* in the interactions of these proteins and the expression of their genes on the plasmid. Figure 4.1 shows a representation of this gene regulatory network.

In the absence of inhibition, each of the three proteins reaches a steady-state concentration resulting from a balance between its production and degradation rates within the bacterium. But with cross-inhibition by the other two repressors, this network architecture is capable of producing oscillations. Imagine that we start with *lacI* at high concentration. It inhibits *TetR* production, so that the concentration of *TetR* soon falls due to degradation. That leaves *cI* free from inhibition, so it increases in concentration and inhibits *lacI* production. Consequently, *lacI* soon falls to low concentration, allowing *TetR* to build up, which inhibits *cI* and eventually allows the concentration of *lacI* to recover. Thus the concentration of each repressor waxes and wanes, out of phase with the other repressors. For this scenario to produce oscillations, it is important that the length of the loop is odd, so that the indirect effect of each repressor on itself is negative: 1 inhibiting 2 which allows 3 to build up and inhibit 1. The same kind of alternation would occur in a sequential inhibition loop of length 5, 7, and so on. But in a loop of length 4, one could have repressors 1 and 3 remaining always at high concentration, inhibiting 2 and 4, which always remain at low concentrations.

However, oscillations are not the only possible outcome in a loop of length 3. Instead, the three repressor concentrations might approach a steady state in which each is present, being somewhat inhibited by its repressor but nonetheless being transcribed at a sufficient rate to balance degradation. To understand the conditions under which one behavior or the other will be present, we need to develop a dynamic model for the network.

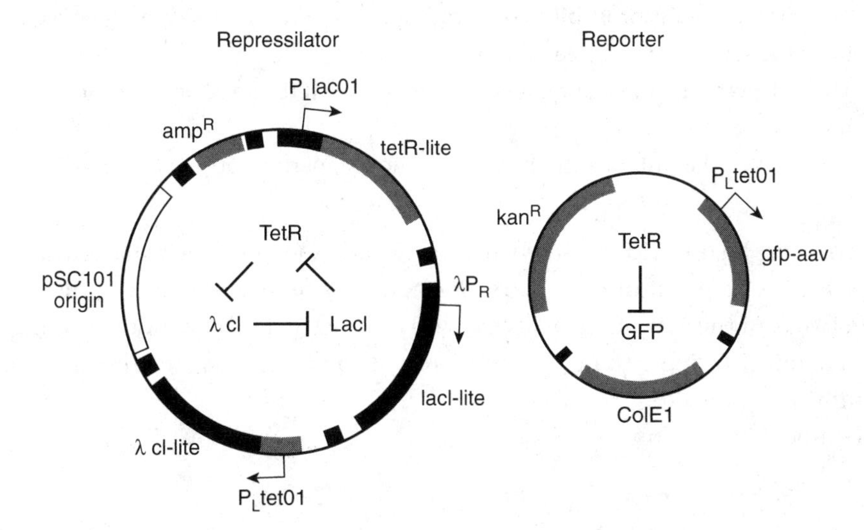

Figure 4.1 A schematic diagram of the repressilator. Genes for the *lacI, TetR* and *cI* repressor proteins together with their promoters are assembled on a plasmid which is inserted into an *E. coli* bacterium. A second plasmid that contains a *TetR* promoter region and a gene that codes for green fluorescent protein is also inserted into the bacterium. In the absence of *TetR*, the bacteria with these plasmids will produce green fluorescent protein (from Elowitz and Leibler 2000).

4.2.1 Formulating a Model

Elowitz and Leibler formulated a model system of differential equations that describes the rates of change for the concentration p_i of each protein repressor and the concentration m_i of its associated mRNA in their network. Here the subscripts i label the three types of repressor: we let i take the values *lacI, tetR, cI*. For each i, the equations give the rates of change of p_i and m_i. If we were interested in building a model of high fidelity that would be quantitatively correct, we would need to measure these rates and determine whether they depend upon other quantities as well. That would be a lot of work, so we settle for less. We make plausible assumptions about the qualitative properties of the production rates of repressors and their associated mRNAs, hoping that analysis of the model will suggest which general properties are important to obtain oscillations in the network. The assumptions are as follows:

- There is a constant probability of decay for each mRNA molecule, which has the same value for the mRNA of all three repressors.
- The synthesis rate of mRNA for each repressor is a decreasing function of the concentration of the repressor that inhibits transcription of that RNA. Again, the three functions for each mRNA are assumed to be the same.

- There is a constant probability of decay for each protein molecule, again assumed to be the same for the three repressors.
- The synthesis rate of each repressor is proportional to the concentration of its mRNA.
- Synthesis of the mRNA and repressors is independent of other variables.

None of these assumptions is likely to be satisfied exactly by the real network, with the possible exception of the fourth. Rather, they represent the main features of the network in the simplest possible way. We construct such a model hoping that the dynamical properties of the network will be insensitive to the simplifying assumptions we make.

The model equations are

$$\dot{m}_i = -m_i + \frac{\alpha}{1+p_j^n} + \alpha_0$$
$$\dot{p}_i = -\beta(p_i - m_i). \qquad [4.1]$$

When i takes one of the values *lacI,tetR,cI* in the equations for $\dot{m}_i$, the corresponding value of j is *cI,lacI,tetR*. That is, j corresponds to the protein that inhibits transcription of i. This model has been called the *repressilator* (a repression-driven oscillator).

The differential equations for the concentrations of the mRNAs m_i and proteins p_i all consist of two components: a positive term representing production rate, and a negative term representing degradation. For mRNA the production rate is $\alpha/(1+p_j^n)+\alpha_0$ and the degradation rate is $-m_i$; for protein the production rate is βm_i and the degradation rate is βp_i. Thus each concentration in the model is a "bathtub" (in the sense of Chapter 1) with a single inflow (production) and outflow (degradation). The dynamics can become complicated because the tubs are not independent: the "water level" in one tub (e.g., the concentration of one repressor) affects the flow rate of another.

The parameters in these equations are α_0, α, β, and n, representing the rate of transcription of mRNA in the presence of a saturating concentration of repressor, the additional rate of transcription in absence of inhibitor, the ratio of the rate of decay of protein to mRNA, and a "cooperativity" coefficient in the function describing the concentration dependence of repression.[2]

[2]Units of time and concentration in the model have been "scaled" to make these equations "nondimensional." This means that the variables m_i, p_i, and time have been multiplied by fixed scalars to reduce the number of parameters that would otherwise be present in the equations. One of the reductions is to make the time variable correspond to the decay rate of the mRNA so that the coefficient of m_i in the equation for $\dot{m}_i$ is -1: when time is replaced by $\tau = at$ in the equation

$$\frac{dx}{dt} = -ax$$

the result is

$$\frac{dx}{d\tau} = \frac{dx}{dt}\frac{dt}{d\tau} = (-ax)\frac{1}{a} = -x.$$

The model uses the function $\alpha/(1 + p_j^n) + \alpha_0$ to represent the repression of mRNA synthesis. As the concentration of the inhibiting protein increases, the synthesis rate falls from $\alpha + \alpha_0$ (when inhibiting protein is absent) to α_0 (when inhibiting protein is at very high concentration). The *Hill coefficient* n reflects the "cooperativity" of the binding of repressor to promoter. This function is "borrowed" from the theory of enzyme kinetics, and is used here as a reasonable guess about how the synthesis rate depends upon repressor concentration. It is not based upon a fit to experimental data that would allow us to choose a rate function that corresponds quantitatively to the real system. This is one more way in which the model is unrealistic.

Exercise 4.1. How would the repressilator model change if the transcription rates of the three genes differed from each other? if the translation rates of the three mRNA differed from each other?

4.2.2 Model Predictions

The model gives us an artificial world in which we can investigate when a cyclic network of three repressors will oscillate versus settling into a steady state with unchanging concentrations. This is done by solving *initial value problems* for the differential equations: given values of m_i and p_i at an initial time t_0, there are unique functions of time $m_i(t)$ and $p_i(t)$ that solve the differential equation and have the specified initial values $m_i(t_0), p_i(t_0)$. Though there are seldom formulas that give the solutions of differential equations, there are good numerical methods for producing approximate solutions by an iterative process. In *one-step* methods, a time step h_1 is selected and the solution estimated at time $t_1 = t_0 + h_1$. Next a time-step h_2 is selected and the solution estimated at time $t_2 = t_1 + h_2$, using the computed (approximate) values of $m_i(t_1), p_i(t_1)$. This process continues, producing a sequence of times t_j and approximate solution values $m_i(t_j), p_i(t_j)$. The result of these computations is a *simulation* of the real network: the output of the simulations represents the dynamics of the model network.

Figure 4.2 illustrates the results of a simulation with initial conditions

$$(m_{lacI}, m_{tetR}, m_{cl}, p_{lacI}, p_{tetR}, p_{cl}) = (0.2, 0.1, 0.3, 0.1, 0.4, 0.5)$$

and parameters $(\alpha_0, \alpha, \beta, n) = (0, 50, 0.2, 2)$ while Figure 4.3 gives the solutions for the same initial conditions and parameters except that $\alpha = 1$. These two simulations illustrate that the *asymptotic* behavior of the system depends upon the values of the parameters. In both cases, the solutions appear to settle into a regular behavior after a *transient* period. In the simulation with the larger value of α, the trajectory tends to an oscillatory solution after a transient lasting approximately 150 time units. When α is small, the trajectory approaches an equilibrium after approximately 50 time units and remains steady thereafter. Approach to a periodic oscillation and approach to an equilibrium are *qualitatively* different

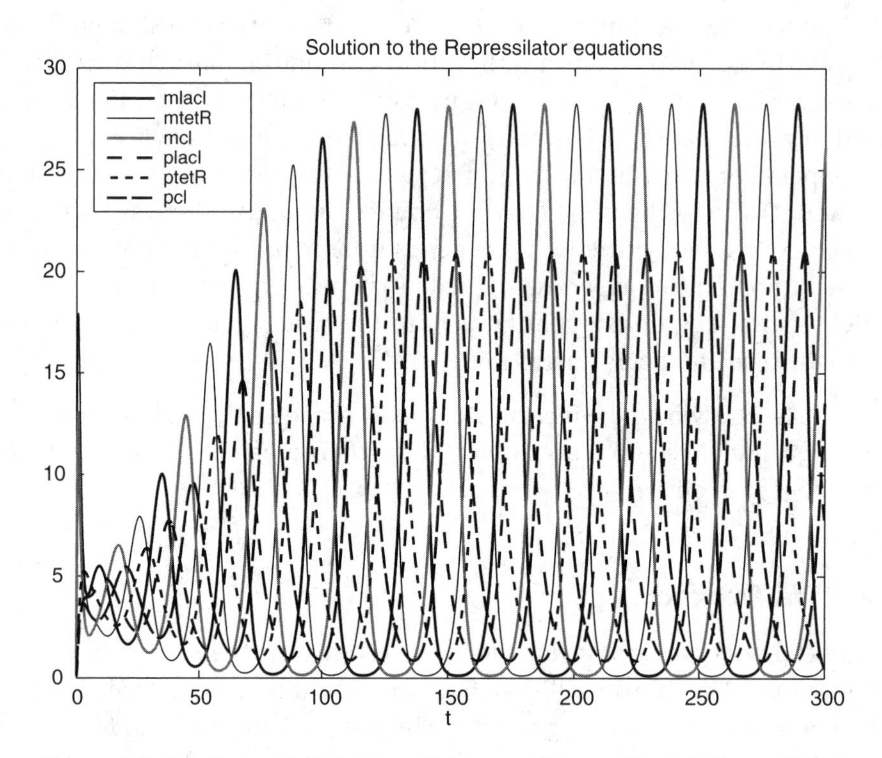

Figure 4.2 Solutions of the repressilator equations with initial conditions $(m_{lacI}, m_{tetR}, m_{cI}, p_{lacI}, p_{tetR}, p_{cI}) = (0.2, 0.1, 0.3, 0.1, 0.4, 0.5)$. The parameters have values $(\alpha_0, \alpha, \beta, n) = (0, 50, 0.2, 2)$.

behaviors, not just differences in the magnitude of an oscillation. When α is small, there are no solutions that oscillate and all solutions with positive concentrations at their initial conditions approach the equilibrium found here. We conclude that there must be a special value of the parameter α where a transition occurs between the regime where the solutions have oscillations and the regime where they do not. Recall that α represents the difference between transcription rates in the absence of repressor and in the presence of high concentrations of repressor. Oscillations are more likely to occur when repressors bind tightly and reduce transcription rates substantially.

Model simulations allow us to answer specific "what if" questions quickly. For example, we can explore the effects of changes in other parameters upon the model behavior. Figure 4.4 shows a simulation for parameters $(\alpha_0, \alpha, \beta, n) = (1, 50, 0.2, 2)$ and the same initial value as above. Here, we have made the "residual" transcription rate in the presence of repressor positive. Again the oscillations die away and the trajectory approaches an equilibrium solution. The rate of approach is slower than when $(\alpha_0, \alpha, \beta, n) = (0, 1, 0.2, 2)$.

So when do trajectories tend to oscillatory solutions and when do they tend to equilibrium (steady-state) solutions? Are other types of long-term behavior

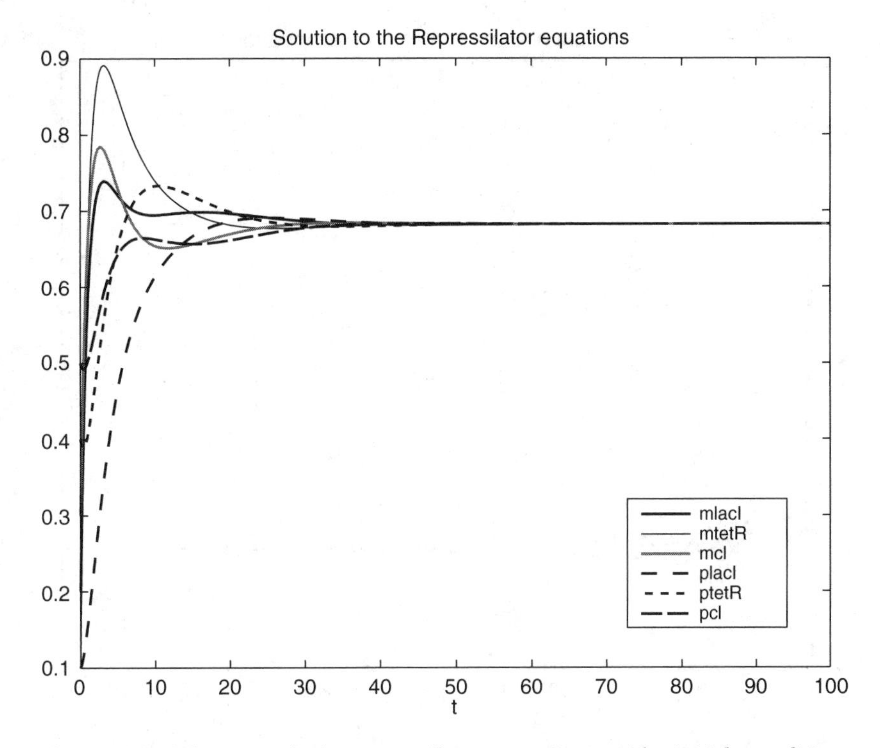

Figure 4.3 Solutions of the repressilator equations with initial conditions $(m_{lacI}, m_{tetR}, m_{cl}, p_{lacI}, p_{tetR}, p_{cl}) = (0.2, 0.1, 0.3, 0.1, 0.4, 0.5)$. The parameters have values $(\alpha_0, \alpha, \beta, n) = (0, 1, 0.2, 2)$.

possible? These are more difficult questions to answer. Chapter 5 gives a general introduction to mathematical theory that helps us answer these questions. Dynamical systems theory provides a conceptual framework for thinking about all of the solutions to systems of differential equations and how they depend upon parameters. Here we preview a few of the concepts while discussing the dynamics of the repressilator.

Equilibrium solutions of a system of differential equations occur where the right-hand sides of the equations vanish. That is, in order for all state variables to remain constant, each of them must have its time derivative equal to 0. In our example, this condition is a system of six nonlinear equations in the six variables m_i, p_i. This may seem daunting, but we can rapidly reduce the complexity of the problem by observing that the equations $\dot{p}_i = -\beta(p_i - m_i)$ imply that $m_i = p_i$ at the equilibria. This leaves only three variables (the three m's or the three p's) to solve for.

Next, we observe that the system of equations [4.1] is symmetric with respect to permuting the repressors and mRNAs: replacing *lacI,tetR,cl* by *cl,lacI,tetR* throughout the equations gives the same system of equations again. A consequence of this symmetry is that equilibrium solutions are likely to occur with all the m_i equal.

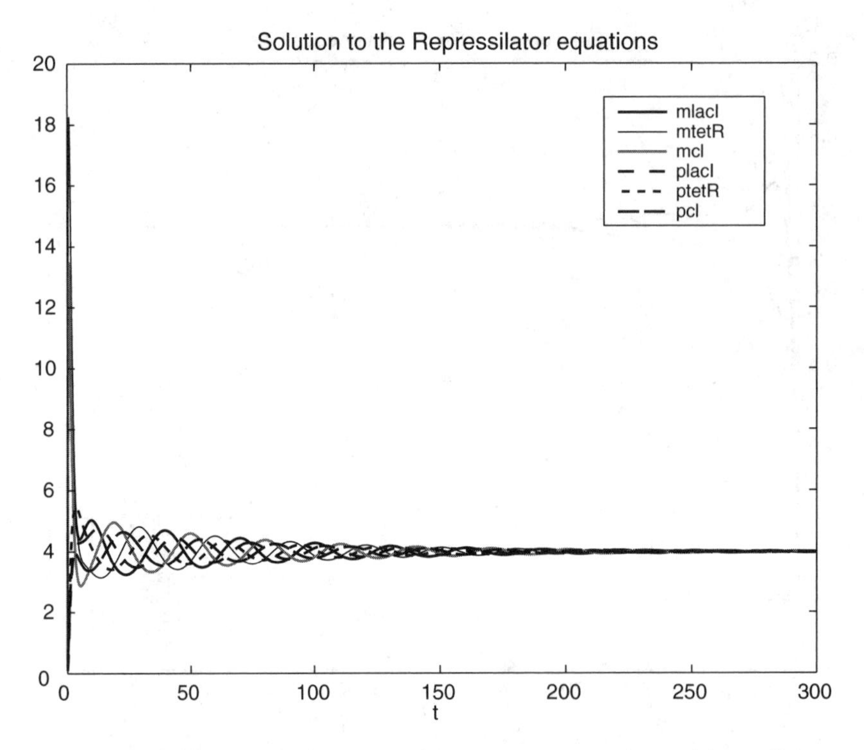

Figure 4.4 Solutions of the repressilator equations with initial conditions $(m_{lacI}, m_{tetR}, m_{cI}, p_{lacI}, p_{tetR}, p_{cI}) = (0.2, 0.1, 0.3, 0.1, 0.4, 0.5)$. The parameters have values $(\alpha_0, \alpha, \beta, n) = (1, 50, 0.2, 2)$.

Thus, if $-p + \alpha/(1 + p^n) + \alpha_0 = 0$, there is an equilibrium solution of the system [4.1] with $m_i = p_i = p$ for each index i. The function $E(p) = -p + \alpha/(1 + p^n) + \alpha_0$, plotted in Figure 4.5, is decreasing for $p > 0$, has value $\alpha_0 + \alpha$ at $p = 0$, and tends to $-\infty$ as $p \to \infty$, so it has exactly one positive root p. This equilibrium solution of the equation is present for all sensible parameter values in the model. A slightly more complicated argument demonstrates that other equilibrium solutions (i.e., ones with unequal concentrations for the three repressors) are not possible in this model.

The stability of the equilibrium determines whether nearby solutions tend to the equilibrium. The procedure for assessing the stability of the equilibrium is to first "linearize" the system of equations and then use linear algebra to help solve the linearized system of equations.[3] The linearized system has the form $\dot{\xi} = A\xi$ where A is the 6×6 *Jacobian* matrix of partial derivatives of the right-hand side of [4.1] at the equilibrium point. The stability of the linear system is determined

[3]We will discuss linearization and the mathematical theory of linear systems of differential equations in Chapter 5.

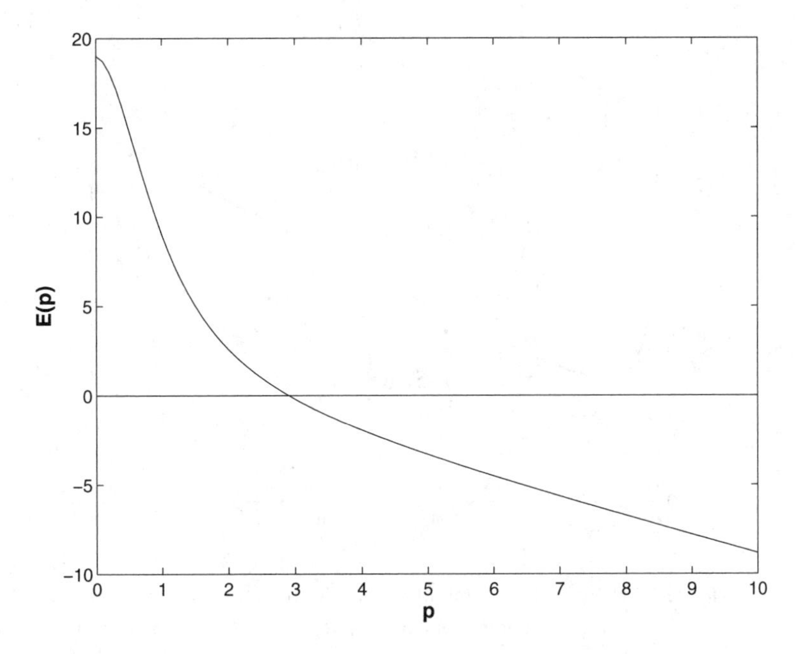

Figure 4.5 Graph of the function $E(p) = -p + 18/(1 + p^2) + 1$.

by the eigenvalues of A. In particular, the equilibrium is stable if all eigenvalues of A have negative real parts.

By analyzing the eigenvalues of the Jacobian matrix, Elowitz and Leibler find that the stability region consists of the set of parameters for which

$$\frac{3X^2}{4 + 2X} < \frac{(\beta + 1)^2}{\beta} \qquad [4.2]$$

where

$$X = -\frac{\alpha n p^{n-1}}{(1 + p^n)^2}$$

and p is the steady state defined by [4.3]

$$p = \frac{\alpha}{1 + p^n} + \alpha_0.$$

The stability region [4.2] is diagrammed in Figure 4.6. If the parameter combination $Y = 3X^2/(4 + 2X)$ lies in the "Stable" region in the diagram, the equilibrium is stable; otherwise it is unstable and the network oscillates. The curve $Y = (\beta + 1)^2/\beta$ that bounds the stability region attains its minimum value 4 at $\beta = 1$. Recall that β is the degradation rate of the proteins, and the model has been scaled so the mRNAs all degrade at rate 1. Thus, the minimum occurs when the proteins and mRNAs have similar degradation rates. This situation gives the widest possible range of values for the other model parameters, which determine the values of X and Y, such that the model equilibrium is unstable. Elowitz and Leibler conclude from this analysis that the propensity for oscillations in

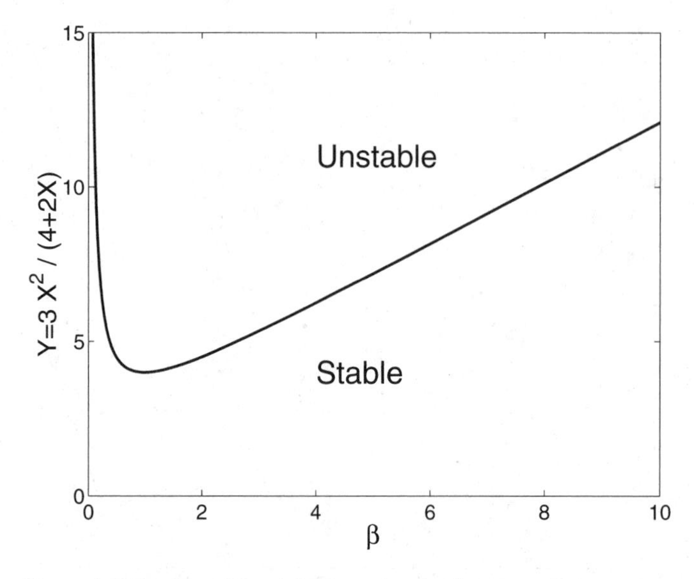

Figure 4.6 Diagram of the stability region for the repressilator system in terms of β and the parameter combination $Y = 3X^2/(4 + 2X)$, with X defined by equations [4.3].

the repressilator is greatest when the degradation rates for protein and mRNA are similar.

If the parameters are such that $Y < 4$, then the equilibrium is stable for any value of β. If $Y > 4$, then there is a range of β values around $\beta = 1$ for which the equilibrium is unstable. So the larger the value of Y, the broader the range of β values giving rise to oscillations. To make use of this result for designing an oscillating network, we have to relate the stability region (defined in terms of Y) to the values of the model parameters (α, α_0, n). This is difficult to do in general because the relationships are nonlinear, but one situation conducive to oscillations can be seen fairly easily. Suppose that $\alpha_0 = 0$; then in [4.3] we have $p = \alpha/(1 + p^n)$ and so

$$\alpha = p(1 + p^n). \quad [4.4]$$

This implies $p \approx 0$ if $\alpha \approx 0$, while p will be large if α is large. Moreover, substituting [4.4] into the definition of X in equation [4.3] we get

$$X = -\frac{np^n}{1 + p^n}.$$

Thus if $p \approx 0$ we will have $X \approx 0$ and therefore stability of the equilibrium. Increasing α (and consequently increasing p and the magnitude of X) gives larger and larger values of Y, moving into the "Unstable" region in Figure 4.6. This suggests that one way to get oscillations is by using repressors that bind tightly ($\alpha_0 \approx 0$) and cause a large drop in the synthesis of mRNA (α large). On the other hand, if α_0 becomes large, then [4.3] implies that $p \geq \alpha_0$ so p is also large. The first line of equation [4.3] then implies that X is small when α_0 is large, giving stability.

So the model suggests three design guidelines to produce an oscillatory network:

- comparable degradation rates,
- tight binding by the repressors, and
- genes that are abundantly expressed in the absence of their repressors.

Exercise 4.2. In the repressilator model, the limit $\beta \to \infty$ corresponds to a situation in which the mRNA and repressor concentrations for each gene and gene product are the same. Why? In this limit, the model reduces to one for just the three mRNA concentrations. Implement this model and explore (1) the differences between its solutions and those of the full repressilator model with $\beta = 10$ and (2) whether the reduced model can also produce oscillations.

4.3 Networks That Act as a Switch

The second example we examine is a simpler network with a toggle switch or "flip-flop" that was designed by Gardner, Cantor, and Collins (2000). The biological systems that they engineered are similar to the ones Elowitz and Leibler utilized: *E. coli* altered by the insertion of plasmids. The primary difference between the two lies in the architecture of the altered gene regulatory networks of the engineered bacteria. Instead of a cyclic network of three repressors that inhibit synthesis of a mRNA, Gardner et al. engineered a network of just two repressors, each of which inhibits the synthesis of the mRNA for the other. Their goal was to produce regulatory systems that were *bistable*. A system is bistable if it has more than one attracting state—in this case two stable equilibria. They successfully engineered several networks that met this goal, all using the *lacI* repressor as one of the two repressors. Figure 4.7 shows a schematic diagram of the network together with data illustrating its function. They also demonstrated that there are networks with the same architecture of mutual inhibition that are not bistable.

To understand which kinetic properties lead to bistability in a network with mutual inhibition, we again turn to differential equations models. The differential equations model that Gardner et al. utilized to study their network is simpler than the system [4.1]. The kinetics of mRNA and protein synthesis for each repressor is aggregated into a single variable, leaving a model with just two state variables u and v representing concentrations of the two repressors. The model

(a)

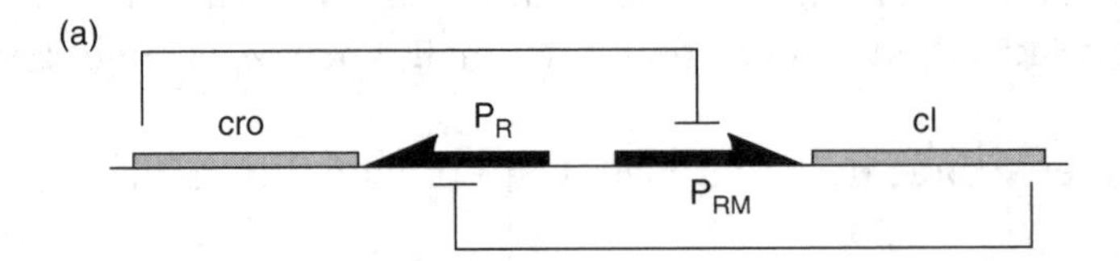

(b)

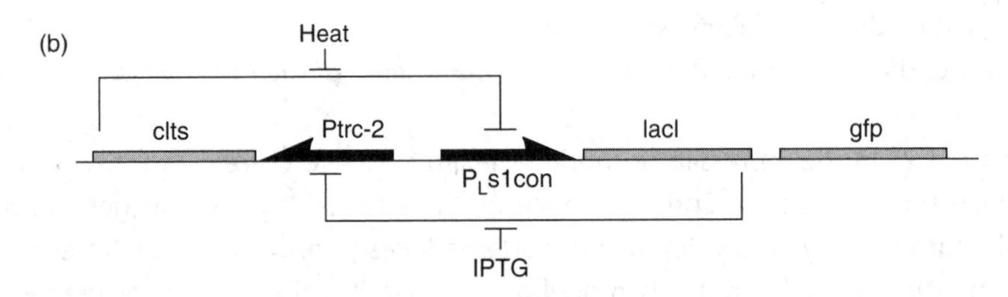

(c)

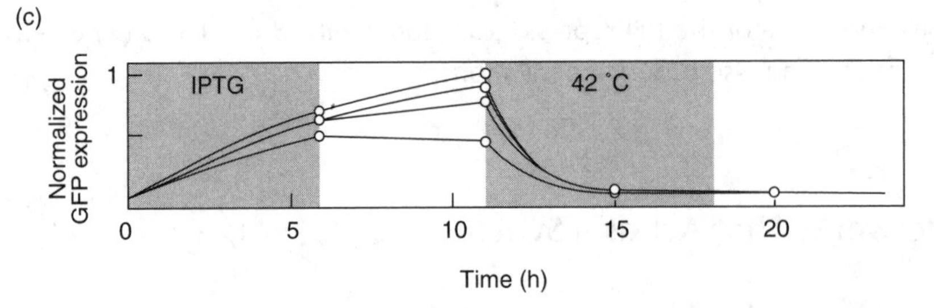

Figure 4.7 A schematic diagram of bistable gene regulatory networks in bacteriophage λ. Two genes code for repressor proteins of the other gene. (a) A natural switch. (b) A switch engineered from *cl* and *LacI* genes in which the repressor of the lacl gene is a temperature sensitive version of the Cl protein. Green fluorescent protein is also produced when the *cl* gene is expressed. (c) In the first gray bar, IPTG eliminates repression of the *clts* gene by *LacI*. The cell continues to produce Cl when IPTG is removed. In the second gray period, the temperature is raised, eliminating repression of the lacl gene by Cl protein. Transcription of the the *cl* gene stops and does not resume when the temperature is reduced (from Hasty et al. 2002).

equations are

$$\begin{aligned} \dot{u} &= -u + \frac{\alpha_u}{1 + v^\beta} \\ \dot{v} &= -v + \frac{\alpha_v}{1 + u^\gamma}. \end{aligned} \qquad [4.5]$$

The system [4.5] is simpler than the repressilator system [4.1]. The repressilator equations [4.1] can be reduced to a smaller system of equations using similar reasoning to that used in Chapter 1 to develop the Michaelis-Menten equation for enzyme kinetics. If β is large in the repressilator model [4.1], then p_i changes rapidly until $p_i \approx m_i$ holds. If we make the approximation that $p_i = m_i$, then the repressilator equations yield a system similar to [4.5] with one differential equation for each of its three mRNA-protein pairs. Biologically, this assumes a situation where protein concentrations are determined by a balance between fast translation and degradation processes. Transcription (mRNA synthesis) happens more slowly than the translation of protein, so the protein concentration remains in balance with the mRNA concentration. Since this approximation in the repressilator model corresponds to increasing β without bound, the stability diagram Figure 4.6 suggests that the oscillations disappear in the reduced system.

Comparing model [4.5] with [4.1], we also see that it assumes $\alpha_0 = 0$ but allows different synthesis rates and cooperativity for the two repressors. If $\alpha_u = \alpha_v$ and $\beta = \gamma$, then the system is symmetric with respect to the operation of interchanging u and v. We make this assumption in investigating the dynamics of the model. We set

$$\alpha_u = \alpha_v = a, \text{and } \beta = \gamma = b. \qquad [4.6]$$

The system [4.5] does not admit periodic solutions.[4] All of its solutions tend to equilibrium points. Thus we determine whether the system is bistable by finding the equilibrium points and their stability. We exploit the symmetry of [4.5] with [4.6] to seek equilibrium solutions that are symmetric under interchange of u and v; that is, $u = v$. We see from [4.5] that a symmetric equilibrium (u, u) is a solution of $f(u) = 0$ where $f(u) = -u + a/(1 + u^b)$. There will always be exactly one symmetric equilibrium because

- $f(u)$ is decreasing: $f'(u) = -1 - bau^{b-1}/(1 + u^b)^2 < 0$,
- $f(0) = a > 0$, and
- $f(a) = -a^{1+b}/(1 + a^b) < 0$.

Looking for asymmetric equilibria takes a bit more work. To do that we have to consider the two *nullclines*—the curves in the (u, v) plane where $\dot{u} = 0$ (the u nullcline) and where $\dot{v} = 0$ (the v nullcline). These are given, respectively, by the equations

$$u = a/(1 + v^b) \qquad [4.7]$$

$$v = a/(1 + u^b). \qquad [4.8]$$

Figure 4.8 shows these curves for selected values of a when $b = 2$. There is always one point of intersection at the symmetric equilibrium, which exists for any values of a and b. For $a = 1$ the symmetric equilibrium is the only intersection and hence the only equilibrium, but for $a = 3$ there are three points of intersection and hence three equilibrium points for the model. The transition between these situations (called a *bifurcation*) occurs when the nullclines are tangent to

[4]For those of you who have studied differential equations, the proof is short: the divergence of the system [4.5] is negative, so areas in the phase plane shrink under the flow. However, the region bounded within a periodic solution would be invariant and maintain constant area under the flow.

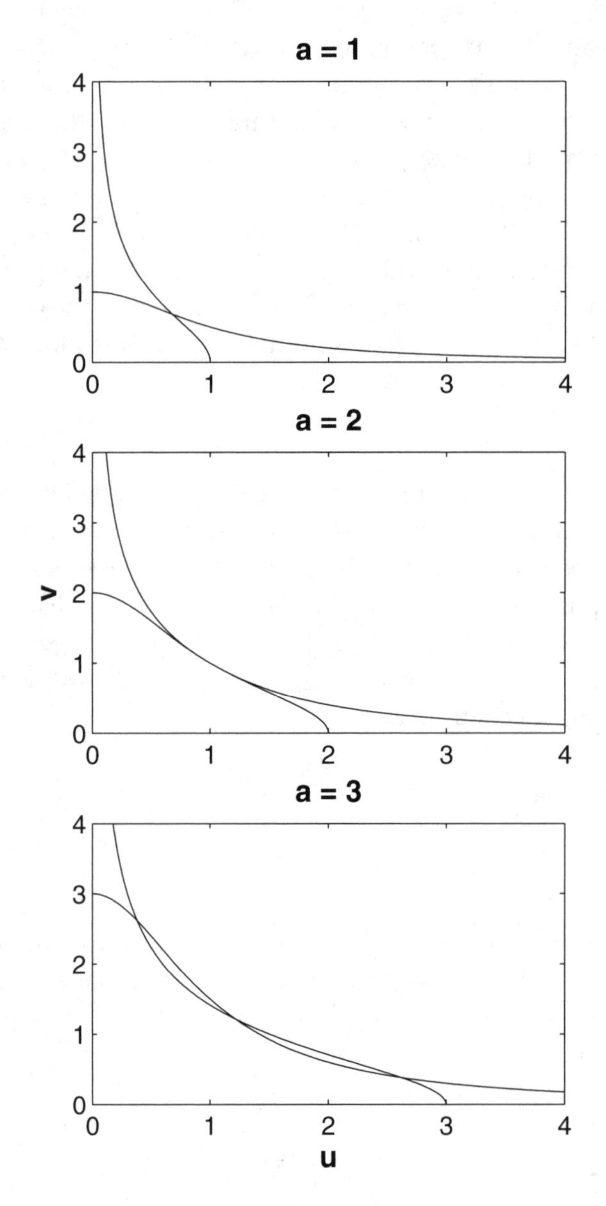

Figure 4.8 Plots of the nullclines for the model [4.5] in the (u, v) plane for values of a below, above, and exactly at the bifurcation between one and three equilibria for the model.

each other at the symmetric equilibrium, that is, they have the same slope (the case $a = 2$ in Figure 4.8).

Exercise 4.3. For what parameter combinations (a, b) do the two nullclines have exactly the same slope at the symmetric equilibrium?

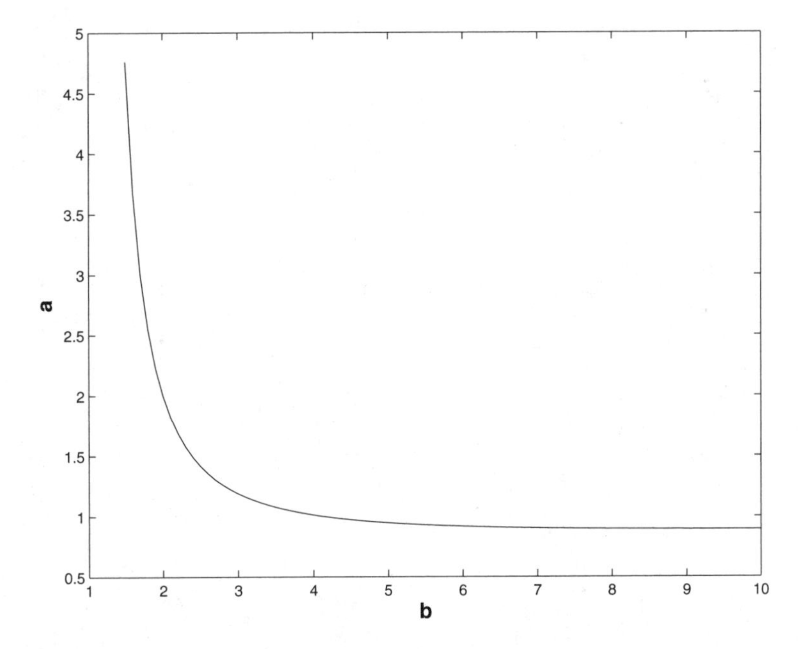

Figure 4.9 Bifurcation diagram of the (b, a) plane for the model [4.5]. For values of (b, a) below the bifurcation curve, there is a single equilibrium point, while for values of (b, a) above the bifurcation curve, there are three equilibrium points.

The conclusion of this exercise is that the bifurcation of the symmetric equilibrium occurs where

$$a = b(b - 1)^{-(1+b)/b}.$$

Figure 4.9 shows the graph of this function. Note that, when $b = 2$, the bifurcation occurs at $a = 2$. For values of (a, b) below the bifurcation curve, there is a single equilibrium point, while for values of (a, b) above the bifurcation curve, there are three equilibrium points.

Figures 4.10 and 4.11 show phase portraits in these two cases. In Figure 4.10 the symmetric equilibrium is stable, and all solutions tend toward it. In Figure 4.11, the symmetric equilibrium has become unstable but the two other equilibria are both stable. Some initial conditions produce trajectories that approach one stable state while other initial conditions produce trajectories that approach the second stable state. The bifurcation that occurs in the model is called a *pitchfork* bifurcation because one equilibrium splits into three as a parameter is varied. Pitchfork bifurcations are typically found at symmetric equilibria in systems that have a symmetry. Figure 4.11 also shows the nullclines of the system. The three equilibrium points lie at the intersections of the nullclines. These figures illustrate the visualization of phase portraits for systems of two differential equations. We

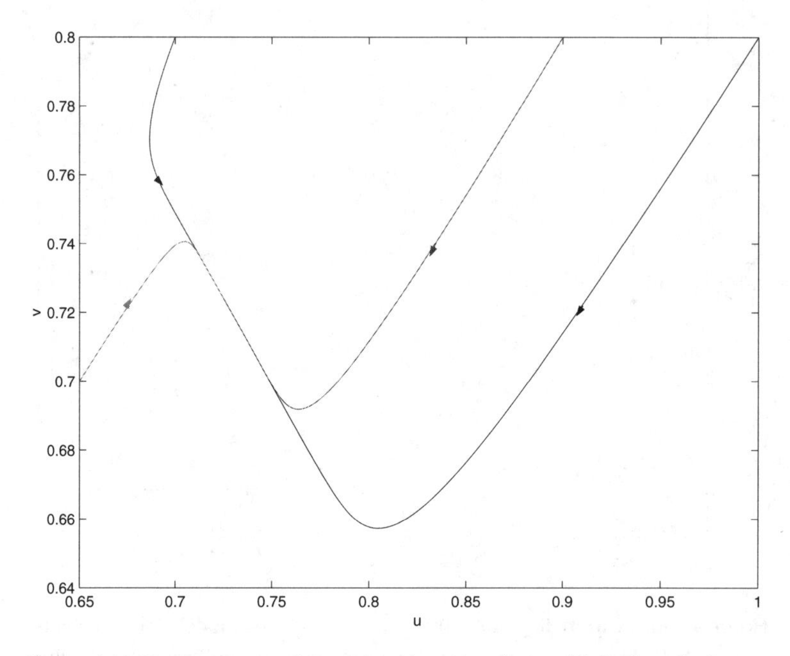

Figure 4.10 Phase portrait of the (u, v) plane for the model [4.5] with $(b, a) = (3, 1)$. There is a single equilibrium point, which is stable and lies on the line $u = v$.

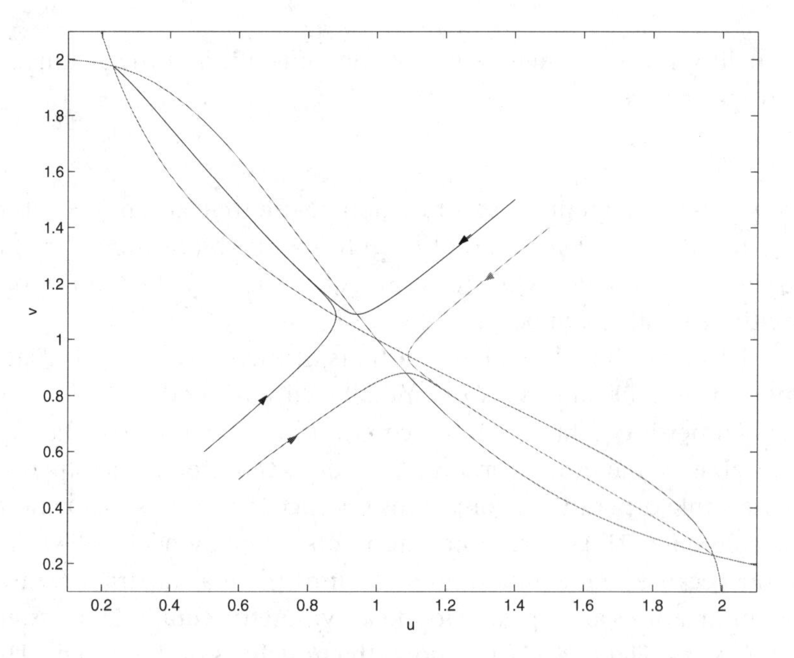

Figure 4.11 Phase portrait of the (u, v) plane for the model [4.5] with $(e, a) = (3, 2)$. There are three equilibrium points, located at the intersections of the nullclines (the curves that reach the boundary of the bounding box).

see the paths taken by points (u, v) as they move along solutions to the system of equations. The techniques that we have applied here,

- nondimensionalizing the model to reduce the number of parameters,
- plotting the nullclines to locate equilibria, and
- determining the local stability of the equilibria as a function of parameter values,

are often all it takes to determine the important properties of a model's solutions.

As with the repressilator, the model [4.5] gives clues about how to choose repressors in this regulatory network that will make a good toggle switch. In particular, increasing protein synthesis rates makes the two stable states of the system move farther apart in the phase plane. The enhanced distance between the repressor concentrations makes it both easier to distinguish the two stable states and harder for perturbations to push the switch from one state to another. Large distance between the stable states makes it hard for "noise" to cause the system to switch spontaneously between them; larger inputs are needed to switch the system when one desires to do so. It is not evident on the phase plane plots, but increasing a also makes the switching between the states faster. Increasing the cooperativity parameter b makes the bifurcation value of the protein synthesis rate a decrease. We conclude that large a and b both work to make the switch more robust but perhaps harder to purposefully switch between states.

Exercise 4.4. Explore the behavior of the solutions to the system [4.5] when the parameters $\alpha_u \neq \alpha_v$ and $\beta \neq \gamma$. Can you find significantly different types of dynamical behavior than in the cases where each pair of parameters is equal?

4.4 Systems Biology

"Systems biology" is a new term, created in the past few years to describe an area of research whose boundaries remain fuzzy. Although organs, organisms, and ecosystems are all systems—in the conventional sense that we can regard them as composed of interacting elements—systems biology tends to focus on processes at the cellular level. The new journal *Systems Biology* has the following introduction:

> Systems biology involves modelling and simulating the complex dynamic interactions between genes, transcripts, proteins, metabolites and cells using integrated systems-based approaches. Encompassing proteomics, transcriptomics, metabolomics and functional genomics, systems biology uses computational and mathematical models to analyze and simulate networks, pathways and the spatial and temporal relationships that give rise to cause and effect in living systems. Such work is of great importance to a better understanding of disease mechanisms, pharmaceutical drug discovery and drug target validation.

The "omics" listed above refer to methods for collecting and analyzing large amounts of data about the molecular biology of cells. Whole genome sequences of many organisms are one striking example of such data. The humane genome has billions of nucleotides and thousands of genes. Computational tools are needed to manage these data, and even tasks like searching for similar sequences in different genomes require efficient algorithms. Systems biology seeks to discover the organization of these complex systems. Cells have intricate internal structure, with organelles like ribosomes and mitochondria and smaller molecular complexes like DNA polymerase that play an essential role in how cells function. Structural information is not enough, however, to understand how cells work. Life is inherently a dynamic process. There are varied approaches that systems biology is taking to model and simulate these dynamics. Systems biology research has been accelerating rapidly, but this is an area of research that is in its infancy and fundamentally new ideas may be needed to close the enormous gap between goals that have been articulated and present accomplishments.

The models of gene regulation in this chapter are simple examples of systems biology models. One simplified view of the cell is to regard it as a soup of reacting chemicals. This leads us to model the cell as sets of chemical reactions as we have done with our simple gene regulation circuits. Conceptually, we can try to "scale up" from systems with two or three genes and their represssors to whole cells with thousands of genes, molecules, and molecular complexes that function as distinct entities in cellular dynamics. Practically, there are enormous challenges in constructing models that faithfully reproduce the dynamical processes we observe in cells. We give a brief overview of some of these challenges for systems biology research in this section, from our perspective as dynamicists.

Much of current systems biology research is directed at discovering "networks" and "pathways." Viewing the cell as a large system of interactions among components like chemical reactions, its dynamics is modeled as a large set of ordinary differential equations. There is a network structure to these equations that comes from mass balance. Each interaction transforms a small group of cellular components into another small group. We can represent the possible transforms as a *directed graph* in which there is a vertex for each component and an arrow from component A to component B if there is an interaction that directly transforms A to B. We can seldom see directly the interactions that take place in the cell, but there are two classes of techniques that enable us to infer some of the network structure. First, we can detect correlations of the concentrations of different entities with "high-throughput" measurements. Example techniques include microarrays that measure large numbers of distinct mRNA molecules and the simultaneous measurements of the concentration of large numbers of proteins with mass spectrometry. Second, genetic modifications of an organism enable us to decrease or enhance expression of individual genes and to modify proteins in specific ways. The effects of these genetic modifications yield insight into the pathways and processes that involve specific genes and proteins.

Exercise 4.5. Draw a graph that represents the interactions between components of the repressilator network.

Construction of networks for gene regulation and cell signaling is one prominent activity for systems biology. Networks with hundreds to thousands of nodes have been constructed to represent such processes as the cell cycle (Kohn 1999) and the initiation of cancer (http://www.gnsbiotech.com/biology.php). For the purposes of creating new drugs, such networks help to identify specific targets in pathways that malfunction during a disease. Can we do more? What can we infer about how a system works from the structure of its network? We can augment the information in the network by classifying different types of entities and their interactions and using labels to identify these on a graph. Feedback in a system can be identified with loops in the graph, and we may be able to determine from the labels whether feedback is positive or negative; i.e., whether the interactions serve to enhance or suppress the production of components within the loop. However, as we have seen with the simple switch and repressilator networks, a single network can support qualitatively different dynamics that depend on quantitative parameters such as reaction rates. Thus, we would like to develop dynamic models for cellular processes that are consistent with the network graphs.

Differential equation models for a reaction network require rate laws that describe how fast each reaction proceeds. In many situations we cannot derive the rate laws from first principles, so we must either measure them experimentally or choose a rate law based on the limited information that we do have. Isolating the components of reactions in large network models and measuring their kinetics is simply not feasible. As in the two models considered earlier in this chapter, the Michaelis-Menten rate law for enzyme kinetics derived in Chapter 1 and mass action kinetics are frequently used in systems biology models. The functional forms chosen for rate laws usually have parameters that still must be given values before simulations of the model are possible. With neural models for the electrical excitability of a neuron, the voltage clamp technique is used to estimate many of the model parameters, but better experimental techniques are needed for estimating parameters in the rate laws of individual cellular interactions. Thus, estimating model parameters is one of the challenges we confront when constructing dynamic models of cellular networks.

Transformation of networks and rate laws into differential equations models is a straightforward but tedious and error-prone process when done "by hand." Moreover, comparison and sharing of models among researchers is much easier when standards are formulated for the computer description of models. Accordingly, substantial effort has been invested in the development of specialized languages for defining systems biology models. One example is SBML, the Systems Biology Markup Language, that is designed to provide a "tool-neutral" computer-readable format for representing models of biochemical reaction networks (http://sbml.org/index.psp).

Exercise 4.6. Describe the types of computations performed by three different systems biology software tools. You can find links to many tools on the homepage of the SBML website.

Once a model is constructed we face the question as to whether the simulations are consistent with experimental observations. Since even a modest-size model is likely to have dozens of unmeasured parameters, we can hardly expect that the initial choices that have been made for all of these will combine to yield simulations that do match data. What do we do to "tune" the model to improve the fit between simulations and data? The simplest thing that can be done is to compute simulations for large numbers of different parameters, observing the effects of parameter changes. We can view the computer simulations as a new set of experiments and use the speed of the computer to do lots of them. As the number of parameters increases, a brute force "sweep" through the parameter space becomes impossible. For example, simulating a system with each of ten different values of k parameters that we vary independently requires 10^k simulations altogether. Thus, brute force strategies to identify parameters can only take us so far.

When simulations have been carried out for differential equation models arising in many different research areas, similar patterns of dynamical behavior have been observed. For example, the way in which periodic behavior in the repressilator arises while varying parameters is similar to the way in which it arises in the Morris-Lecar model for action potentials. Dynamical systems theory provides a mathematical framework for explaining these observations that there are "universal" qualitative properties that are found in simulation of differential equations models. We use this theory to bolster our intuition about how properties of the simulations depend upon model parameters. Tyson, Chen, and Novak (2001) review model studies of the cell cycle in yeast, illustrating how dynamical systems theory helps interpret the results of simulations of these models.

Mathematics can also help estimate parameters in a model in another way. If we are interested in the *quantitative* fit of a model to data, then we can formulate parameter estimation as an *optimization* problem. We define an objective function that will be used to measure the difference between a simulation and data. A common choice of objective function is *least squares*. If $x_1, \ldots, x_n$ are observations and $y_1, \ldots, y_n$ are the corresponding values obtained from simulation, then the least squares objective function is $\sum(y_i - x_i)^2$. It is often useful to introduce weights $w_i > 0$ and consider the weighted least squares objective function $\sum w_i(y_i - x_i)^2$.

The value of this objective function is non-negative and the simulation fits the data perfectly if and only if the value of the objective function is zero. The value of any objective function f will depend upon the parameters of a model. The optimization problem is to find the parameters that give the minimum value of the objective function. There are several well-developed computational strategies for solving this optimization problem, but it would take us far afield to pursue them here. Some of the computational tools that have been developed for systems biology (http://www.gepasi.org/,http://www.copasi.org) attempt to solve parameter estimation problems using some basic optimization algorithms. The methods have been tested successfully on systems of a few chemical reactions, but we still do not know whether they will make significant improvements in estimating parameters of large models. Some of the difficulties encountered in application of optimization methods to estimating parameter of network models are addressed by Brown and Sethna (2003).

4.4.1 Complex versus Simple Models

Finding basic "laws" for systems biology is a compelling challenge. There is still much speculation about what those principles might be and how to find them. There is wide agreement that computational models are needed to address this challenge, but there is divergence of opinion as to the complexity of the models we should focus upon.

Taking advantage of experimental work aimed at identifying interaction networks in cells, systems biologists typically formulate and study models with many state variables that embrace the complexity of the system being modeled, rather than trying to identify and model a few "key" variables. Nobody denies the necessity of complex models for applied purposes, where any and all biological details should be incorporated that improve a model's ability to make reliable predictions, though as we discuss in Chapter 9, putting in too much biological detail can sometimes be bad for forecasting accuracy. For example, in a human cancer cell model developed to help design and optimize patient-specific chemotherapy (Christopher et al. 2004), the diagram for the cell apoptosis module includes about forty interlinked state variables and links to two other submodels of undescribed complexity ("cell cycle module" and "p53 module").

But for basic research, there is a long-standing and frequently heated debate over the value of simple "general" models versus complex "realistic" models as tools. On one side Lander (2004, p. 712) dismisses both simple models and the quest for generality:

> ... starting from its heyday in the 1960s, [mathematical biology] provided some interesting insights, but also succeeded in elevating the term "modeling" to near-pejorative status among many biologists. For the most part mathematical biologists

> sought to fit biological data to relatively simple models, with the hope that fundamental laws might be recognized .... This strategy works well in physics and chemistry, but in biology it is stymied by two problems. First, biological data are usually incomplete and extremely imprecise. As new measurements are made, today's models rapidly join tomorrow's scrap heaps. Second, because biological phenomena are generated by large, complex networks of elements there is little reason to expect to discern fundamental laws in them. To do so would be like expecting to discern the fundamental laws of electromagnetism in the output of a personal computer.

In order to "elevate investigations in computation biology to a level where ordinary biologists take serious notice", the "essential elements" include "network topologies anchored in experimental data" and "fine-grained explorations of large parameter spaces" (Lander 2004, p. 714).

The alternative view is that there are basic "pathways," "motifs," and "modules" in cellular processes that are combined as building blocks in the assembly of larger systems, and that there are definite principles in the way in which they are combined. In contrast to Lander, we would say that biological phenomena are generated by evolution, which uses large complex networks for some tasks; this distinction is crucial for the existence of fundamental laws and the levels at which they might be observed. Evolution has the capacity to dramatically alter the shape and behavior of organisms in a few generations, but there are striking similarities in the molecular biology of all organisms, and cross-species comparisons show striking similarities in individual genes. Natural selection has apparently preserved—or repeatedly discovered—successful solutions to the basic tasks that cells face. Biological systems are therefore very special networks which may have general organizational principles.

We build complex machines as assemblies of subsystems that interact in ways that do not depend upon much of the internal structure of the subsystems. We also create new devices using preexisting components—consider the varied uses of teflon, LCDs, and the capacitor. Many biologists (e.g., Alon 2003) have argued that designs produced by evolution exhibit the same properties—modularity and recurring design elements—because good design is good design, whether produced by human intelligence or the trial-and-error of mutation and natural selection. Alon (2003) notes in particular that certain network elements that perform particular tasks, such as buffering the network against input fluctuations, occur repeatedly in many different organisms and appear to have evolved independently numerous times, presumably because they perform the task well. The existence of recurring subsystems with well-defined functions opens the possibility of decomposing biological networks in a hierarchical fashion: using small models to explore the interactions within each kind of subsystem, and different small models to explore the higher-level interactions among subsystems that perform different functions. How well this strategy works depends on the extent to which whole-system behavior is independent of within-subsystem details, which still remains to be seen.

Proponents of simpler models often question the utility of large computer simulations, viewing them as a different type of complex system that is far less interesting than the systems they model. In the context of ecological models, mathematical biologist Robert May (1974, p. 10) expressed the opinion that many "massive computer studies could benefit most from the installation of an on-line incinerator" and that this approach "does not seem conducive to yielding general ecological principles." In principle one can do "fine-grained explorations of large parameter spaces" to understand the behavior of a computational model, as Lander (2004) recommends; in practice it is impossible to do the 10^{50} simulation runs required to try out all combinations of 10 possible values for each of 50 parameters. Similarly Gavrilets (2003) argued that the theory of speciation has failed to advance because of over-reliance on complex simulation models:

> The most general feature of simulation models is that their results are very specific. Interpretation of numerical simulations, the interpolation of their results for other parameter values, and making generalizations based on simulations are notoriously difficult ... simulation results are usually impossible to reproduce because many technical details are not described in original publications.

According to Gavrilets (2003) the crucial step for making progress is "the development of simple and general dynamical models that can be studied not only numerically but analytically as well." Ginzburg and Jensen (2004) compare complex models with large parameter spaces to the Ptolemaic astronomy based on multilayered epicycles: adaptable enough to fit any data, and therefore meaningless.

Of course the quotes above were carefully selected to showcase extreme positions. Discussions aiming for the center are also available—indeed, the quotes above from May (1974) were pulled from a balanced discussion of the roles of simple versus complex models, and we also recommend Hess (1996), Peck (2004), and the papers by Hall (1988) and Caswell (1988) that deliberately stand at opposite ends of the field and take aim at each other. The relative roles that simple and complex dynamic models will play in systems biology are uncertain, but it seems that both will be used increasingly as we accumulate more information about the networks and pathways of intracellular interactions.

4.5 Summary

This chapter has explored the dynamics of differential equation models for two synthetic gene regulation networks. The models were chosen to illustrate how simulation is used to study nonlinear differential equations. The focus is upon long-time, asymptotic properties of the solutions. We distinguish solutions that tend to an equilibrium, or steady state, from solutions that tend to periodic orbits. In the next chapter, we give an introduction to the mathematics of dynamical systems theory as a conceptual framework for the study of such questions. The theory emphasizes qualitative properties of solutions to differential equations like the distinction between solutions that tend to equilibria and those that tend to periodic orbits. The results of simulations are fitted into a geometric picture that shows the asymptotic properties of all solutions.

The application of dynamical systems methods to gene regulatory networks is a subject that is still in its infancy. To date, systems biology has been more effective in developing experimental methods for determining networks of interactions among genes, proteins, and other cellular components than in developing differential equation models for these complex systems. Simulation is essentially the only general method for determining solutions of differential equations. Estimating parameters in large models that lead to good fits with data is an outstanding challenge for systems biology. We do not know whether there are general principles underlying the internal organization of cells that will help us to build better models of intracellular systems. The mathematics of dynamical systems theory and optimization theory do help us in investigations of large models of these systems.

4.6 References

Alon, U. 2003. Biological networks: the tinkerer as an engineer. Science 301: 1866–1867.

Atkinson, M., M. Savageau, J. Myers, and A. Ninfa. 2003. Development of genetic circuitry exhibiting toggle switch or oscillatory behavior in *Escherichia coli*. Cell 113: 597–607.

Brown, K., and J. Sethna. 2003. Statistical mechanics approaches to models with many poorly known parameters. Physical Review E 68: 021904.

Caswell, H. 1988. Theory and models in ecology: A different perspective. Ecological Modelling 43: 33–44.

Christopher, R., A. Dhiman, J. Fox, R. Gendelman, T. Haberitcher, D. Kagle, G. Spizz, I. G. Khalil, and C. Hill. 2004. Data-driven computer simulation of human cancer cell. Annals of the NY Academy of Science 1020: 132–153.

Elowitz, M., and S. Leibler. 2000. A synthetic oscillatory network of transcriptional regulators. Nature 403: 335–338.

Ferber, D. 2004. Microbes made to order. Science 303: 158–161.

Gardner, T., C. Cantor, and J. Collins. 2000. Construction of a genetic toggle switch in *Escherichia coli*. Nature 403: 339–342.

Gavrilets, S. 2003. Models of speciation: What have we learned in 40 years? Evolution 57: 2197–2215.

Ginzburg, L. R., and C.X.J. Jensen. 2004. Rules of thumb for judging ecological theories. Trends in Ecology and Evolution 19: 121–126.

Hall, C.A.S. 1988. An assessment of several of the historically most influential theoretical models in ecology and of the data provided in their support. Ecological Modelling 43: 5–31.

Hasty, J., D. McMillen, and J. Collins. 2002. Engineered gene circuits. Nature 402: 224–230.

Hess, G. R. 1996. To analyze, or to stimulate, is that the question? American Entomologist 42: 14–16.

Kohn, K. 1999. Molecular interaction map of the mammalian cell cycle control and DNA repair system. Molecular Biology of the Cell 10: 2703–2734.

Lander, A. D. 2004. A calculus of purpose. PLoS Biology 2: 712–714.

May, R. M. 1974. *Stability and Complexity in Model Ecosystems*, 2nd edition. Princeton University Press, Princeton, NJ.

Peck, S. L. 2004. Simulation as experiment: A philosophical reassessment for biological modeling. Trends in Ecology and Evolution 19: 530–534.

Tyson, J. J., K. Chen, and B. Novak. 2001. Network dynamics and cell physiology. Nature Reviews: Molecular Cell Biology 2: 908–916.

5 Dynamical Systems

This chapter is a mathematical interlude about dynamical systems theory of ordinary differential equations. The goal of this theory is to describe the solutions of these systems in geometric terms. The theory classifies patterns that are found in simulations of many models and gives a mathematical justification for why these particular patterns are observed. We emphasize the descriptive language developed by the theory and the associated concepts, but this chapter can only give the briefest of introductions to this rich theory. The second half of the chapter explores the dynamics of the Morris-Lecar model introduced in Chapter 3 as a case study for application of the theory.

The conceptual picture at the center of dynamical systems theory is that of points moving in an abstract *phase space* according to well-defined rules. For deterministic systems, we assume that where points go is determined by their current position in the phase space. The path through the phase space that a point takes is called its *trajectory*. Usually, we will be interested in the *asymptotic* or limit behavior of the trajectories, where they go after long times. A common behavior is for regions of trajectories to approach the same limit set. This limit set is called an *attractor* for the system. In the case of the bistable switch considered in Chapter 4, the phase space is the positive quadrant of the plane, and the system has two attractors, both *equilibrium points* or *steady states*. The number and types of attractors that exist may vary with changes in system parameters. These changes are called *bifurcations* of the system and its attractors.

This chapter restricts its attention to systems with one- and two-dimensional phase spaces. The phase spaces of these systems are easier to visualize than those of higher-dimensional systems, and their attractors have a simpler structure than the chaotic attractors which are possible in higher dimensions. Pictures that divide the phase spaces into regions with different limit sets are *phase portraits*. *Generic* planar dynamical systems have attractors that are equilibrium points and periodic orbits. We will describe how to systematically compute the attractors of generic planar systems and the regions of trajectories tending toward them, using the Morris-Lecar system as a case study. Computer algorithms are used to find equilibrium points and compute trajectories in determining these phase portraits. Qualitative differences in the phase portraits for two sets of parameter values signal that a bifurcation occurs if one follows a path in parameter space joining them. We describe the bifurcations that one expects to encounter in such investigations—there are only a small number of different types that occur as a single parameter is varied. In this analysis, we shall encounter the algebra of eigenvalues and eigenvectors once again.

The chapter begins with a description of one-dimensional systems. The next section introduces general terminology and considers mathematical foundations. The third section discusses linear systems of differential equations and how these are used to study properties of a system near an equilibrium point. This is followed by a section that gives guidelines for analyzing nonlinear two-dimensional systems. These guidelines are applied to the Morris-Lecar example in the fifth section. The sixth section describes several bifurcations and gives illustrations from the Morris-Lecar model. The final section of the chapter gives a brief discussion of some of the issues that arise in numerically computing trajectories.

5.1 Geometry of a Single Differential Equation

We begin our discussion of dynamical systems in one dimension. A single differential equation

$$\dot{x} = f(x) \tag{5.1}$$

defines the motion of a point on the line. The function $x(t)$ describes the position x of the point at each time, and $f(x)$ gives the velocity at which the point is moving. We want to determine the limit of $x(t)$ as $t \to \infty$. Geometrically, what can happen is pretty simple. The values of x at which $f(x) = 0$ are *equilibrium* points: if $f(x_0) = 0$ the constant function $x(t) \equiv x_0$ is a solution to the equation. The equilibrium points divide the line into intervals on which $f(x)$ is positive and intervals on which it is negative; see Figure 5.1. In an interval on which $f(x)$ is positive, $x(t)$ increases, while in an interval on which $f(x)$ is negative, $x(t)$ decreases. If f is itself a differentiable function of x, trajectories do not cross the equilibrium points. Each trajectory that is not an equilibrium then either increases or decreases. Moreover, each trajectory will tend either to an equilibrium or to $\pm\infty$ as $t \to \infty$. This is a simple example of a phase portrait for a dynamical system. The line is the phase space of the system and the solution trajectories move along the line. The equilibrium points identify the asymptotic behavior of trajectories: where they go as $t \to \pm\infty$. In slightly more technical terms, the

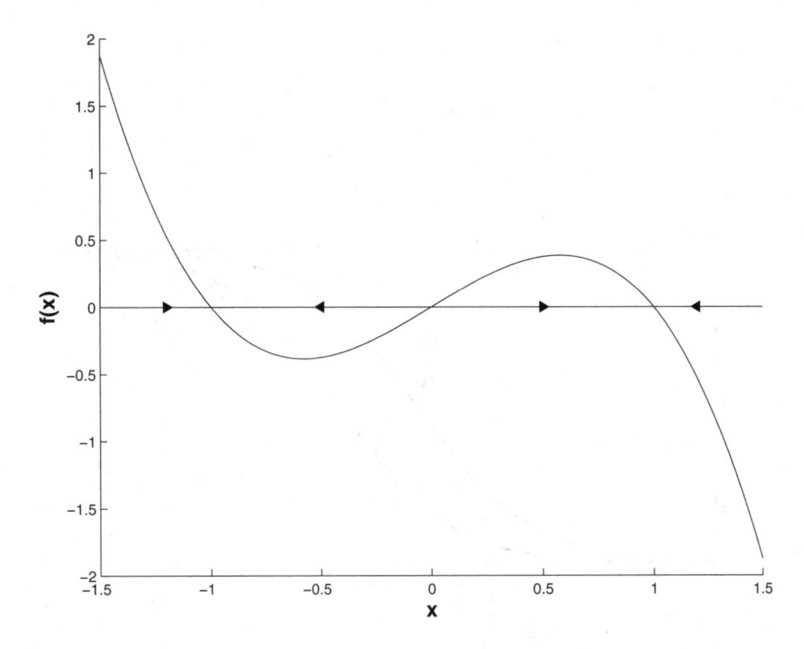

Figure 5.1 The phase line of a one-dimensional differential equation with three equilibrium points. The values of the vector field are plotted as the graph of a function and arrows show the direction of the vector field along the line.

limit set of a bounded trajectory is an equilibrium point. Similarly, the backward limit set of a trajectory that is bounded as $t \to -\infty$ is an equilibrium point.

The classical logistic model for density-dependent population growth is defined by the one-dimensional vector field

$$\dot{x} = rx\left(1 - \frac{x}{K}\right). \quad [5.2]$$

Here x represents the size of a population whose per capita growth rate declines with population size, decreasing linearly from a value r at very small population sizes to 0 at population size K. This equation can be solved explicitly, but we choose to represent its solutions graphically in Figure 5.2. This figure gives a different perspective on the phase line of this equation as a dynamical system. The vertical direction is the phase line of the system. The graphs of solutions show how they move up or down along the phase line as a function of time. As $t \to \infty$, the trajectories that are shown approach the equilibrium point at $x = K$. As $t \to -\infty$, the two lower trajectories approach the equilibrium point 0. The upper trajectory tends to ∞ as t decreases, "reaching" ∞ in finite time. Observe that x increases when $\dot{x}$ as given by the equation is positive and decreases when $\dot{x}$ is negative.

Exercise 5.1. Compare the graph of $f(x) = rx(1 - x/K)$ with Figure 5.2. What is the relationship of the maximum values of $f(x)$ to the trajectories of the differential equa-

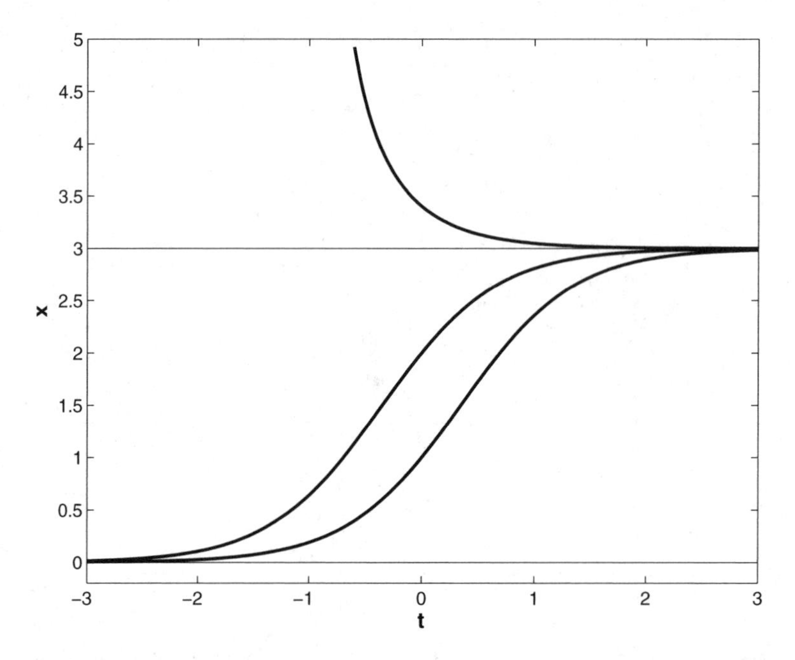

Figure 5.2 Plots of solutions $x(t)$ to the differential equation $\dot{x} = 2x(1 - x/3)$. The equilibrium points are at $x = 0$ and $x = 3$. Solutions in the interval $(0, 3)$ approach 3 as $t \to \infty$ and 0 as $t \to -\infty$. Solutions with $x > 0$ also approach 3 as $t \to \infty$ but tend to ∞ as t decreases.

tion plotted in Figure 5.2? How will Figure 5.2 change if the values of r and K are changed in the equation?

Exercise 5.2. Imagine a population in which deaths exceed births when the population is small, but per capita reproductive rate *increases* with population size. This might happen, for example, if it is easier to find a mate in a larger population. If the differential equation describing growth rate of this population is

$$\dot{x} = rx\left(-1 + \frac{x}{K}\right), \tag{5.3}$$

analyze what can happen to trajectories with $x(0) > 0$.

5.2 Mathematical Foundations: A Fundamental Theorem

Dynamical systems are defined by systems of ordinary differential equations that express the rates at which a collection of dependent variables $(x_1, x_2, \ldots, x_n)$ vary in time. The dependent variables can represent any quantities, and we write $x_i(t)$ to denote the function that describes how the ith dependent variable changes in time. We will also collect the dependent variables into an n-dimensional vector $\mathbf{x} = (x_1, x_2, \ldots, x_n)$ and write $\mathbf{x}(t)$ for the function that describes how the vector changes in time. We use R^n to denote the set of all n-dimensional vectors. When $n = 2$ or 3, we think of the vector $\mathbf{x}(t)$ as a point moving in the plane or in space. When $n > 3$, visualization of the vector $\mathbf{x}$ strains our imagination and we often resort to looking at simultaneous plots of the functions $x_i(t)$. Nonetheless, we still regard $\mathbf{x}(t)$ as a moving point that sweeps out a curve. The differential equations themselves take the form

$$\dot{x}_i = f_i(x_1, x_2, \ldots, x_n), \qquad [5.4]$$

expressing the *rates* of change of each dependent variable in terms of the current values of all the dependent variables. In this equation, we write $\dot{x}_i$ for the derivative of x_i with respect to time.[1] We also gather the equations together in the vector form $\dot{\mathbf{x}} = \mathbf{f}(\mathbf{x})$.

The equations [5.4] define a *vector field* that assigns the vector $\mathbf{f}(\mathbf{x})$ to the point $\mathbf{x}$.[2] In this setting, the set of all $\mathbf{x}$ at which $\mathbf{f}$ is defined is called the *phase space* of the vector field and the solutions are called *trajectories*. This is most readily visualized in the case $n = 2$ when the phase space is the plane, and we can interpret the vector field as assigning an arrow to each point of the plane. Figure 5.3 shows an example of a two-dimensional vector field, the arrows giving the values of the vector field on a grid of points and the heavy solid curves showing three solutions to the system of differential equations

$$\begin{aligned} \dot{x} &= 1/2, \\ \dot{y} &= x. \end{aligned} \qquad [5.5]$$

Most of this chapter will deal with vector fields in the plane.

A vector field "points the way" for trajectories of [5.4], giving the tangent vectors to the solution curves. It is plausible (and true if $\mathbf{f}$ is itself differentiable) that there is exactly one solution beginning at each location of phase space at time $t = 0$. This is the content of the *existence and uniqueness* theorem for ordinary differential equations.

> *If $\mathbf{f} : R^n \to R^n$ is a differentiable function defined on the domain U and $\mathbf{x}_0 \in U$, then there is an $a > 0$ and a unique differentiable function $\mathbf{x} : [-a, a] \to R^n$ such that $\mathbf{x}(0) = \mathbf{x}_0$ and $\dot{\mathbf{x}}(t) = \mathbf{f}(\mathbf{x}(t))$ for $-a \le t \le a$.*

Solving the differential equations produces the curve $\mathbf{x}(t)$. The vector field determines trajectories, but an initial point $\mathbf{x}(0)$ must be chosen to specify a

[1] The equations [5.4] are called *autonomous* because the functions f_i do not depend upon t. One can study *nonautonomous* equations in which the f_i may depend upon t, but the geometric interpretations of their solutions are more complicated.

[2] The terms *vector* and *point* are sometimes used interchangeably and sometimes in distinct ways. When both refer to x, we call $f(x)$ a *tangent vector* to distinguish it from x. Some authors call the quantities x points and the quantities $f(x)$ vectors. We have tried to avoid confusion by making minimal use of the term vector to describe either x or $f(x)$.

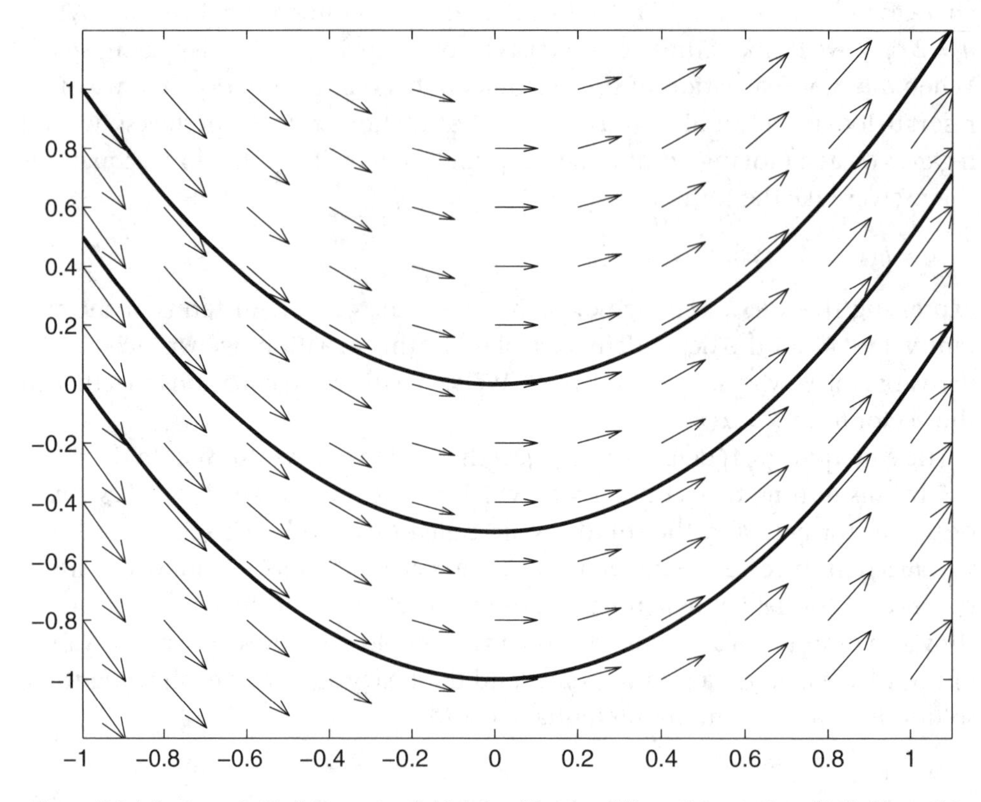

Figure 5.3 A "quiver plot" of the vector field defined by the differential equations [5.5]. The solid curves are the solutions of the equations passing through the points $(0,0)$, $(0,-0.5)$, and $(-1,0)$. The vector field arrows are tangent to the solution curves passing through the tail of the arrows.

unique trajectory. The solutions fit together: if the solution $\mathbf{x}(t)$ arrives at the point $\mathbf{y}_0$ at time s, then the solution $\mathbf{y}$ beginning at $\mathbf{y}_0$ follows the same path as the trajectory that began at $\mathbf{x}(0)$. In a formula, the trajectories satisfy $\mathbf{y}(t) = \mathbf{x}(s+t)$.

The resulting picture is that the phase space is partitioned into curves. Through each point there is a curve, and different curves never intersect. These curves constitute the *phase portrait* of the system and they are the main object that we want to study. *Dynamical systems theory* seeks to describe the phase portraits geometrically and determine how the phase portraits change as the underlying vector field varies. The theory has developed a language to describe the patterns displayed by phase portraits and a mathematical perspective that explains the ubiquity of some of these patterns. The complexity possible in these pictures depends upon the dimension of the phase space. One dimension allows few possibilities; two dimensions provide room for a richer set of alternatives. For systems with more than two dependent phase space variables, *chaotic* trajectories that combine unpredictability with surprising order may occur. We shall restrict attention in this chapter to models with just two dependent variables.

5.3 Linearization and Linear Systems

5.3.1 Equilibrium Points

Equilibrium points play a special role in the analysis of one-dimensional vector fields. This is also true in higher dimensions. The location and analysis of equilibrium points is a starting point for the mathematical description of phase portraits. In this section, we study the properties of a vector field near an equilibrium point by linearization of the vector field at the equilibrium. We describe how to compute the linearization as the Jacobian matrix of derivatives at the equilibrium, and how to solve the linear system with elementary functions in terms of its eigenvalues and eigenvectors. More complete developments of this theory can be found in many texts on differential equations, for example, Blanchard et al. (2002); Hirsch et al. (2004). At the end of the section, we discuss the relationship between the flow of a vector field near an equilibrium and its linearization.

We begin our discussion of linearization with dimension one. A one-dimensional vector field is defined by a single differential equation $\dot{x} = f(x)$. This equation has an equilibrium point at x_0 if $f(x_0) = 0$. The derivative $f'(x_0)$ is the slope of the best linear approximation to f at x_0. The *linearization* of the differential equation at x_0 is defined to be the equation $\dot{y} = ay$ with $a = f'(x_0)$. The solutions of the linearized equation have the form $y(t) = c\exp(at)$ where the constant c is determined by initial conditions. If $a < 0$, then the solutions of the linearized equation tend to 0 as $t \to \infty$, while if $a > 0$, the solutions of the linearized equation tend to 0 as $t \to -\infty$. This behavior carries over to the original equation in the following way. If $f'(x_0) < 0$, then f is positive in an interval immediately to the left of x_0 and negative in an interval immediately to the right of x_0. Thus, there is an interval containing x_0 so that all solutions in the interval tend to x_0 as $t \to \infty$. Similarly, if $f'(x_0) > 0$, there is an interval containing x_0 so that all solutions in the interval tend to x_0 as $t \to -\infty$. Thus, the stability of the equilibrium point is determined by the linearized vector field when $f'(x_0) \neq 0$.

Exercise 5.3. Can anything be said in geneneral about the stability of an equilibrium x_0 where $f'(x_0) = 0$? Consider the behavior of the systems defined by $\dot{x} = x^2$, $\dot{x} = x^3$, and $\dot{x} = -x^3$ in the vicinity of the origin.

We turn now to the n-dimensional vector field defined by $\dot{\mathbf{x}} = \mathbf{f}(\mathbf{x})$. *Equilibrium* solutions are solutions $\mathbf{x}(t) \equiv \mathbf{x}_0$ with $\mathbf{f}(\mathbf{x}_0) = 0$. Finding equilibria is one of the first steps that we undertake in computing phase portraits. Stable equilibria can be located as the limits of trajectories, but we desire methods that will directly locate all the equilibria. Iterative *root-finding* algorithms are used for this purpose. *Newton's method* is the most frequently used root-finding algorithm, and one of the simplest. We describe here how it works. Newton's method uses the *Jacobian* of $\mathbf{f}$, defined as the matrix $D\mathbf{f}$ whose (i, j) entry is

$$D\mathbf{f}_{i,j} = \left(\frac{\partial f_i}{\partial x_j}\right). \quad [5.6]$$

If $\mathbf{u}$ is a vector at which $\mathbf{f}(\mathbf{u})$ is pretty small, the linear approximation of $\mathbf{f}$ near $\mathbf{u}$ is $\mathbf{f}(\mathbf{u}) + D\mathbf{f}(\mathbf{x} - \mathbf{u})$ with $D\mathbf{f}$ evaluated at $\mathbf{u}$. Assuming that $D\mathbf{f}$ has a matrix inverse, we solve the system of equations $\mathbf{f}(\mathbf{u}) + D\mathbf{f}(\mathbf{x} - \mathbf{u}) = 0$ to obtain $\mathbf{x} = \mathbf{u} - (D\mathbf{f})^{-1}\mathbf{f}(\mathbf{u})$. Newton's method uses this value of $\mathbf{x}$ as the next guess for a solution of $\mathbf{f}(\mathbf{x}) = 0$. It iterates this procedure by defining the discrete map $\mathbf{N}(\mathbf{u}) = \mathbf{u} - (D\mathbf{f})^{-1}\mathbf{f}(\mathbf{u})$. Beginning with an initial vector $\mathbf{u}_0$, one defines $\mathbf{u}_1, \mathbf{u}_2, \ldots$ by $\mathbf{u}_1 = \mathbf{N}(\mathbf{u}_0)$, $\mathbf{u}_2 = \mathbf{N}(\mathbf{u}_1)$, and generally $\mathbf{u}_{j+1} = \mathbf{N}(\mathbf{u}_j)$. If $\mathbf{f}(\mathbf{x}_0) = 0$, $D\mathbf{f}(\mathbf{x}_0)$ has a matrix inverse, and $\mathbf{u}_0$ is close enough to $\mathbf{x}_0$, then the iterates $\mathbf{u}_i$ of the Newton map converge to $\mathbf{x}_0$ very rapidly. The method doesn't always work, either because the Jacobian has no inverse or because $\mathbf{u}_0$ was not close enough to a root of $\mathbf{f}$. Newton's method is fast enough that it can be tried repeatedly with many randomly chosen initial vectors $\mathbf{u}_0$. As a cautionary note, it is always a good idea to check your answer to a problem—even if the answer was produced by a computer. The most direct way to check whether you have found an equilibrium is to evaluate $\mathbf{f}(\mathbf{x})$. No computer method is guaranteed to find all of the equilibria of all systems, so do not be surprised if a software package fails sometimes in this task.

Exercise 5.4. Compute the equilibrium points of the repressilator and toggle switch models of Chapter 4 with Newton's method: write a script that takes an initial point as input, iterates the Newton map N, and checks to see whether each iterate has converged to a specified tolerance. If it has, then the script should return this value. If convergence is not obtained after a chosen number of iterates, the script should return an "error" message.

5.3.2 Linearization at Equilibria

Having located an equilibrium $\mathbf{x}_0$ for the vector field defined by $\dot{\mathbf{x}} = \mathbf{f}(\mathbf{x})$, its *linearization* at $\mathbf{x}_0$ is the system $\dot{\mathbf{y}} = \mathbf{A}\mathbf{y}$ where A is the Jacobian matrix $D\mathbf{f}$ at $\mathbf{x}_0$:

$$(A_{ij}) = \left.\frac{\partial f_i}{\partial x_j}\right|\mathbf{x}_0. \quad [5.7]$$

Here, $\mathbf{y}$ represents the displacement $\mathbf{x} - \mathbf{x}_0$ from equilibrium. Computing the matrix (A_{ij}) of the linearization is an exercise in differentiation. On large systems, this may take too long to carry out "by hand" and people are error-prone, so automated methods are often used. The simplest method is to use the finite-difference approximation

$$\left.\frac{\partial f_i}{\partial x_j}\right|\mathbf{x}_0 \approx \frac{1}{h}(f_i(\mathbf{x}_0 + h\mathbf{e}_j) - f_i(\mathbf{x}_0)) \quad [5.8]$$

where $\mathbf{e}_j$ is the jth unit vector, in which all components are 0 except for a 1 as the jth component. However, there is a subtlety in obtaining highly accurate values of the Jacobian through suitable choices of the increment h. The issue involves the balance between "round-off" and "truncation" errors. *Truncation error* refers to the fact that the difference quotient on the right-hand side of [5.8] is only an approximation to the derivative, with an error that is comparable in magnitude to h. *Round-off* error occurs because computer arithmetic is not exact. To limit memory requirements, computers normally round numbers to a fixed number of leading binary "digits." Call this number k. So, when we compute products or more complex arithmetic expressions, there is usually a round-off error that occurs after k significant digits. This is likely to happen when we compute $\mathbf{f}(\mathbf{x}_0)$ and $\mathbf{f}(\mathbf{x}_0 + h\mathbf{e}_j)$. The difference $f_i(\mathbf{x}_0 + h\mathbf{e}_j) - f_i(\mathbf{x}_0)$ subtracts two numbers that are almost equal, each of which has only k significant digits. When we perform the subtraction, leading significant digits cancel and the result has fewer significant digits than either term. When we divide by h, the error is amplified. If h is small enough, this amplified round-off error may be larger than the truncation error. The best approximation achievable with finite-difference approximation using [5.8] is typically half the number of digits of precision used in the computer arithmetic. A better approximation can be obtained from a *centered* finite difference

$$\frac{\partial f_i}{\partial x_j}\bigg|_{\mathbf{x}_0} \approx \frac{1}{h}\left(f_i\left(\mathbf{x}_0 + \frac{h}{2}\mathbf{e}_j\right) - f_i\left(\mathbf{x}_0 - \frac{h}{2}\mathbf{e}_j\right)\right) \qquad [5.9]$$

but this requires more function evaluations.

Example.

We illustrate the accuracy of finite-difference calculations with a simple example. Consider the quadratic function $f(x) = 1 + x + 3x^2$. The derivative of f is $f'(x) = 1 + 3x$. Let us approximate $f'(1)$ with finite-difference approximations

$$g(h) = \frac{f(1+h) - f(1)}{h} \qquad [5.10]$$

and evaluate the residual $r(h) = g(h) - f'(1) = g(h) - 7$. A bit of algebra gives the exact value of the residual as $r(h) = 3h$. This is the truncation error in calculating the derivative with the finite-difference formula. It suggests that the smaller we take h, the more precise our approximation of the derivative. In practice, that's not what happens when we use a computer to evaluate the finite difference formula [5.8]! Figure 5.4 shows a log-log plot of the absolute value of $r(h)$ versus h, calculated with Matlab by substituting h into the formula $(1/h)(1 + (1 + h) + 3(1 + h)^2 - 5) - 7$. The figure shows that the best accuracy is obtained for values of h that are approximately 10^{-8} and that the accuracy which is achieved is also about 10^{-8}. What is happening is that for smaller values of h, the round-off error in calculating the numerator $(f(1 + h) - f(1)$ of g has magnitude roughly

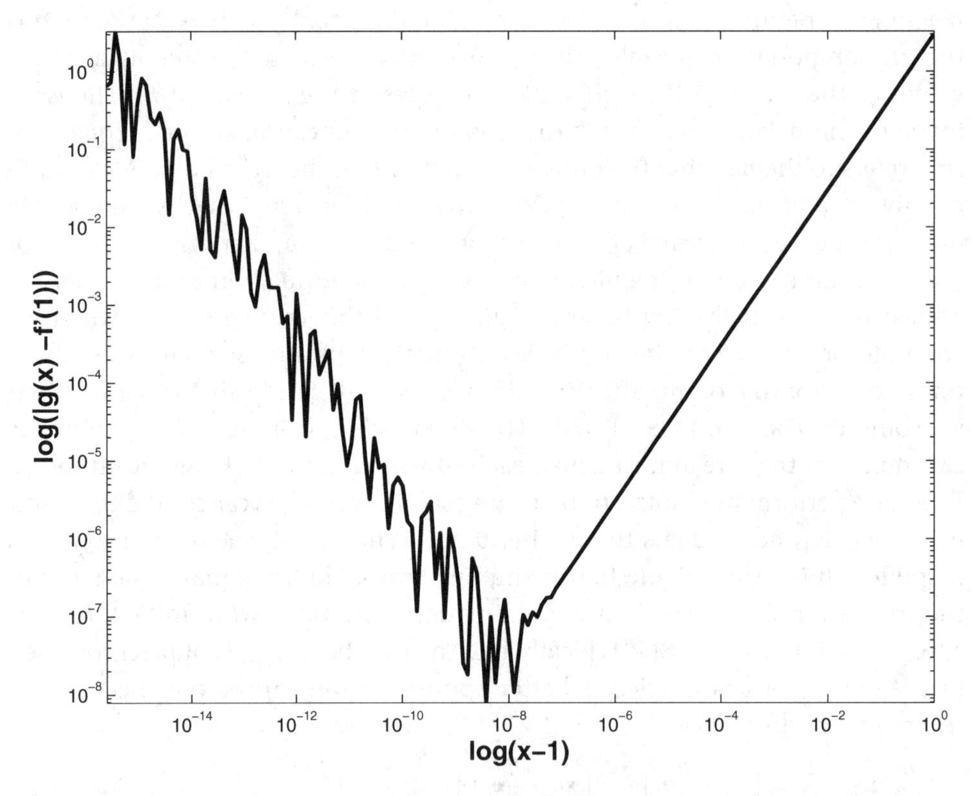

Figure 5.4 A log-log plot of the residual (error) $|r(h)|$ obtained in calculating the derivative of the function $f(x) = 1 + x + 3x^2$ with a finite-difference formula.

10^{-16}.[3] Dividing this round-off error by h, makes the error grow. If $h = 10^{-k}$, then the error has magnitude roughly 10^{k-16}. When $k > 8$, the round-off error becomes larger than the truncation error. Bottom line: If it is easy to use an explicit formula for the derivative of a function f, do so.

Exercise 5.5. Compute the linearization of the equilibrium points of the toggle switch model of Chapter 4 in two ways: with analytical formulas for the derivatives and with finite differences. Do this twice, once for parameter values for which there is a single equilibrium and once for parameter values for which there are three equilibria.

5.3.3 Solving Linear Systems of Differential Equations

We want to extend results from one-variable calculus to solve the linear system $\dot{\mathbf{y}} = \mathbf{A}\mathbf{y}$. Recall that the solutions of the scalar equation $\dot{y} = ay$ are $y(t) = c\exp(at)$. The value of c is typically determined by an initial condition. If $y_0 = y(t_0)$ is the initial condition, then $c = \exp(-at_0)y_0$. These formulas can be extended to work

[3]The smallest number larger than 1 that can be represented in "double-precision" floating-point arithmetic is approximately $1 + 10^{-16}$.

with matrices. If $\mathbf{v}$ is an eigenvector of $\mathbf{A}$ with real eigenvalue λ, recall that then $\mathbf{Av} = \lambda\mathbf{v}$. This implies that $\mathbf{y}(t) = c\exp(\lambda t)\mathbf{v}$ is a solution of the linear system $\dot{\mathbf{y}} = \mathbf{Ay}$ for any value of c, as is verified by differentiating $\mathbf{y}(t)$ to obtain

$$\dot{\mathbf{y}} = c\exp(\lambda t)\lambda\mathbf{v} = c\exp(\lambda t)\mathbf{Av} = \mathbf{Ay}(t). \quad [5.11]$$

The system $\dot{\mathbf{y}} = \mathbf{Ay}$ is linear because sums and scalar multiples of solutions are also solutions. In formulas, if $\mathbf{y}(t)$ and $\mathbf{w}(t)$ are solutions and c is a scalar, then $\mathbf{y}(t) + \mathbf{w}(t)$ and $c\mathbf{y}(t)$ are also solutions. This, too, is verified by substitution into the equation. Some matrices $\mathbf{A}$ have n distinct real eigenvalues λ_i. If $\mathbf{v}_i$ are their eigenvectors, then every vector $\mathbf{y}$ can be written in exactly one way as a sum $\mathbf{y} = \sum c_i\mathbf{v}_i$ for suitable constants c_i. (This says that the $\mathbf{v}_i$ are a *basis* of R^n.) Using the linearity of the differential equation, we find that $\mathbf{y}(t) = \sum c_i \exp(t\lambda_i)\mathbf{v}_i$ is a solution. In the case of n distinct real eigenvalues, all solutions can be written this way. To solve the initial value problem with $\mathbf{y}_0 = \mathbf{y}(t_0)$ specified, we solve $\mathbf{y}_0 = \sum c_i \exp(t_0\lambda_i)\mathbf{v}_i$ for the c_i. This is a system of n linear equations in n unknowns that has a unique solution since the eigenvectors $\mathbf{v}_i$ are linearly independent.

The signs of the eigenvalues have a large impact on the qualitative properties of the solutions to the linear system $\dot{\mathbf{y}} = \mathbf{Ay}$. In the directions of eigenvectors having negative eigenvalues, the solutions tend toward the origin with increasing time, while in the directions of eigenvectors having positive eigenvalues, the solutions tend away from the origin. Let us examine the possibilities for $n = 2$:

- Two negative eigenvalues—*stable node*: The origin is an attractor. All solutions tend toward it.
- One negative, one positive eigenvalue—*saddle*: Solutions tend toward the origin along one eigenvector and away from it along the other.
- Two positive eigenvalues—*unstable node*: The origin is a repellor. All solutions tend away from it.

Figure 5.5 shows plots of phase portraits of a stable node and a saddle. The phase portrait of an unstable node looks exactly like the phase portrait of a stable node except that the direction of motion is away from the origin. In each case, the matrix $\mathbf{A}$ is diagonal and the equations for the coordinates x and y have the form $\dot{u} = \lambda u$. The eigenvalues for the stable node are -1 for x and -5 for y. The function $\exp(-5t)$ decreases *much* more rapidly than $\exp(-t)$, so the ratio $y(t)/x(t) \to 0$ and trajectories approach the origin along the x-axis. The y-axis is the exception since $x(t)$ remains 0 along this line. For the saddle shown in Figure 5.5b, the eigenvalues are 1 for x and -2 for y. The coordinate axes are invariant under the flow. The trajectories on the y-axis approach the origin, the ones on the x-axis tend away from the origin. All of the other trajectories approach the x-axis as $t \to \infty$ and the y-axis as $t \to -\infty$.

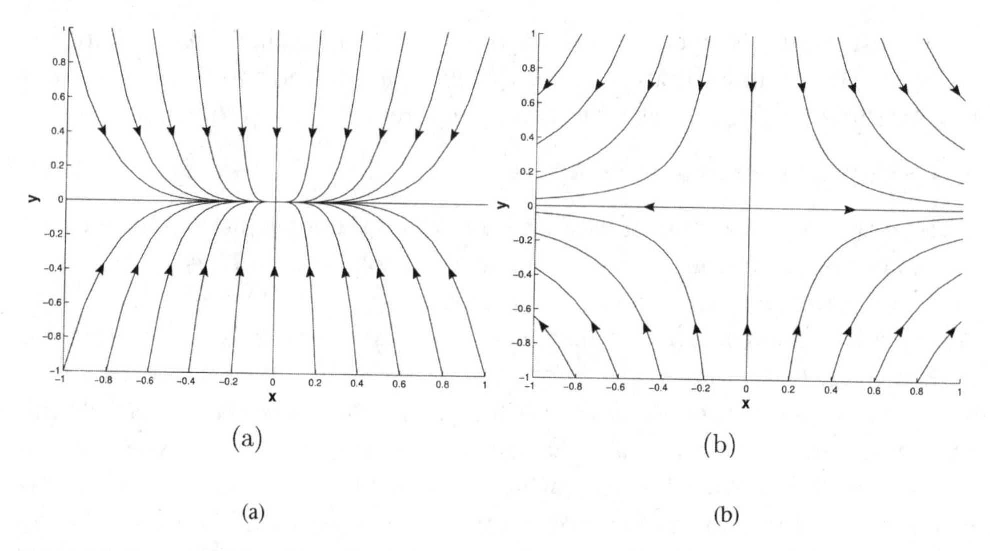

Figure 5.5 Phase portraits of two-dimensional linear vector fields with (a) a stable node and (b) a saddle.

If the eigenvectors of a linear system are not orthogonal as in the previous example, then the phase portraits shown above can be significantly distorted. Figure 5.6 shows phase portraits of vector fields with the matrices

$$\mathbf{A} = \begin{bmatrix} -5 & 4 \\ 0 & -1 \end{bmatrix}, \qquad \mathbf{B} = \begin{bmatrix} 1 & -3 \\ 0 & -2 \end{bmatrix}. \tag{5.12}$$

The matrix A has eigenvalues -1 and -5 with eigenvectors $(1, 1)$ and $(1, 0)$. Its phase portrait is displayed in Figure 5.6a. All trajectories converge to the origin, but for many the distance to the origin *increases* for some time while the trajectory approaches the line of slope 1. As in the previous example, the trajectories flow roughly parallel to the eigenvector of the "fast" eigenvalue -5 and then approach the origin along the eigenvector of the "slow" eigenvalue -1. The matrix B has eigenvalues 1 and -2 with the same eigenvectors $(1, 0)$ and $(1, 1)$ as A. Its phase portrait is displayed in Figure 5.6b. The trajectories flow toward the x-axis while moving away from the line of slope 1. Points on the line of slope 1 approach the origin, while all other trajectories are unbounded as $t \to \infty$.

These pictures illustrate the flow of linear two-dimensional vector fields with two real, nonzero eigenvalues. However, many matrices have complex eigenvalues and we want to solve these systems as well. A fundamental example is the matrix

$$\mathbf{A} = \begin{bmatrix} a & -b \\ b & a \end{bmatrix}. \tag{5.13}$$

By direct substitution into the equation $\dot{\mathbf{y}} = \mathbf{A}\mathbf{y}$, we find that the functions

$$\mathbf{y}(t) = \exp(at)\begin{pmatrix} \cos(bt) \\ \sin(bt) \end{pmatrix}, \qquad \mathbf{y}(t) = \exp(at)\begin{pmatrix} -\sin(bt) \\ \cos(bt) \end{pmatrix} \tag{5.14}$$

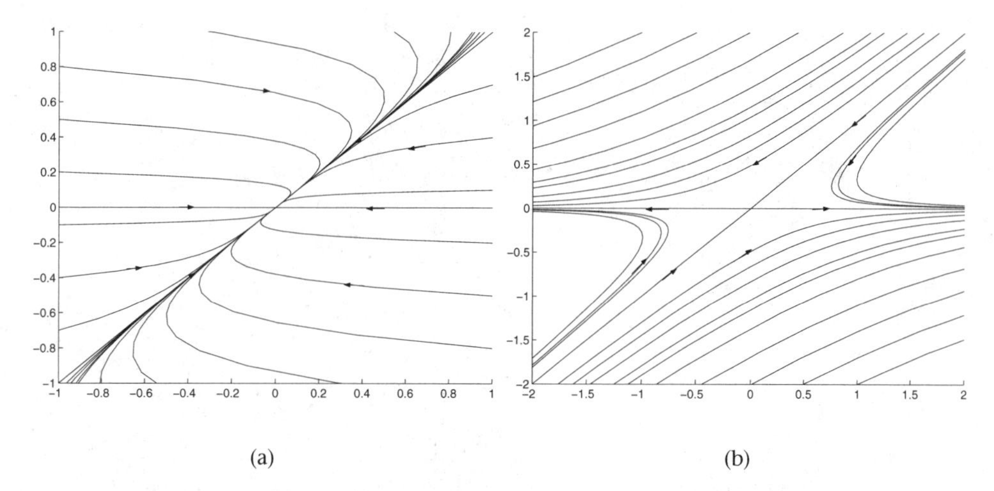

Figure 5.6 Phase portraits of two-dimensional linear vector fields with (a) a stable node and (b) a saddle with eigenvectors that are not orthogonal.

are solutions. The general solutions are linear combinations of these two:

$$\mathbf{y}(t) = c_1 \exp(at) \begin{pmatrix} \cos(bt) \\ \sin(bt) \end{pmatrix} + c_2 \exp(at) \begin{pmatrix} -\sin(bt) \\ \cos(bt) \end{pmatrix}. \qquad [5.15]$$

The behavior of the phase portraits depends upon the value of a. There are three cases:

- $a < 0$—*stable focus*: The origin is an attractor. All solutions spiral toward it.
- $a = 0$—*center*: Solutions lie on circles, and points in the phase plane rotate around the origin at uniform velocity.
- $a > 0$—*unstable focus*: The origin is a repellor. All solutions spiral away from it.

Figure 5.7 displays phase portraits for $(a, b) = (-0.1, 1)$ and for $(a, b) = (0, 1)$. The general case of a 2×2 matrix with complex eigenvalues can be reduced to this one by linear changes of coordinates. When the eigenvalues are purely imaginary, the trajectories are ellipses that need not be circles.

The remaining cases of 2×2 matrices that we have not yet discussed are those with a zero eigenvalue and those with a single eigenvalue of multiplicity 2. Along the eigenvector of a zero eigenvalue, there is an entire line of equilibrium points. If the second eigenvalue is negative, all trajectories approach the line of equilibria along a trajectory that is parallel to the second eigenvector.

Cases with a single eigenvalue are somewhat more complicated. Consider the matrix

$$\mathbf{A} = \begin{bmatrix} \lambda & a \\ 0 & \lambda \end{bmatrix}. \qquad [5.16]$$

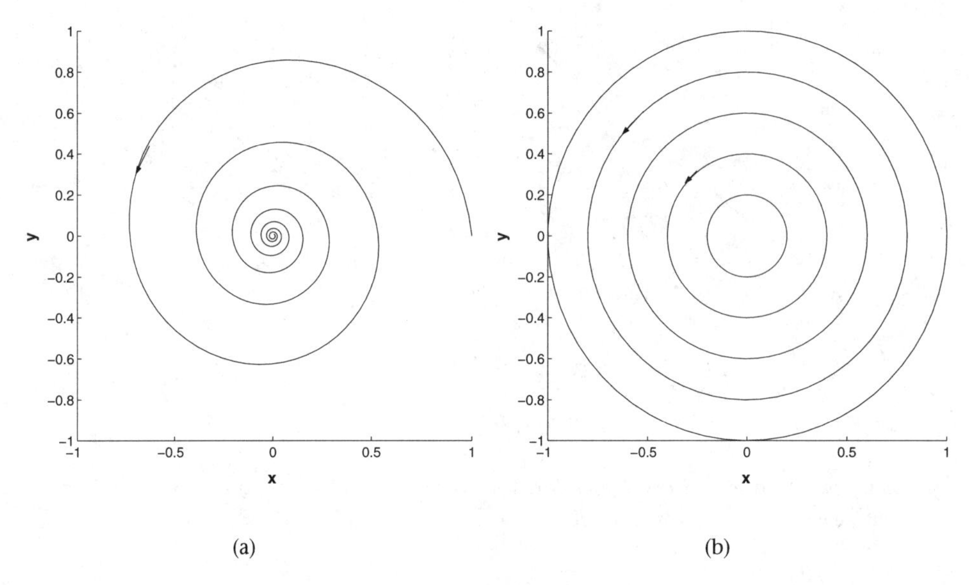

Figure 5.7 Phase portraits of two-dimensional linear vector fields with (a) a stable focus and (b) a center.

When $a \neq 0$, this matrix has only one eigenvalue, namely, λ, and a single eigenvector along the x-axis. The general solution of the equation $\dot{\mathbf{y}} = \mathbf{A}\mathbf{y}$ is

$$\mathbf{y}(t) = (c_1 + c_2 a t)\exp(\lambda t)\begin{pmatrix} 1 \\ 0 \end{pmatrix} + c_2 \exp(\lambda t)\begin{pmatrix} 0 \\ 1 \end{pmatrix}. \qquad [5.17]$$

When $a \neq 0$ and $\lambda < 0$, these curves all tend toward the origin tangent to the x-axis. This is typical of what happens for systems with a single negative eigenvalue that has only one eigenvector. Contrast this with the behavior of the system when $a = 0$ and $\lambda < 0$. Then *every* vector is an eigenvector and all trajectories flow radially toward the origin. Similarly, when $\lambda > 0$, all of the solutions flow from the origin tangent to the x-axis ($a \neq 0$) or radially ($a = 0$). When $\lambda = 0$ and $a \neq 0$ trajectories are lines parallel to the x-axis, while if $\lambda = 0$ and $a = 0$, $\mathbf{A} = 0$ and all points are equilibria.

Exercise 5.6. Draw phase portraits for the following matrices that illustrate the three cases discussed in the previous paragraph:

$$\mathbf{A} = \begin{bmatrix} 0 & 0 \\ 1 & -1 \end{bmatrix}, \quad \mathbf{B} = \begin{bmatrix} -1 & 5 \\ 0 & -1 \end{bmatrix}, \quad \mathbf{B} = \begin{bmatrix} -1 & 0 \\ 0 & -1 \end{bmatrix}. \qquad [5.18]$$

This completes our discussion of linear vector fields in dimension two. We comment briefly on the solution of linear systems in dimensions larger than two. The first step is to find the eigenvalues of the matrix $\mathbf{A}$, both real and complex. Theoretically, the eigenvalues are roots of the *characteristic polynomial* of $\mathbf{A}$, a polynomial of degree n. Complex eigenvalues therefore come in complex conjugate pairs $a \pm bi$. Each pair of complex eigenvalues has a two-dimensional plane and a basis for this plane in which the origin is a focus ($a \neq 0$) or a center ($a = 0$). When there are no multiple roots of the characteristic polynomial, R^n has a basis consisting of eigenvectors, including the basis vectors in the planes of complex eigenvalues. When there are multiple eigenvalues, a more refined analysis that determines the *Jordan normal form* of the matrix is required. In all cases, the solutions can be written explicitly as elementary functions and the solution of the initial value problem can be found with linear algebra.

5.3.4 Invariant Manifolds

We continue our discussion of the general linear system $\dot{\mathbf{y}} = \mathbf{A}\mathbf{y}$. A linear subspace $V \subset R^n$ is a set that is closed under scalar multiplication and vector addition: if $\mathbf{v}, \mathbf{w} \in V$ and $a, b \in R$, then $a\mathbf{v} + b\mathbf{w} \in V$. The subspace V is an *invariant subspace* of $\mathbf{A}$ if $\mathbf{v} \in V$ implies $\mathbf{A}\mathbf{v} \in V$. In addition, trajectories with initial point in an invariant subspace V remain entirely within V. One dimensional invariant subspaces are just lines through the origin in the direction of eigenvectors. Finding the invariant subspaces of $\mathbf{A}$ is a problem in linear algebra, one whose solution we do not describe here. Nonetheless, we will make use of particular invariant subspaces that have dynamical meaning. The *stable manifold* E^s of $\mathbf{A}$ is the largest invariant subspace so that all the eigenvalues of $\mathbf{A}$ restricted to E^s have negative real parts; i.e, the eigenvalues are $a < 0$ or $a + bi$ with $a < 0$. Similarly the *unstable manifold* E^u is the largest invariant subspace so that all the eigenvalues of $\mathbf{A}$ restricted to E^u have positive real parts. The *center manifold* E^c of $\mathbf{A}$ is the largest invariant subspace so that all the eigenvalues of $\mathbf{A}$ restricted to E^c have zero real parts. It is a theorem of linear algebra that R^n is the direct sum of E^s, E^u, and E^c; that is, every vector $\mathbf{v} \in R^n$ can be written in a unique way as $\mathbf{v} = \mathbf{v}_s + \mathbf{v}_u + \mathbf{v}_c$ with $\mathbf{v}_s \in W^s, \mathbf{v}_u \in W^u$ and $\mathbf{v}_c \in W^c$. If R^n has a basis of eigenvectors of A, then each lies in the stable, unstable, or center manifold depending on whether its real part is negative, positive, or zero. Each of these manifolds then has a basis consisting of eigenvectors of A.

The dynamical significance of the stable manifold E^s is that the trajectories in E^s are the ones that approach the origin as $t \to \infty$. Moreover, the convergence happens at an exponential rate: if there are no eigenvalues with real parts in the interval $[-a, 0), a > 0$, then there is a constant c so that

$$|\mathbf{y}(t)| \leq c \exp(-at)|\mathbf{y}(0)|$$

for all trajectories in E^s and $t > 0$. The trajectories in E^u approach the origin in negative time as $t \to -\infty$. Trajectories in E^c do not tend to the origin in either forward or backward time. Thus the "splitting" of R^n into stable, unstable, and center subspaces contains the stability information about trajectories. In particular, the origin is an asymptotically stable equilibrium of a linear system if and only if all of its eigenvalues have negative real parts.

For a nonlinear system with an equilibrium $\mathbf{x}_0$, we would like to use the linearization at $\mathbf{x}_0$ to deduce stability information about the nonlinear system in a region around $\mathbf{x}_0$. This can be done successfully except in the case of the center manifold directions. An equilibrium that has no center directions is called *hyperbolic*. For hyperbolic equilibria, the *stable manifold theorem* asserts the existence of nonlinear counterparts of the stable and unstable manifolds of the linearized system. Specifically, it proves that there are maps $h^s : E^s \to R^n$ and $h^u : E^u \to R^n$ whose images are invariant under the flow of the nonlinear system and tangent to E^s and E^u. The nonlinear stable manifold of $\mathbf{x}_0$ consists of points that flow to $\mathbf{x}_0$ as $t \to \infty$, while the nonlinear unstable manifold of $\mathbf{x}_0$ consists of points that flow to $\mathbf{x}_0$ as $t \to -\infty$. In particular, equilibria whose eigenvalues all have negative real parts are asymptotically stable. Equilibria whose eigenvalues all have positive real parts are sources, with all nearby trajectories tending to the equilibria as $t \to -\infty$.

In two-dimensional systems, an equilibrium with one positive and one negative equilibrium is a saddle with one-dimensional stable and unstable manifolds. Each of these manifolds is formed by a pair of trajectories, called *separatrices*. Trajectories near the stable manifold, but not on it, approach the equilibrium and then depart near one of the two separatrices comprising the unstable manifold. The separatrices play a central role in dividing the phase plane into regions with similar asymptotic behavior as $t \to \pm\infty$.

The linearization of an equilibrium does not determine stability of the nonlinear system in center directions. Consider the one-dimensional system $\dot{x} = x^2$. The origin is an equilibrium point, and its eigenvalue is 0. Except at 0, $\dot{x} > 0$, so trajectories with negative initial conditions approach 0 as $t \to \infty$ while trajectories with positive initial conditions tend to ∞. Thus, the origin is an equilibrium that is stable from one side of its central direction and unstable from the other.

5.3.5 Periodic Orbits

The technique of linearization can be applied to periodic orbits as well as equilibria. Recall that a periodic orbit of period T is a trajectory with $\mathbf{x}(t+T) = \mathbf{x}(t)$ for all times t. The *flow map* ϕ_T is defined by following trajectories for T time units from each point: $\phi_T(\mathbf{y})$ is $\mathbf{x}(T)$ where $\mathbf{x}(t)$ is the trajectory with initial condition $\mathbf{x}(0) = \mathbf{y}$. A periodic orbit with period T consists of fixed points of the flow map ϕ_T. To study the periodic orbit containing the point $\mathbf{y}$, we compute the Jacobian derivative $D\phi_T$ of ϕ_T at $\mathbf{y}$. The map ϕ_T is normally computed by numerical integrating trajectories, so we expect to use numerical integration to compute $D\phi_T$ as well. The eigenvalues of $D\phi_T$ give stability information about the periodic orbit. There is always an eigenvalue 1 with eigenvector in the flow direction. This direction is also tangent to the periodic orbit. If there are $n-1$ eigenvalues with magnitude smaller than one, then the periodic orbit is an attractor with nearby trajectories tending toward it. If there is an eigenvalue with magnitude larger than one, then the periodic orbit is unstable and some nearby trajectories tend away from it. There is a version of the stable manifold theorem for periodic orbits, but in two dimensional systems the situation is quite simple since there is only one direction transverse to the periodic orbit. Unless 1 is a double eigenvalue of $D\phi_T$, the periodic orbit is either an attractor (eigenvalue with magnitude smaller than one) or repellor (eigenvalue with magnitude larger than one).

5.4 Phase Planes

We turn now to two-dimensional vector fields with phase space the plane. Chapter 4 introduced the example

$$\begin{aligned} \dot{u} &= -u + \frac{\alpha_u}{1+v^\beta} \\ \dot{v} &= -v + \frac{\alpha_v}{1+u^\gamma} \end{aligned} \qquad [5.19]$$

and discussed some of its properties. In particular, we found parameters for the model for which there were two stable nodes and one saddle. Our goal here is to learn how to draw the phase portraits of systems like this in a systematic way, relying upon numerical methods for three basic tasks: finding equilibrium points, computing individual trajectories, and computing eigenvalues of matrices. While individual trajectories can be approximated with numerical methods, we want more than individual trajectories. We want enough information that we can predict qualitatively what all the trajectories look like, after computing only a few of them. This task is easy for one-dimensional systems: nonequilibrium trajectories are increasing or decreasing functions of time that tend to equilibria or $\pm\infty$. When the phase space is two dimensional, we can still give a pretty complete "recipe" for determining phase portraits with rather mild assumptions.

The key to drawing phase portraits of two-dimensional systems is to determine where trajectories go as $t \to \pm\infty$. This idea is embodied in the concept of the limit sets of a trajectory. The (forward or ω) limit set of a trajectory is the set of points that the trajectory repeatedly gets closer and closer to as $t \to \infty$. The (backward or α) limit set of a trajectory is the set of points that the trajectory repeatedly gets closer and closer to as $t \to -\infty$. Periodic orbits and equilibrium points are their own forward and backward limit sets. One objective in determining a phase portrait is to find the limit sets, and for each limit set to find the points with that limit set. The plane is divided into subsets, so that all of the trajectories in each subset have the same forward limit set and the same backward limit set. Once we know these subsets, when we select an initial point for a trajectory, we know where the trajectory is going and where it came from. An especially important type of system is one in which all trajectories have a single equilibrium point x_0 as forward limit. In this case, x_0 is said to be *globally attracting*. The origin is globally attracting for a linear system in which all of the eigenvalues have negative real part. Figures 5.5a and 5.7a show examples.

Limit sets of flows in the plane are highly restricted because trajectories do not cross each other. The key result is the *Poincaré-Bendixson theorem* which states that a limit set of a bounded trajectory is either a periodic orbit or contains an equilibrium point. Figure 5.8 gives an example of a limit set in a two-dimensional vector field that is more complicated than a single equilibrium point or a periodic orbit. The Poincaré-Bendixson theorem leads to a systematic procedure for finding the phase portrait of the vector field

$$\begin{aligned} \dot{x} &= f(x,y) \\ \dot{y} &= g(x,y) \end{aligned} \qquad [5.20]$$

in the plane. There are three steps.

1. ***Locate the equilibrium points and determine their stability by linearization.*** A graphical procedure for finding the equilibria is to draw the nullclines. The x nullcline is the curve $f(x,y) = 0$ on which the vector field points vertically up or down; the y nullcline is the curve $g(x,y) = 0$ on which the vector field points horizontally. The equilibrium points are the intersections of the nullclines. The matrix of partial derivatives

$$\mathbf{A} = \begin{bmatrix} \frac{\partial f}{\partial x} & \frac{\partial f}{\partial y} \\ \frac{\partial g}{\partial x} & \frac{\partial g}{\partial y} \end{bmatrix}$$

evaluated at an equilibrium point is its Jacobian. When the eigenvalues of A are neither zero nor pure imaginary, the equilibrium is hyperbolic. As we saw in Section 5.3.4 the eigenvalues determine the qualitative features of trajectories near the equilibrium.

2. ***Compute the stable and unstable manifolds of any saddle equilibria.*** To compute the unstable manifold, we numerically compute two trajectories with initial conditions that are slightly displaced from the equilibrium along the direction of the eigenvector with positive eigenvalue. For the second trajectory, we start on the opposite side of the equilibrium point from the first initial point. To compute the stable manifold, we compute two trajectories backward in time, starting with initial conditions slightly displaced from the equilibrium along the direction of

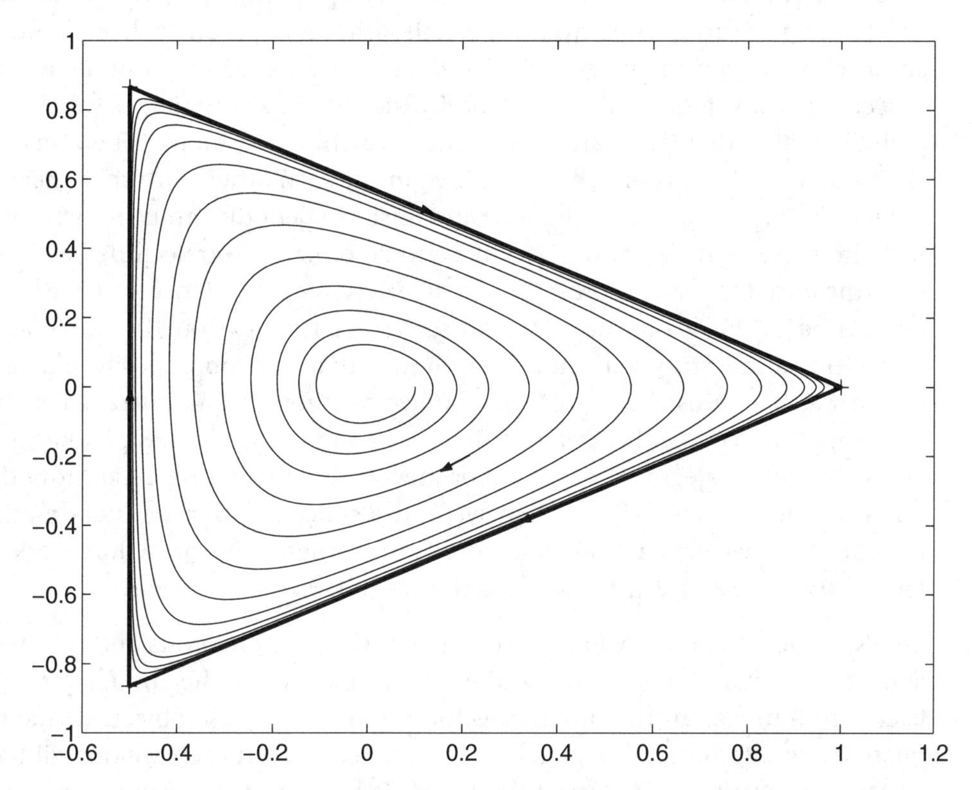

Figure 5.8 The heavy triangle is the forward limit set of the spiraling trajectory. There are three saddles at the vertices of the triangle. Each side of the triangle is a *heteroclinic* trajectory that lies in the stable manifold of one vertex and the unstable manifold of another.

the eigenvector with negative eigenvalue. The forward limit set of each of the two trajectories in the unstable manifold will be a stable equilibrium, a stable limit cycle, a saddle point (perhaps the same one!) or the trajectory will be unbounded. Similarly, the backward limit set of each of the two trajectories in the stable manifold will be an unstable equilibrium, an unstable limit cycle, a saddle or the trajectory will be unbounded in backward time. Knowing this, we make sure to integrate for long enough that the limit behavior of the trajectories is apparent.

3. ***Search for periodic orbits.*** Every periodic orbit must contain an equilibrium point in its interior. Continuing to assume that no eigenvalues at equilibria are zero or pure imaginary, there must be an interior equilibrium point that is not a saddle.[4] This prompts us to look for periodic orbits by numerically integrating

[4]This statement is proved by studying the *index* of a vector field along closed curves. The index of the vector field on a curve without equilibrium points measures the number of times that the vector field rotates around the origin while traversing the curve.

trajectories that start near attracting or repelling equilibrium points. We compute trajectories forward in time from the neighborhood of an unstable equilibrium and backward from the neighborhood of a stable equilibrium. If none of these trajectories converges to a periodic orbit, then there are none. If we do find periodic orbits, then we want to continue searching for more. There may be two or more periodic orbits that are nested inside each other, with the inner and outer ones forming a ring. The stable and unstable periodic orbits in such a nest will alternate. Between two adjacent periodic orbits, the trajectories will flow from the unstable periodic orbit to the stable periodic orbit. Once we have found one periodic orbit γ in a nest, we can search for the next one by numerically integrating a trajectory with initial conditions that start on opposite side of γ, but near it. If γ is unstable, we compute trajectories forward in time; if γ is stable, we compute trajectories backward in time. Each time we find a new periodic orbit as the limit of a trajectory, we search for the next by integrating a trajectory that starts on the opposite side. When we find a trajectory that does approach a periodic orbit, we have found all the orbits in the nest. This procedure works so long as there are a finite number of periodic orbits.[5]

That's it. Once we have found the equilibria, the periodic orbits and the stable and unstable manifolds of the saddles, we can see what the the forward and backward limits of all the other trajectories must be. These objects divide the phase plane into different regions, and the trajectories in each region will have the same forward and backward limit sets. Since trajectories cannot cross one another, there is no ambiguity about what are the limit sets for each region. To make this discussion concrete, we now analyze the dynamics of a model system.

Exercise 5.7. Draw phase portraits of the toggle switch model of Chapter 4. There are no periodic orbits, so the main task beyond those of previous exercises is to compute the stable and unstable manifolds of the saddle point, when there is a saddle point.

5.5 An Example: The Morris-Lecar Model

Recall from Chapter 5 that the Morris-Lecar equations are a model for the membrane potential of a barnacle muscle, defined by the following equations:

$$\begin{aligned} C\frac{dv}{dt} &= i - g_{Ca}m_\infty(v)(v - v_{Ca}) - g_K w(v - v_K) - g_L(v - v_L) \\ \tau_w(v)\frac{dw}{dt} &= \phi(w_\infty(v) - w) \end{aligned} \qquad [5.21]$$

[5] Some systems have continuous families of periodic orbits, something that is impossible to verify rigorously by numerical integration.

$$m_\infty(v) = 0.5\left(1 + \tanh\left(\frac{v - v_1}{v_2}\right)\right)$$

$$w_\infty(v) = 0.5\left(1 + \tanh\left(\frac{v - v_3}{v_4}\right)\right)$$

$$\tau_w(v) = \frac{1}{\cosh((v - v_3)/2v_4)}$$

The variables are the membrane potential v and a gating variable w that represents activation of a potassium current. Here we have made the assumption that the calcium activation variable m is always at its voltage-dependent steady state. This makes m an explicit function of voltage rather than a phase space variable and reduces the dimension of the system from three to two.

The Morris-Lecar system displays a variety of dynamical phenomena (Rinzel and Ermentrout 1998). We examine two sets of parameter values, chosen for illustrative purposes rather than their biological significance. The two sets of parameters give qualitatively different phase portraits and are listed in Table 5.1.[6] We follow the procedure described in the previous section for constructing the phase portraits. Figure 5.9 shows the nullclines for these two sets of parameter values. For parameter set 1, the nullclines have a single point of intersection and the Jacobian at this point has complex eigenvalues whose values are approximately $-0.009 \pm 0.080i$. Thus, this equilibrium is a stable focus. For parameter set 2, there are three intersections of the nullclines. From lower left to upper right, the eigenvalues at these equilibria are approximately $(-0.025 \pm 0.114i)$, $(-0.046, 0.274)$, and $(0.076 \pm 0.199i)$. The equilibria are a stable focus, a saddle, and an unstable focus, respectively. The computation of the equilibria and their stability completes the first step in determining their phase portraits.

Figure 5.10 is a phase portrait for the set of parameters with a single equilibrium point. Since there are no saddle points, we proceed to look for periodic trajectories. We begin by computing a backward trajectory with initial point near the equilibrium. This is seen to converge to a periodic orbit. We next select an initial point on the outside of this unstable periodic orbit and compute its trajectory forward. This converges to another (stable) periodic orbit. Trajectories outside the stable limit cycle tend to ∞, as seen by computing backward from an initial point just outside the second periodic orbit. Five trajectories are plotted in Figure 5.10. The triangle is located at the stable equilibrium point. The two bold trajectories are the periodic orbits: the small periodic orbit is unstable and the large periodic is stable. The figure also shows two trajectories, one that flows from the outside of the unstable periodic orbit to the stable periodic orbit and one that flows from the unstable periodic orbit to the equilibrium point.

[6] The two parameters that differ between the two sets are g_{Ca} and ϕ. Experimentally, g_{Ca} can be reduced with substances that block calcium channels, while ϕ changes with temperature.

Parameter	Set 1	Set 2
g_{Ca}	4.4	5.5
g_K	8	8
g_L	2	2
v_{Ca}	120	120
v_K	−84	−84
v_L	−60	−60
C	20	20
ϕ	0.04	0.22
i	90	90
v_1	−1.2	−1.2
v_2	18	18
v_3	2	2
v_4	30	30

Table 5.1 Parameter values for the Morris-Lecar system.

Figure 5.11 overlays graphs of $v(t)$ and $w(t)$ along the large periodic orbit. The graph of v shows v rising from a membrane potential of approximately −50 mV to a threshold near −20 mV and then abruptly increasing to a membrane potential of approximately 30 mV. The membrane then repolarizes with its potential returning to its minimum of about −50 mV. The gating variable w oscillates in response to these changes in membrane potential. As the membrane potential v rises and falls, w evolves toward its "steady-state" value $w_\infty(v)$. However, the rate at which it does this is slow enough that it seldom reaches its instantaneous steady state value before v changes substantially. Thus, the changes in w "lag" behind the changes in v, with w reaching its minimum and maximum values after v has reached its minimum and maximum values.

We can use the phase portrait Figure 5.10 to determine the limit set of all other trajectories for parameter set 1. Initial points that lie inside the unstable periodic orbit tend to the equilibrium point as time increases, spiraling as they do so because the equilibrium is a focus. As time decreases, these initial points tend to the unstable periodic orbit. Initial points between the two periodic orbits converge to the stable periodic orbit as time increases and the unstable periodic orbit as time decreases. Initial points outside the stable periodic orbit tend to the stable periodic orbit as time increases and tend to ∞ as time decreases.

Figure 5.12 shows the phase portrait of the Morris-Lecar model for the parameters with three equilibrium points. The two parameters g_{Ca} and ϕ have changed their values from the previous set. We have determined that there are three equilibrium points, each with different stability. The next step in computing the phase portrait is to compute the stable and unstable manifolds of the saddle.

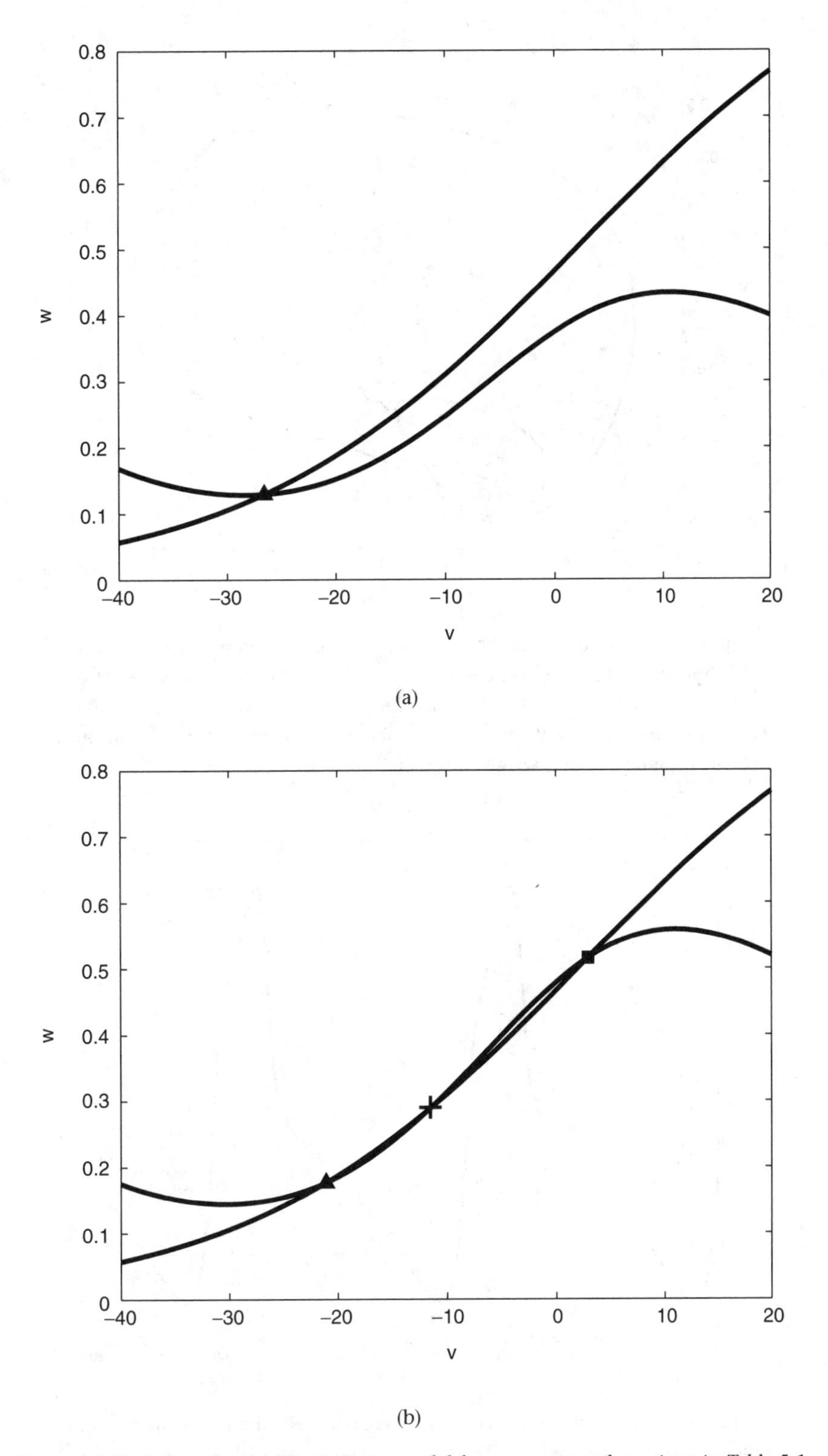

Figure 5.9 Nullclines for the Morris-Lecar model for parameter values given in Table 5.1.

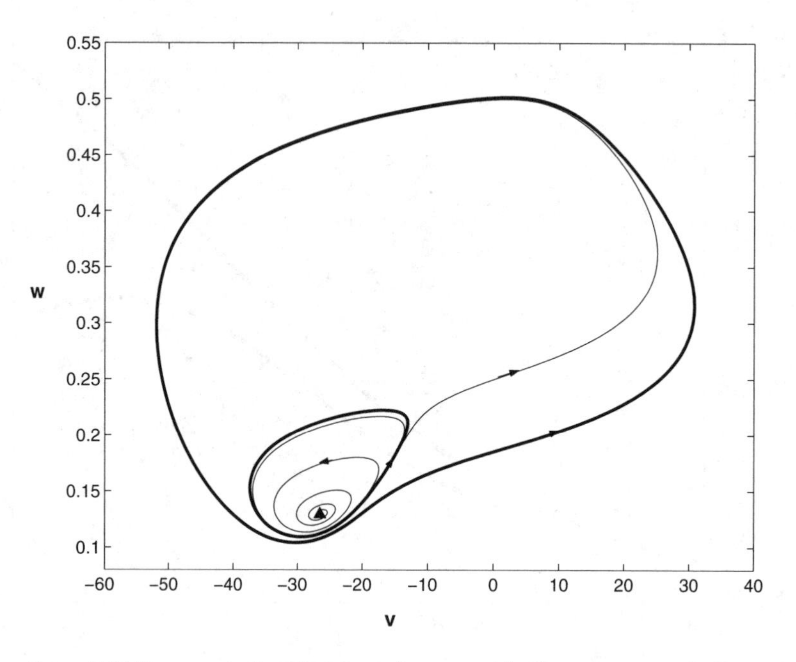

Figure 5.10 Phase portrait of the Morris-Lecar model. There are two periodic orbits (bold curves) and a stable equilibrium (triangle). One trajectory flows from the small unstable periodic orbit to the large stable periodic orbit; the remaining trajectory flows from the small unstable periodic orbit to the equilibrium point. Parameter values are given by Set 1 in Table 5.1.

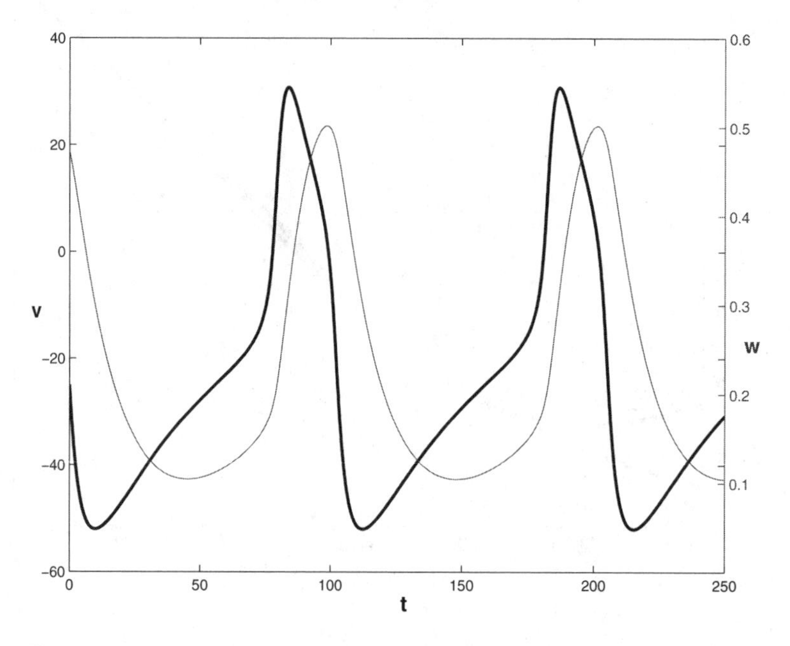

Figure 5.11 Time series of (v, w) for the large periodic orbit of the Morris-Lecar model shown in Figure 5.10. The graph of v is drawn bold.

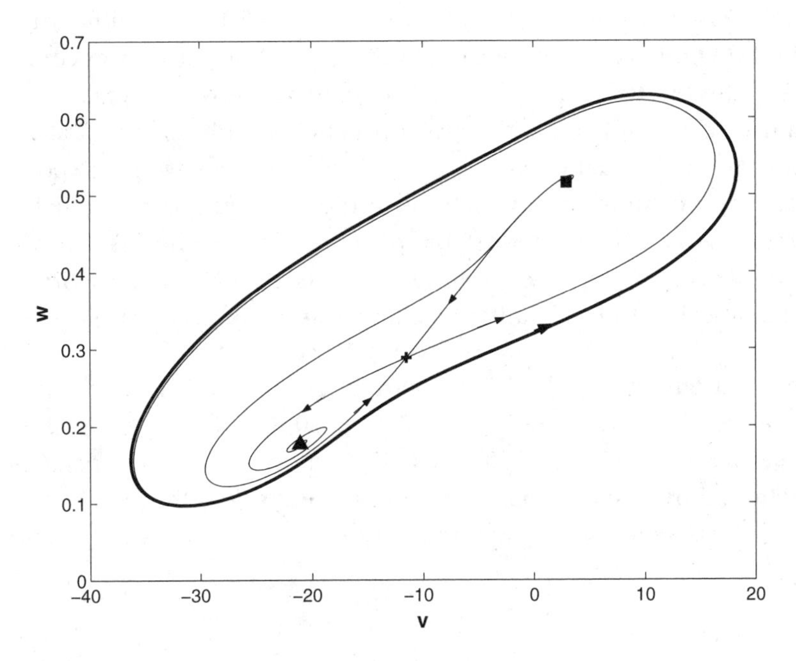

Figure 5.12 Phase portrait of the Morris-Lecar model. There is one periodic orbits (bold curve) and a three equilibrium points, one stable (triangle), one saddle (plus) and one source (square). The stable and unstable manifolds of the saddle are shown. Parameter values are given by Set 2 in Table 5.1

This is done by choosing four initial conditions near the saddle along the eigenvectors of the linearization. We compute two trajectories backward from the initial conditions that lie on opposite sides of the saddle in the direction of the stable eigenvector, and we compute two trajectories forward from the initial conditions that lie on opposite sides of the saddle in the direction of the unstable eigenvector. We observe that one of the trajectories in the unstable manifold tends to the stable equilibrium but that the other accumulates at a stable periodic orbit, shown in bold. Both branches of the stable manifold tend to the unstable equilibrium point as $t \to -\infty$. The final step in determining the phase portrait is to look for additional periodic orbits. Periodic orbits cannot occur inside the periodic orbit we have found because every closed curve inside the periodic orbit either (1) does not surround an equilibrium point or (2) intersects the stable or unstable manifold of the saddle. Thus, the only possible location for another periodic orbit is outside the one we have found. However, backward trajectories with initial points outside this periodic orbit tend to ∞, as we verify by integration. We conclude that there is a single periodic orbit.

Exercise 5.8. Estimate the ratio by which the distance to the unstable equilibrium point increases each time a trajectory spirals once around the equilibrium point in Figure 5.10.

The two phase portraits shown in Figures 5.10 and 5.12 have different numbers of equilibrium points and periodic orbits. If we vary the parameters continuously from the first set in the Morris-Lecar example to the second set, i.e., g_{Ca} from 4.4 to 5.5 and ϕ from 0.04 to 0.22, then we must encounter bifurcations at which the phase portraits make qualitative changes. Comparing the two phase portraits, we must encounter parameters at which the number of equilibrium points increases and parameters at which the unstable periodic orbit of the first phase portrait disappears. The next section discusses rudiments of bifurcation theory, a subject that systematically studies qualitative changes in phase portraits.

Exercise 5.9. Compute another phase portrait for the Morris-Lecar model with parameter values $v_1 = -1.2$, $v_2 = 18$, $v_3 = 2$, $v_4 = 30$, $g_{Ca} = 5.5$, $g_K = 8$, $g_L = 2$, $v_K = -84$, $v_L = -60$, $v_{Ca} = 120$, $C = 20$, $\phi = 0.22$, and $i = 91$. Only the parameter i has changed from those used in Figure 5.12. In what ways does the phase portrait differ from those displayed in Figures 5.10 and 5.12?

5.6 Bifurcations

The previous two sections discussed how to compute the phase portraits of two-dimensional vector fields. We implicitly emphasized the properties of structurally stable vector fields. A vector field is *structurally stable* if all small enough perturbations of the vector field have qualitatively similar phase portraits. For planar vector fields, structurally stable vector fields are characterized by the following properties:

1. Equilibrium points are hyperbolic, i.e., their linearizations have no zero or pure imaginary eigenvalues.
2. Periodic orbits are hyperbolic, i.e., their linearizations each have an eigenvalue different from 1.
3. There are no saddle connections: trajectories that lie in both the stable manifold of a saddle and the unstable manifold of a saddle (possibly the same saddle).

When a system depends upon parameters, like the the Morris-Lecar model, we expect to find regions in the parameter space with structurally stable vector fields separated by boundaries yielding vector fields that are not structurally stable. The two different parameter sets we examined in the Morris-Lecar model are each structurally stable, but they are qualitatively dissimilar, with different numbers of equilibria and periodic orbits. Here we investigate how we get from one phase portrait to another as we vary parameters in a model system. This is the subject of *bifurcation theory*.

Bifurcation theory looks at the typical behavior of *families* of vector fields that depend upon parameters. We designate a certain number of parameters as *active parameters* and examine how the phase portraits of the system change as the active parameters are varied. For example, there are thirteen parameters in the Morris-Lecar system, but we might designate g_{Ca} and ϕ as active parameters and look just at variations of these. We think in terms of determining phase portraits for the system across ranges of values for the active parameters. In a laboratory experiment, imagine running a series of experiments with different values of the active parameters. For each set of parameters, we do one or more experiments in which we allow the system time to reach its limit state. We may use initial conditions from the final state of a previous experiment or reset them. Numerically, we compute trajectories for different values of the active parameters as well as different initial points. In applying root finding algorithms to locate equilibrium points, we can try to track the position of the equilibria as continuous functions of the parameters.

Experience with many experiments and computations of this type suggests that there are modes of bifurcation that occur repeatedly in different systems. These modes have been mathematically analyzed and classified by ways in which vector fields may fail to be structurally stable. For two dimensional vector fields with one active parameter, the list of typical bifurcations is rather short, with just five types:

1. *Saddle-node bifurcation:* The Jacobian at an equilibrium point has a zero eigenvalue.
2. *Hopf bifurcation:* The Jacobian at an equilibrium point has a pair of pure imaginary eigenvalues.
3. *Saddle-node of limit cycle bifurcation:* A flow map Jacobian at a periodic orbit has double eigenvalue 1.
4. *Homoclinic bifurcation:* There is a trajectory in both the stable and unstable manifold of a single saddle.
5. *Heteroclinic bifurcation:* There is a trajectory in both the stable manifold of one saddle and the unstable manifold of another saddle.

The changes in phase portraits that occur with each type of bifurcation have also been characterized. We use examples to illustrate these patterns of bifurcation, but do not fully discuss their generality. We begin with the saddle-node bifurcation. The family of two-dimensional vector fields defined by

$$\begin{aligned} \dot{x} &= \mu + x^2 \\ \dot{y} &= -y \end{aligned} \qquad [5.22]$$

has a saddle-node bifurcation when $\mu = 0$. We are interested in how the phase portraits of this system change as μ varies. The system is separable: the equation for $\dot{x}$ is independent of y and the equation for $\dot{y}$ is independent of x. Moreover, the behavior of the y variable is always the same, approaching 0 as $t \to \infty$. Thus we restrict our attention to x and its dynamical behavior. The x-axis is invariant: trajectories with initial points on the x axis remain on the axis. There are equilibrium points along the curve $\mu = -x^2$. When $\mu < -x^2$, $\dot{x} < 0$ and x decreases, and when $\mu > -x^2$, $\dot{x} > 0$ and x increases. The equilibrium points and trajectory directions are shown in Figure 5.13. You should think of this figure as a "stack" of phase lines for the x-axis, one for each value of μ. When $\mu < 0$, there are two equilibrium points at $x = \pm\sqrt{-\mu}$. The negative equilibrium is stable and the positive equilibrium is unstable. When $\mu > 0$, there are no equilibrium points, $x \to \infty$ as $t \to \infty$ and $x \to -\infty$ as $t \to -\infty$. When $\mu = 0$, there is a single equilibrium point at $(x, y) = (0, 0)$. It is the forward limit set of trajectories that start with $x < 0$ and the backward limit set of trajectories that start with $x > 0$ and $y = 0$. At a general saddle-node bifurcation, a pair of equilibrium points coalesce and disappear, producing a qualitative change in the phase portrait. In order for saddle-node bifurcation to happen, there must be a zero eigenvalue at the bifurcating equilibrium point.[7]

Let us examine saddle-node bifurcation in the Morris-Lecar model with active parameter g_{Ca}. The equations that locate the saddle-node bifurcation are $\dot{v} = 0, \dot{w} = 0, \det(A) = 0$ where A is the Jacobian of the vector field at (v, w). The dependence of these equations on v is complicated and messy, but the dependence on w and g_{Ca} is linear. We can exploit this observation to solve $\dot{v} = 0, \dot{w} = 0$ for w and g_{Ca}, obtaining

$$\begin{aligned} w(v) &= w_\infty(v) = 0.5\left(1 + \tanh\left(\frac{v - v_3}{v_4}\right)\right) \\ g_{Ca}(v) &= \frac{i - g_K w(v - v_K) - g_L(v - v_L)}{m_\infty(v)(v - v_{Ca})} \end{aligned} \qquad [5.23]$$

at an equilibrium point. Note that we allow ourselves to vary the parameter g_{Ca} to find an equilibrium at which v has a value that we specify. We substitute these values into $\det(A)$. With the help of the computer algebra system Maple, we compute the value of $\det(A)$ as a function of v, using these substitutions for w and g_{Ca}. Figure 5.14 plots these values. It is evident that there are two values of v for which $\det(A) = 0$. The approximate values of g_{Ca} at these saddle-node points are 5.32 and 5.64. Figure 5.15 shows the nullclines for these two parameter values. This figure illustrates that the saddle-node bifurcations of a

[7] The *implicit function theorem* implies that an equilibrium with no zero eigenvalues varies smoothly with respect to parameters. The determinant of a matrix is the product of the eigenvalues, so a *defining equation* for a saddle-node bifurcation is that the determinant of the Jacobian vanishes. Notice that the curve of equilibria in (x, μ) space is a smooth parabola, defined by $\mu + x^2 = 0$. However, at the bifurcation, we cannot solve for the equilibrium point as a function of the parameters. Numerical *continuation* methods designed to track the curve of equilibria follow the curve around its "turning point" rather than incrementing the parameter and searching for equilibrium points at each fixed parameter value.

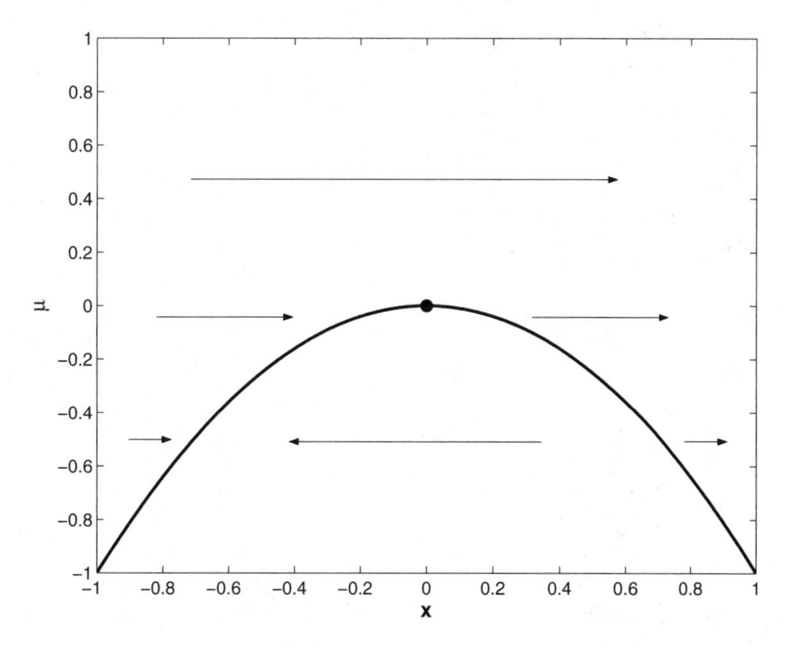

Figure 5.13 The dynamics of system [5.22] along the x-axis for varying μ. The equilibrium points are the parabola, with the saddle-node point at the origin. The trajectories are horizontal, with their directions shown by the arrows. Above the equilibrium curve trajectories flow right, while below the equilibrium curve trajectories flow left.

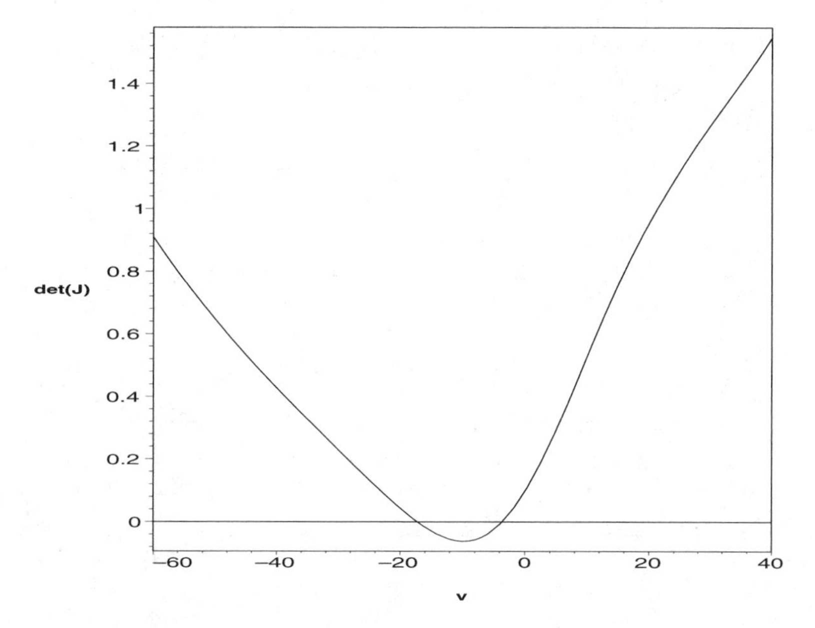

Figure 5.14 The values of w and g_{Ca} have been determined when there is an equilibrium point at a specified value of v. The determinant of the Jacobian at the equilibrium is plotted as a function of v. Saddle-node bifurcations occur when this function vanishes.

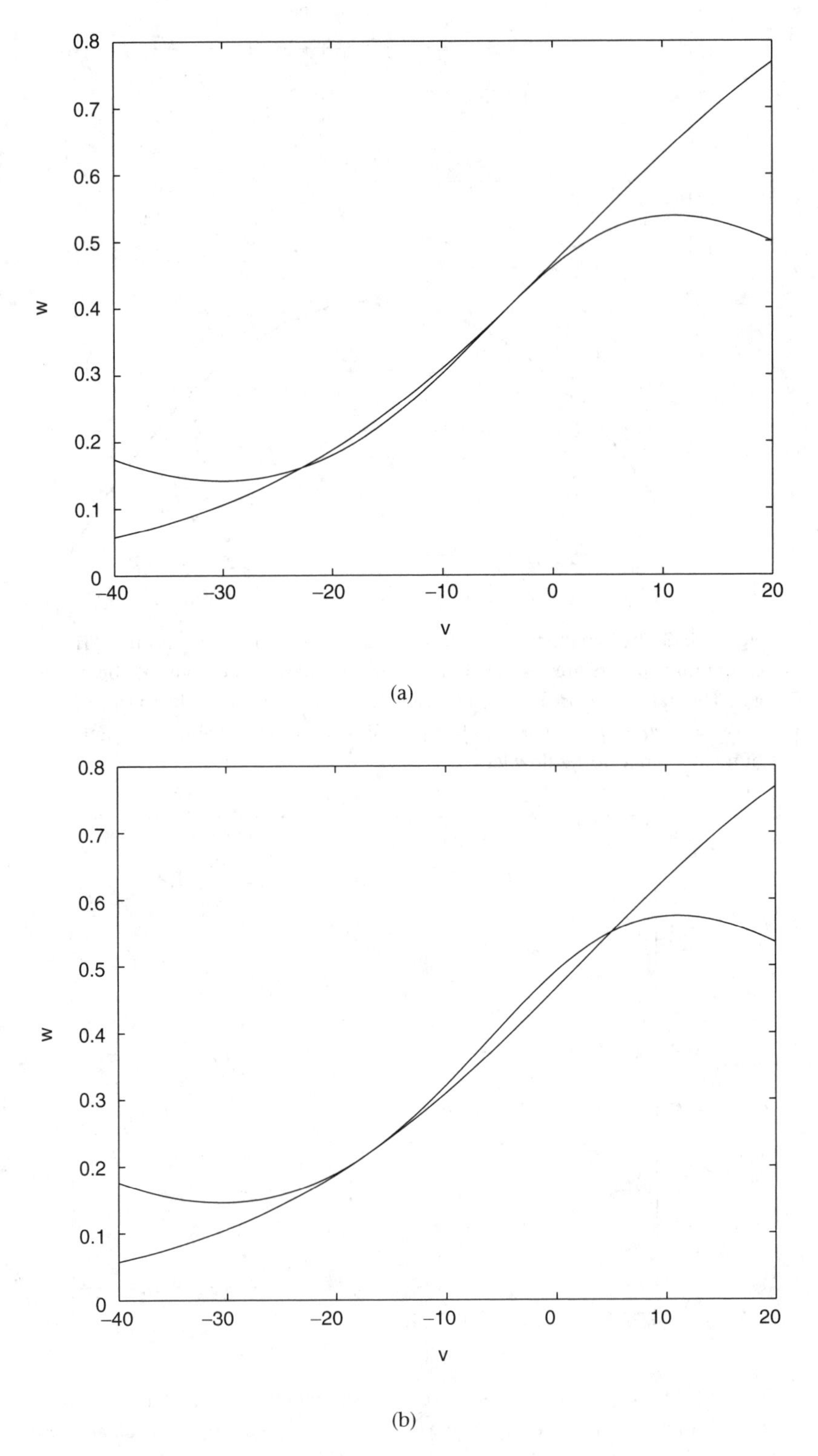

Figure 5.15 Nullclines for the Morris-Lecar system for saddle-node parameter values.

two-dimensional vector field occur when nullclines have a point of tangency. As the parameter g_{Ca} decreases from 5.32, the nullclines of v and w separate near the upper intersection where they are tangent. Simliarly, when g_{Ca} increases from 5.64, the nullclines of v and w separate near the lower, tangential intersection. When g_{Ca} is between the bifurcation values, the nullclines have three points of intersection.

Exercise 5.10. Compute phase portraits of the Morris-Lecar system for the saddle-node parameter values. Pay particular attention to the region around the saddle-node points.

Hopf bifurcation occurs when an equilibrium of a vector field depending upon a single active parameter has a pair of eigenvalues that cross the imaginary axis as the parameter changes. An example of a vector field which undergoes Hopf bifurcation is

$$\begin{aligned} \dot{x} &= (\mu - (x^2 + y^2))x - y \\ \dot{y} &= (\mu - (x^2 + y^2))y + x. \end{aligned} \qquad [5.24]$$

This system has an equilibrium point at the origin for all μ, and its Jacobian is

$$\begin{bmatrix} \mu & -1 \\ 1 & \mu \end{bmatrix}. \qquad [5.25]$$

When $\mu = 0$, this system has a pair of purely imaginary eigenvalues. To analyze the dynamics of the system, we investigate how the function $\rho(x, y) = x^2 + y^2$ varies along trajectories. Differentiating with the chain rule, we obtain

$$\dot{\rho} = 2(x\dot{x} + y\dot{y}) = 2(\mu - (x^2 + y^2))(x^2 + y^2) = 2\rho(\mu - \rho). \qquad [5.26]$$

When $\mu < 0$, $\dot{\rho}$ is negative everywhere but the origin and all trajectories approach the origin. When $\mu > 0$, the circle $\rho = \mu$ is a stable periodic orbit that is the forward limit set of all trajectories except the equilibrium at the origin. The family of periodic orbits emerges from the origin as μ increases from 0. In general, the emergence of a family of periodic orbits, with amplitude $\sqrt{x^2 + y^2}$ growing like $\sqrt{\mu - \mu_c}$, is characteristic of Hopf bifurcation occurring at μ_c. This Hopf bifurcation is *supercritical*: the periodic orbits emerging from the equilibrium point are stable (Figure 5.16). The family

$$\begin{aligned} \dot{x} &= (\mu + (x^2 + y^2))x - y \\ \dot{y} &= (\mu + (x^2 + y^2))y + x \end{aligned} \qquad [5.27]$$

has a *subcritical* Hopf bifurcation (Figure 5.17). The periodic orbits are circles $-\mu = x^2 + y^2$ which exist for $\mu < 0$, and they are unstable. Linearization does not determine whether a Hopf bifurcation is subcritical or supercritical. A definitive quantity can be expressed in terms of the degree-3 Taylor series expansion of

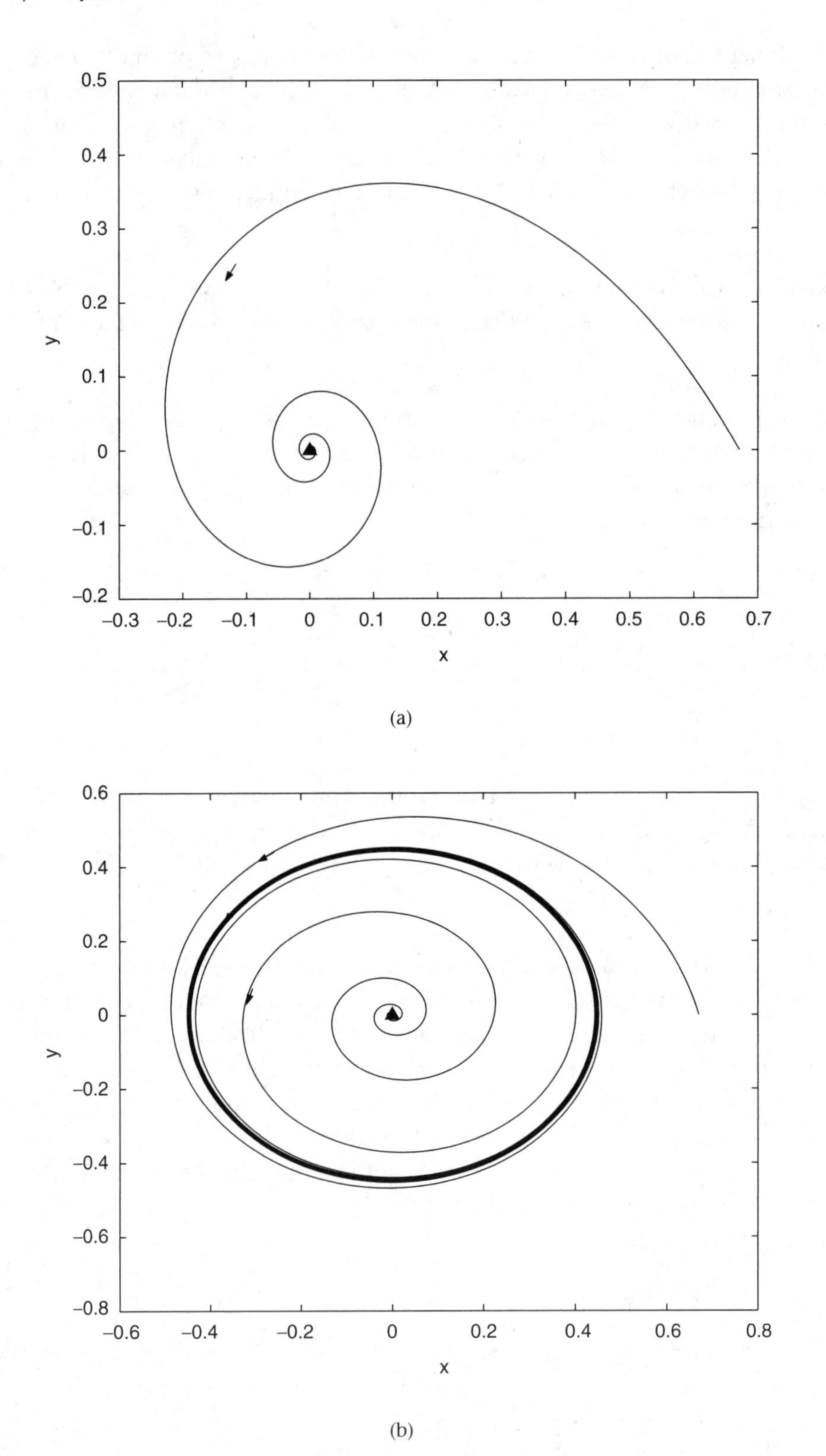

Figure 5.16 Phase portraits of the supercritical Hopf bifurcation for (a) $\mu = -0.1$ and (b) $\mu = 0.2$.

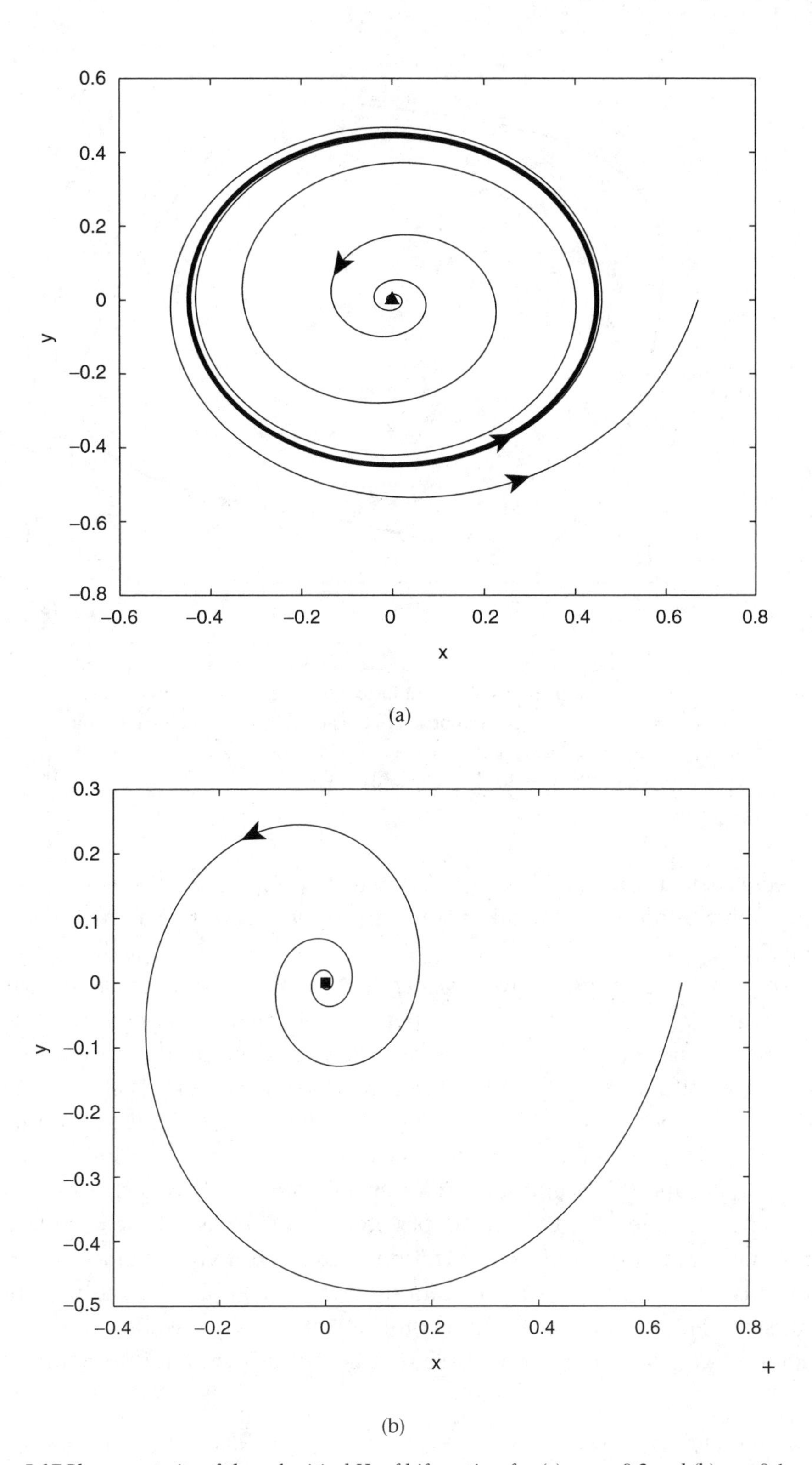

Figure 5.17 Phase portraits of the subcritical Hopf bifurcation for (a) $\mu = -0.2$ and (b) $\mu = 0.1$.

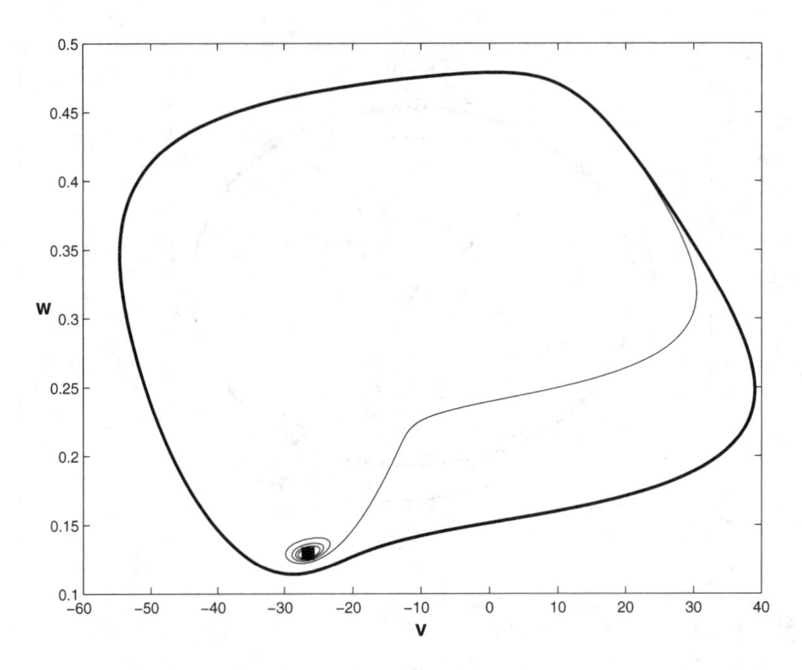

Figure 5.18 Phase portrait of the Morris-Lecar model. There is a single periodic orbit (bold curve) and an unstable equilibrium (square). Trajectories inside the periodic orbit flow from the equilibrium to the periodic orbit. Parameter values are $v_1 = -1.2,\ v_2 = 18,\ v_3 = 2,\ v_4 = 30,\ g_{Ca} = 4.4,\ g_K = 8,\ g_L = 2,\ v_K = -84,\ v_L = -60,\ v_{Ca} = 120,\ C = 20,\ \phi = 0.02$ and $i = 90$.

the vector field at the equilibrium point, but the simplest way to assess whether a Hopf bifurcation is subcritical or supercritical is to compute a few trajectories numerically.

In the Morris-Lecar system, we investigate Hopf bifurcation with active parameter ϕ when $g_{Ca} = 4.4$. For these parameter values, there is a single equilibrium point near $(v, w) = (-26.6, 0.129)$. As ϕ varies, the equilibrium does not move, but its eigenvalues change. A Hopf bifurcation occurs when ϕ is approximately 0.0231. For larger values of ϕ the eigenvalues have negative real part and the equilibrium is a stable focus. For smaller values of ϕ the eigenvalues have positive real part and the equilibrium is an unstable focus. As ϕ decreases from its value 0.04 in Figure 5.10, the smaller periodic orbit shrinks. At the Hopf bifurcation value of ϕ, this unstable periodic orbit collapses onto the equilibrium point. Thus, this is a subcritical Hopf bifurcation in which a family of unstable periodic orbits surrounds a stable equilibrium point, shrinking as the equilibrium becomes unstable. Figure 5.18 shows the phase portrait of the system for $\phi = 0.02$.

Exercise 5.11. Investigate the periodic orbits of the Morris-Lecar model in the vicinity of the bifurcating equilibrium. Plot how their amplitude (diameter) varies with ϕ.

Saddle node of limit cycle and homoclinic bifurcations involve changes in the number of periodic orbits. These bifurcations are *global* in that we must integrate trajectories in order to locate the bifurcations, as contrasted with the *local* saddle-node and Hopf bifurcations that can be determined from locating equilibrium points and their linearizations. At saddle node of limit cycle bifurcations, a pair of periodic orbits coalesce and disappear. In homoclinic bifurcations, there is an equilibrium point $\mathbf{x}_0$ with a trajectory $\mathbf{x}(t)$ that lies in both its stable and unstable manifolds. Thus, $\mathbf{x}(t) \to \mathbf{x}_0$ both as $t \to \infty$ and as $t \to -\infty$, In a system undergoing homoclinic bifurcation, there is a family of periodic orbits that terminates at the homoclinic orbit. As it does so, its period becomes unbounded.[8]

We use the Morris-Lecar system to illustrate saddle-node of limit cycle and homoclinic bifurcations. Starting with the parameters in Figure 5.10 ($g_{Ca} = 4.4$), the two periodic orbits move toward each other as we increase the parameter ϕ from 0.04. When ϕ reaches a value slightly larger than 0.52, the periodic orbits coalesce with one each other and disappear. For larger values of ϕ the stable equilibrium point is a global attractor: it is the limit set of all trajectories (Figure 5.19).

To look for homoclinic orbits in the Morris-Lecar system, we need to choose parameter values for which there is a saddle point. Thus we set $g_{Ca} = 5.5$ and vary ϕ. We find that there is a homoclinic orbit that forms a loop "below" the saddle when ϕ is approximately 0.202554. Figure 5.20 shows phase portraits of the system when $\phi = 0.202553$ and $\phi = 0.202554$. Observe that for $\phi = 0.202553$, there are two periodic orbits, one an unstable orbit that almost forms a loop with a corner at the saddle. One branch of the stable manifold of the saddle comes from this periodic orbit, while both branches of the unstable manifold of the saddle tend to the large stable periodic orbit. For $\phi = 0.202554$, there is a single periodic orbit and both branches of the stable manifold of the saddle come from the unstable equilibrium point while one branch of the unstable manifold tends to the stable equilibrium point. In between these parameter values there is a homoclinic bifurcation that occurs at parameter values where the lower branches of the stable and unstable manifolds of the saddle coincide.

Exercise 5.12. There are additional values of ϕ that give different homoclinic bifurcations of the Morris-Lecar model when $g_{\text{Ca}} = 5.5$. Show that there is one near $\phi = 0.235$ in which the lower branch of the stable manifold and the upper branch of the unstable manifold cross, and one near $\phi = 0.406$ where the upper branches of the stable and unstable manifold cross. (Challenge: As ϕ varies from 0.01 to 0.5, draw a consistent set of pictures showing how periodic orbits and the stable and unstable manifolds of the saddle vary.)

---

[8]A similar phenomenon happens at some saddle-node bifurcations of an equilibrium. After the equilibria disappear at the bifurcation, a periodic orbit that passes through the region where the equilibrium points were located may appear. The term *snic* (saddle node on invariant circle) was used by Rinzel and Ermentrout to describe these saddle-node bifurcations.

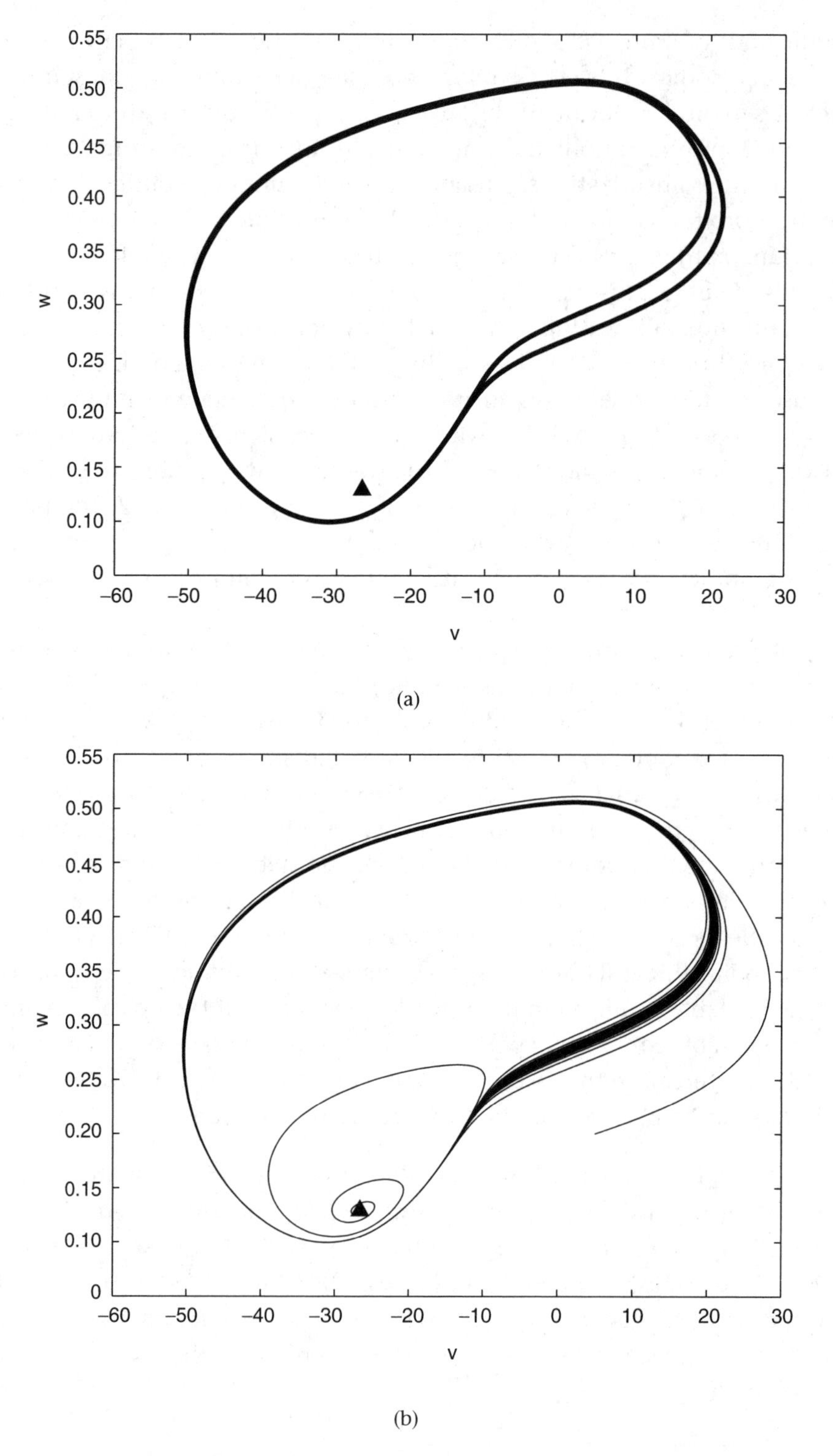

Figure 5.19 Phase portraits of the Morris-Lecar system close to a saddle-node of periodic orbits. (a) Here $\phi = 0.05201$ and there are two nearby periodic orbits. (b) Here $\phi = 0.05202$ and a single trajectory is plotted. It passes very slowly through the region where the periodic orbits were located.

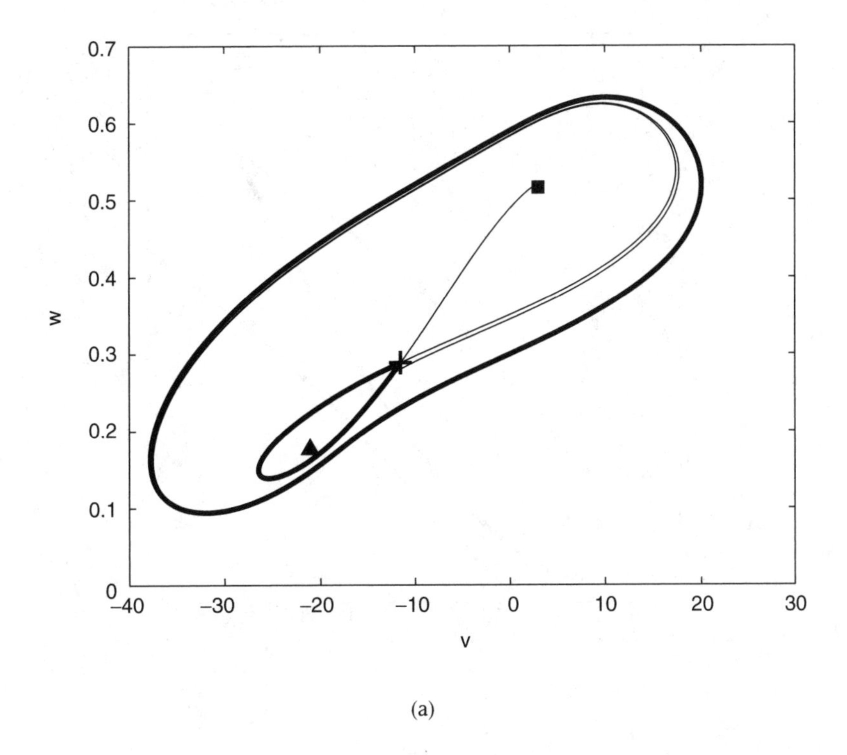

(a)

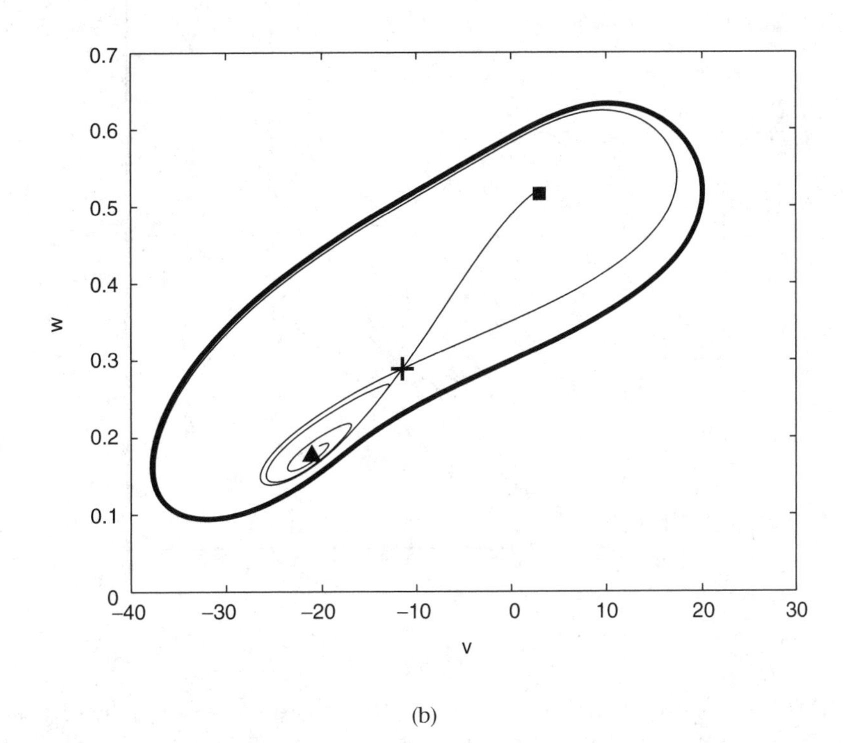

(b)

Figure 5.20 Phase portraits of the Morris-Lecar system close to a homoclinic orbit. (a) and (c) $\phi = 0.202553$ (b) and (d) $\phi = 0.202554$.

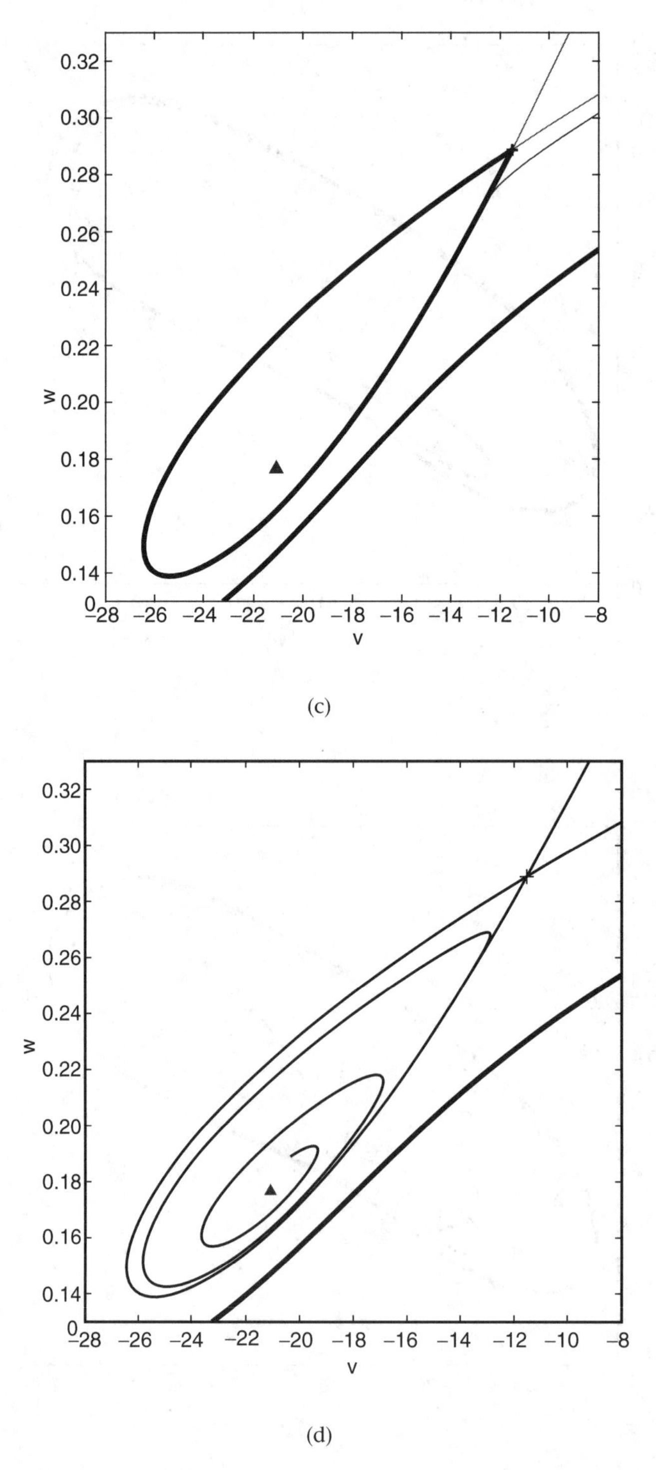

Figure 5.20 (*continued*) The phase portraits (c) and (d) show an expanded view of those in figures (a) and (b). Observe carefully the relative position of the stable and unstable manifolds of the saddle in (c) and (d).

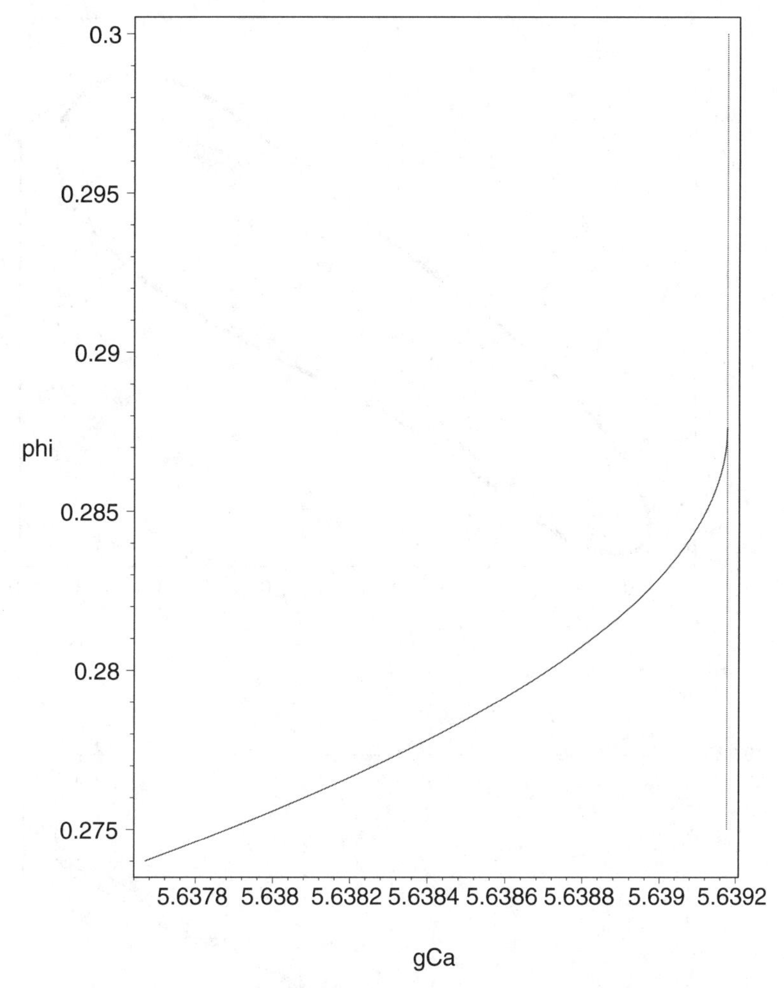

Figure 5.21 Bifurcation curves of saddle-node (vertical) and Hopf bifurcation (curved) in the plane of the parameters (g_{Ca}, ϕ). The curves intersect tangentially at the end point of the Hopf bifurcation curve, a Takens-Bogdanov point.

More complicated bifurcations than the ones described above can be expected to be found in a two-dimensional vector field with more than a single active parameter. The bifurcations encountered in generic systems with k active parameters are called *codimension-k* bifurcations. As k increases, the classification of codimension-k bifurcations becomes increasingly difficult. Even for two-dimensional vector fields, the largest value of k for which the classification is reasonably complete is $k = 3$. We give here two examples of codimension-2 bifurcations in the Morris-Lecar system. When both ϕ and g_{Ca} are varied at the same time, we can find parameter values for which 0 is the only eigenvalue of the Jacobian at an equilibrium point. These parameter values satisfy the defining equations for both saddle-node and Hopf bifurcations. Using

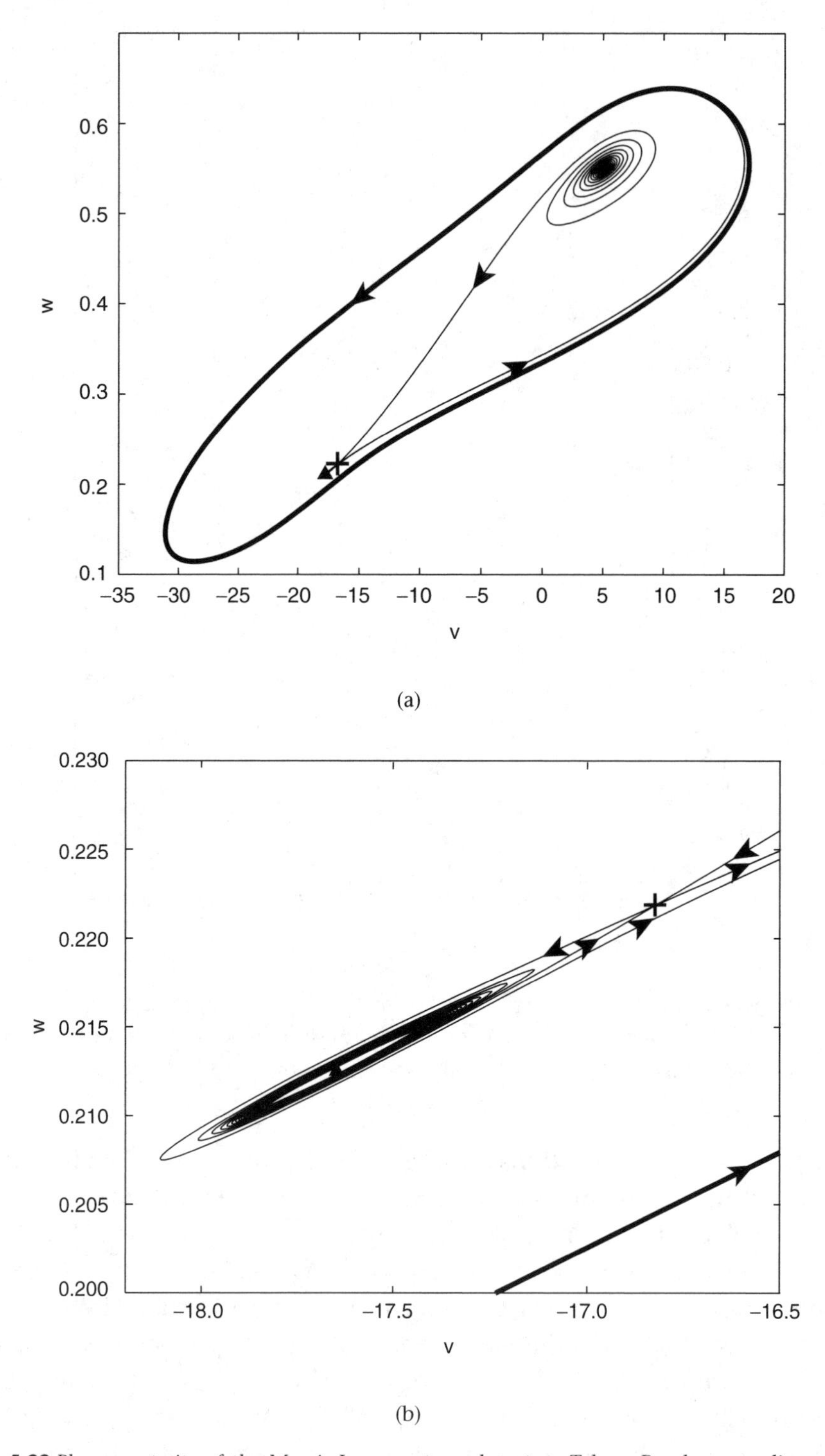

Figure 5.22 Phase portraits of the Morris-Lecar system close to a Takens-Bogdanov codimension-2 bifurcation. (a) There are three equilibrium points (a stable focus, a saddle and an unstable focus) and two periodic orbits. The small periodic orbit is not visible here, but can be seen in the small scale plot (b) of the phase portrait near the two lower equilibrium points.

Maple, we compute an approximate set of parameters for which this happens: $(g_{Ca}, \phi) \approx (5.6392, 0.2876)$. See Figure 5.21. This codimension-2 bifurcation is called a *Takens-Bogdanov* bifurcation, after the two mathematicians who first studied the properties of these bifurcations around 1970 (Guckenheimer and Holmes, 1983). Three of the bifurcations we have studied in the Morris-Lecar model—saddle nodes, Hopf bifurcation, and homoclinic bifurcation—come together at the Takens-Bogdanov bifurcation. The curve of homoclinic bifurcations lies above the curve of Hopf bifurcations in Figure 5.21, but is more difficult to compute. Figure 5.22 shows a phase portrait with $(g_{Ca}, \phi) = (5.638, 0.276)$, parameters near the Takens-Bogdanov parameter values. This phase portrait has three equilibrium points, two of which are close together, and two periodic orbits, one of which is small. Figure 5.22b shows more detail of the region around the two nearby equilibrium points. One of the aspects of the Takens-Bogdanov bifurcation, evident in this figure, is that the eigenvectors of the saddle lie almost in the same direction. As the parameter ϕ is decreased or g_{Ca} is increased, the small unstable periodic orbit shrinks, meeting the equilibrium point at a subcritical Hopf bifurcation. As ϕ is increased or g_{Ca} is decreased, the small unstable periodic orbit becomes a homoclinic orbit and then disappears. When the parameter g_{Ca} increases, the two nearby equilibrium points coalesce with one another in a saddle-node bifurcation.

A second type of codimension-2 bifurcation is the *cusp*. This bifurcation occurs in two-dimensional vector fields when the nullclines intersect with a third-order tangency, like the tangent line to a function at a point of inflection. The parameters that give the saddle-node bifurcation in Figure 5.9a appear to be close to such a point of tangency. We let i be a second active parameter with g_{Ca} and then find a cusp point near $(g_{Ca}, i) = (4.97, 107.1)$. A pair of saddle-node curves emanate from the cusp in the parameter plane. Figure 5.23 shows the nullcline for the cusp values of the parameters along side a plot of the saddle-node curve in the (g_{Ca}, i) parameter plane.

5.7 Numerical Methods

Throughout the past two chapters, we have assumed that we have computer algorithms that reliably determine trajectories with specified initial conditions. This assumption is a good one, but there are pitfalls that must be avoided to obtain reliable results. Since the Morris-Lecar equations cannot be "solved" by finding explicit analytic formulas that express the trajectories as elementary functions, numerical methods that compute approximate solutions proceed step by step in time. There are many methods for solving these initial value problems. The results can vary substantially as we switch from one algorithm to another and as we change algorithmic parameters, so we discuss the fundamental ideas employed in

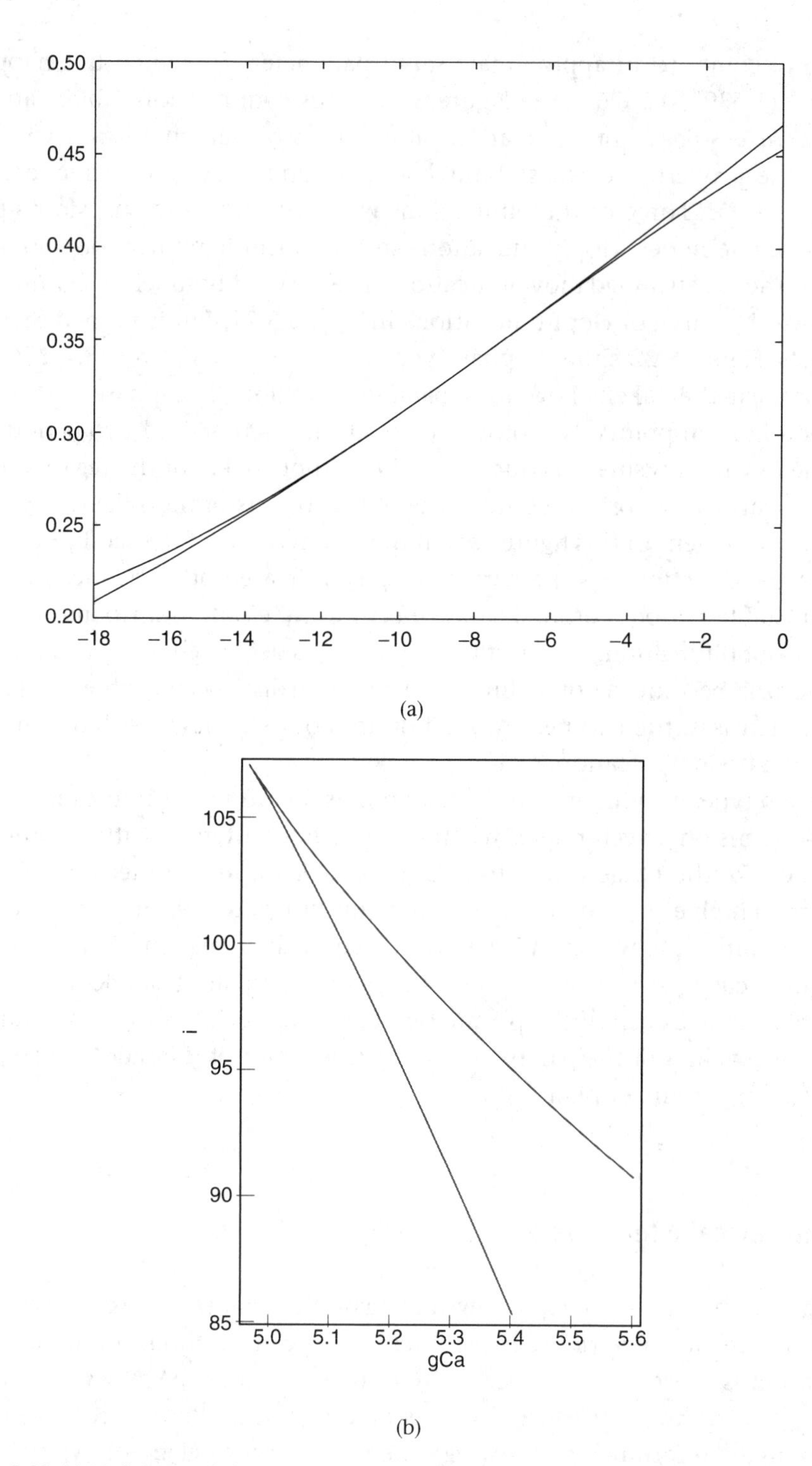

Figure 5.23 (a) Nullclines of the Morris-Lecar system at a cusp point. The two nullclines intersect at a single point to third order. The parameters are as in Table 5.1 except that $(g_{Ca}, i) = (4.9666, 107.06655)$. The location of the cusp point is independent of the parameter ϕ. (b) The saddle-node curve in the (g_{Ca}, i) parameter plane. The cusp bifurcation parameters are at the cusp of the saddle-node curve in this parameter plane.

a few of the simplest numerical methods and display some of their shortcomings. We begin with the simplest method of all, called the (explicit) *Euler* method.

In calculus, we learn that the derivative of a function gives the slope of its best linear approximations. If $g(x)$ is a differentiable function and $\Delta x = x - a$, then the residual $r(x) = g(x) - (g(a) + g'(a)(\Delta x))$ has the property that $r(x)/(\Delta x) \to 0$ as $\Delta x \to 0$. For x close to a, $g(a) + g'(a)(\Delta x)$ gives a good approximation to $g(x)$. The same thing is true for solutions of the system of differential equations $\dot{\mathbf{x}} = \mathbf{f}(\mathbf{x})$. If $\mathbf{x}(t)$ is the trajectory with $\mathbf{x}(t_0) = \mathbf{x}_0$, then $\mathbf{x}(t_0 + h)$ is well approximated by $\mathbf{x}_0 + h\mathbf{f}(\mathbf{x}_0)$ when the time increment $h = t - t_0$ is small. The line traced by $\mathbf{x}_0 + h\mathbf{f}(\mathbf{x}_0)$ as h varies is the one that the trajectory would follow if the vector field were constant, with the same value $\mathbf{f}(\mathbf{x}_0)$ everywhere. Since the vector fields we study are not constant, we fix a (small) positive value of h and use $\mathbf{x}_1 = \mathbf{x}_0 + h\mathbf{f}(\mathbf{x}_0)$ as an approximation to the point we reach on the trajectory at time $t_1 = t_0 + h$. We then take another step to time $t_2 = t_1 + h$ and position $\mathbf{x}_2 = \mathbf{x}_1 + h\mathbf{f}(\mathbf{x}_1)$. We continue in this fashion, using the iteration $t_{n+1} = t_n + h$ and $\mathbf{x}_{n+1} = \mathbf{x}_n + h\mathbf{f}(\mathbf{x}_n)$ to determine $\mathbf{x}_{n+1}, t_{n+1}$ from $\mathbf{x}_n, t_n$. To compute an approximation to $\mathbf{x}(T)$, we pick a *time-step* $h = (T - t_0)/N$ for some integer N, and use N steps of this iteration. This is the Euler method. At each step, the Euler method makes an error that shrinks as $h \to 0$. We hope that, as $h \to 0$, the method will give values that converge to $\mathbf{x}(T)$. Since the number of steps grows as the step length h gets smaller, it is not evident whether we obtain this convergence or how good the approximation is. To study the approximations further, let us look at an example where we can derive formulas for the values of $\mathbf{x}(T)$ and and the points produced by the Euler method.

Example.

Consider the equation $\dot{x} = x$. From calculus, we know that the solutions to this equation are $x(t) = x(0)e^t$. Take $T = 1$ and $x(0) = 1$. Then $x(T) = e \approx 2.718281828$ and we want to see how the values calculated by the Euler method vary when we use N steps of length $h = 1/N$ starting at $x_0 = 1$. We compute the first three steps of the method to be $x_1 = 1 + h$, $x_2 = x_1 + hx_1 = (1 + h)x_1 = (1 + h)^2$, and $x_3 = x_2 + hx_2 = (1 + h)x_2 = (1 + h)^3$. Continuing, we obtain $x_N = (1 + h)^N = (1 + 1/N)^N$. Now $(1 + 1/N)^N \to e$ as $N \to 0$, but it does so quite slowly. Table 5.2 gives some data for this example, showing values obtained by the Euler method when N is a power of 10. The table indicates that ten thousand time-steps are required to obtain four digits of accuracy on this simple equation, a very poor performance. We also observe that each additional digit of accuracy requires about ten times the number of time-steps. In contrast to this method of calculating e and the solution to the differential equation $\dot{x} = x$, consider the Taylor series for

$$e = \sum_{n=0}^{\infty} \frac{1}{n!}.$$

10	2.593742460
100	2.704813829
1000	2.716923932
10000	2.718145927
100000	2.718268237
1000000	2.718280469

Table 5.2 Approximate values for e obtained from the Euler method. The first column gives the number of time steps, the second column, the approximate value of $e \approx 2.718281828$.

If we truncate this series after eleven terms (the last term we give has $n = 10$), it gives an approximate value of e as 2.718281801. Extrapolating from our observations about the Euler method, comparable accuracy would require roughly 100 million time steps! Thus the Euler method is a very poor way of obtaining highly accurate approximations to solutions of differential equations. We want methods that require fewer calculations to achieve a given accuracy.

One path to finding more accurate numerical methods for solving ordinary differential equations is to focus upon the order of the numerical methods. The order is defined by the relationship between the step length h and the error the method makes in computing $\mathbf{x}(t+h)$ from $\mathbf{x}(t)$. If the error in computing one step is proportional to h^{d+1}, then the method has *order d*. When we use N steps of length $h = T/N$ to compute an approximate solution at time $t_0 + T$, $N = T/h$ has order $1/h$. Thus, the error of a method of order d in computing $x(T)$ tends to zero as h^d as $h \to 0$.

In our example of computing e by solving the differential equation $\dot{x} = x$, replacing the Euler method by the degree d truncation of the Taylor series of the exponential function gives increasingly accurate order-d approximations to the solution. For example, a single step of length 1 with $d = 10$ gives two more digits of accuracy than a million steps of the Euler method. Even "low"-degree methods give an enormous improvement: ten steps of length $1/10$ with $d = 4$ give a value of 2.718279744, with accuracy comparable to the Euler method with a million steps.

There are varied approaches to the construction of higher-order methods for solving differential equations. We mention two, one to illustrate the principle and the second because it is frequently used. The second-order *Heun's method* has time-step map $H_h(\mathbf{x}) = \mathbf{x} + (h/2)(\mathbf{f}(\mathbf{x}) + \mathbf{f}(x + h\mathbf{f}(\mathbf{x})))$. Note that producing each time-step requires evaluating $\mathbf{f}$ twice, once at x and once at $x + h\mathbf{f}(\mathbf{x})$. To see that the method has second order, we first compute the Taylor expansion of $H_h(\mathbf{x})$ in h, obtaining $\mathbf{x} + \mathbf{f}(\mathbf{x})h + \frac{1}{2}\mathbf{f}'(\mathbf{x})\mathbf{f}(\mathbf{x})h^2$. Next we differentiate the equation $\dot{\mathbf{x}}(t) = \mathbf{f}(\mathbf{x}(t))$ with respect to t, to get $\ddot{\mathbf{x}} = \mathbf{f}'(\mathbf{x})\mathbf{f}(\mathbf{x})$ using the chain rule.

These equations imply that the degree-1 and -2 terms of the Taylor expansion of $\mathbf{x}(t)$ have coefficients $\mathbf{f}(\mathbf{x})$ and $\mathbf{f}'(\mathbf{x})\mathbf{f}(\mathbf{x})$, agreeing with the Taylor expansion of Heun's method. Similar, but longer, calculations lead to still higher-order methods. For example, the *Runge-Kutta method* with time-step map R_h defined by

$$
\begin{aligned}
\mathbf{k}_1 &= \mathbf{f}(\mathbf{x}) \\
\mathbf{k}_2 &= \mathbf{f}\left(\mathbf{x} + \frac{1}{2}h\mathbf{k}_1\right) \\
\mathbf{k}_3 &= \mathbf{f}\left(\mathbf{x} + \frac{1}{2}h\mathbf{k}_2\right) \\
\mathbf{k}_4 &= \mathbf{f}\left(\mathbf{x} + h\mathbf{k}_3\right) \\
R_h(\mathbf{x}) &= \mathbf{x} + \frac{h}{6}(\mathbf{k}_1 + \mathbf{k}_2 + \mathbf{k}_3 + \mathbf{k}_4)
\end{aligned}
\qquad [5.28]
$$

has fourth order. When accuracy of a numerical simulation is important, use higher-order methods if at all possible.

In addition to accuracy, the *stability* of numerical methods for solving initial value problems is also important. To illustrate the issue, consider Euler's method applied to the equation $\dot{x} = -ax$ with $a > 0$ large. The solution of this equation with $x(0) = x_0$ is $x(t) = e^{-at}x_0$. As $t \to \infty$, the solution approaches 0 very, very quickly. The time-step map from the Euler method is $E_h(x) = x(1 - ah)$. N iterations of E_h send x to $x(1 - ah)^N$. If $ah > 2$, then $|1 - ah| > 1$ and $|1 - ah|^N$ increases geometrically with N. Thus the behavior of the iterates of E_h is completely different from that of the solution of the differential equation. Instead of monotonically approaching 0, the iterates of E_h increase in magnitude with alternating signs. To avoid this unstable behavior of the numerical integration, the time-step is limited by $ah < 2$. If $1 < ah < 2$, the iterates of E_h approach 0, but they oscillate in sign as they do so.

Constraints on the time-step h of a method are a big issue in applications of biological interest. If time-steps are limited by the fastest rates of exponential decay in a system, then we will have difficulty working with models that include fast processes that quickly approach equilibrium. The model for enzyme kinetics introduced in Chapter 1 is already an example of this phenomenon. Simulations of the three-dimensional motions of proteins provide a somewhat different challenge in handling multiple time scales in systems of differential equations. The frequencies of oscillations in individual bonds within a molecule are many orders of magnitude faster than the time required for the folding of the molecule. Attempts to predict three-dimensional structure from simulation of the molecular dynamics must encompass this range of time scales, a task that is still not feasible on today's fastest computers. Within the numerical analysis literature, systems with multiple time scales are called *stiff*. Numerical methods have been created for integrating stiff systems. The simplest of these is the *implicit Euler* method.

Implicit methods require that one solves a system of equations to determine the point at the end of a step. The simplest example is the implicit Euler method defined by the equations $I_h(\mathbf{x}) = \mathbf{x} + h\mathbf{f}(I_h(\mathbf{x}))$. To determine $I_h(\mathbf{x})$, we need to solve these equations at each step. Newton's method is typically used to find the solution. As before, we can analyze the method completely for the linear equation $\dot{x} = -ax$. The equation to be solved is $I_h(x) = x - haI_h(x)$ whose solution is $I_h(x) = x/(1 + ah)$. This implies that after n steps, x is mapped to $x/(1 + ah)^n$. As n increases, this tends to 0 in a monotone way for all $a, h > 0$. Thus the method avoids the instability and step length limitations of the explicit Euler method. In a stiff system, we can usually use step lengths based upon the slower time scales in the system without causing instability due to rapidly decaying components of a solution. Even though the implicit Euler method is more stable than the explicit Euler method, it still has first order. The second-degree Taylor expansion of $x/(1 + ah)$ in h is $x(1 - ah + a^2h^2 - \cdots) = x - ahx + a^2h^2x - \cdots$. The second-degree term in the Taylor expansion of $e^{-ah}x$ is $\frac{1}{2}a^2x$, not a^2x, so the method is only first-order accurate. In general, implicit methods are slower than explicit methods when used with the same step size due to the time required to solve the implicit equations. Their advantage is that it is often possible to take much larger steps and still maintain stability.

The limitations on step size associated with accuracy and instability may change substantially along a trajectory. To deal with this, it is common practice to use methods that change step sizes during the calculation of trajectories. Procedures have been developed for estimating the error of a tentative step and adjusting the step length based upon this information. The simplest procedure is to do the following. If the estimated error exceeds a desired threshold, then a smaller step is chosen. If the estimated error is smaller than a lower threshold, then a larger step is attempted. The effect of such a strategy is to utilize the largest step sizes that are consistent with specified error tolerances. In practice, the additional computational cost in estimating error tolerances is more than offset by the smaller number of steps required to compute trajectories. On occasion, the step length selection procedures can be overly optimistic and give results that do become unstable.

We depend upon thoroughly tested computer algorithms to solve solutions to initial value problems for systems of differential equations, just as we depend upon algorithms to compute eigenvalues and eigenvectors of matrices. However, there is a wider range of performance and more choices to be made in solving differential equations. There are three things to keep in mind when using numerical integration:

- It is easy to get spurious results by misusing an algorithm. So be skeptical and check your results, for example by using more than one step size to solve the same problem, and by comparing solutions to the same problem computed with different methods.

- There are problems, like molecular simulations of protein folding, that remain beyond the capability of current methods. Be realistic in your expectations of what computers can do, and creative in dealing with problems that seem beyond the pale.
- There is a mathematical framework that is helpful in developing a consistent interpretation of simulation results. If numerical results appear to be at odds with the theory, work to resolve the discrepancies.

5.8 Summary

This chapter is the most mathematical in this book. Nonetheless, it gives only a brief introduction to dynamical systems theory. Systems of ordinary differential equations are perhaps the most common type of dynamic models in all of the sciences, so it is worthwhile to learn how to work with them. That is a task that involves lots of computation and lots of mathematics. The language that has been developed to describe the patterns formed by these systems and the logic of concentrating attention on generic systems can be intimidating when we first encounter them. As with most skills, proficiency with this type of analysis comes with experience and practice. We have used a single example, the Morris-Lecar model, to illustrate phenomena described in much greater generality by the theory. We hope that you will explore additional examples of differential equation models with complex dynamics and study the mathematics in more depth so that you can use it with confidence as a tool in your explorations.

The next chapter applies what we have learned here to differential equation models of infectious disease. In the simplest of these models, analytical techniques take us a long way and the dynamics are simpler than those of the Morris-Lecar model. However, sustained oscillations of childhood endemic diseases like chicken pox and measles have been observed, and we will use dynamic models to gain insight into the causes of these oscillations. Subsequent chapters will consider still more complex dynamic models of biological systems where the mathematical foundations for interpreting the model results are weaker than is the case for ordinary differential equations.

5.9 References

Blanchard, P., R. Devaney, and G. Hall. 2002. *Differential Equations*, 2nd edition. Brooks/Cole.

Guckenheimer, J., and P. Holmes. 1983. *Nonlinear Oscillations, Dynamical Systems, and Bifurcation of Vector Fields*. Springer-Verlag.

Hirsch, M. W., S. Smale, and R. L. Devaney. 2004. *Differential Equations, Dynamical Systems, and an Introduction to Chaos*. Elsevier, Amsterdam.

Rinzel J. and B. Ermentrout. 1998. Analysis of neural excitability and oscillations. Pages 251–292 in C. Koch and I. Seger (eds), *Methods in Neuronal Modeling*. MIT Press, Cambridge.

Strogatz, S. 1994. *Nonlinear Dynamics and Chaos: With Applications to Physics, Biology, Chemistry and Engineering*. Perseus Publishing.

6 Differential Equation Models for Infectious Disease

> To those familiar with the manifold complexities of real infections in real populations, our "basic models" may seem oversimplified to the point of lunacy.
>
> *Roy Anderson and Robert May (1992, p. 9)*

The aims of this chapter are the same as those stated by Anderson and May (1992) for their monograph *Infectious Diseases of Humans*: to show how relatively simple models can help to interpret data on infectious diseases, and to help design programs for controlling them. We begin with some classic examples, but end with some very recent developments: models for the emergence and management of drug-resistant disease strains, and models for disease progression within the body with particular reference to HIV/AIDS. The study of infectious diseases is one of the most mature applications of dynamic models in biology, so we can present some real and important "success stories" for simple dynamic models. We limit ourselves here to human diseases, but very similar models are also used for animal and plant disease dynamics (see, e.g., Hudson et al. 2002, Campbell and Madden 1990, and the papers discussed in the Preface), an issue of increasing importance as climate change and other anthropogenic stressors render natural populations increasingly susceptible to disease.

This chapter introduces no new mathematics. Rather, following our general approach, it serves to indicate the enormous scope of potential applications for differential equation models. Previous chapters have examined differential equation models at the within-cell (enzyme kinetics, gene regulation) and whole-cell (neuron excitation) levels. Now we go up to the level of human populations, but the model structures and the tools for their analysis—rescaling, eigenvalues, bifurcations, and so on—are the same.

6.1 Sir Ronald Ross and the Epidemic Curve

Sir Ronald Ross (1857–1932) received the 1902 Nobel Prize in medicine for determining the life cycle of the malaria parasite, in particular the role of mosquitos in the parasite life cycle and as vectors for its transmission between humans. From that humble beginning he went on to found the modern application of dynamic models to the study of infectious diseases.

Ross (1916) gave two motivations for modeling epidemic dynamics. First, he noted that infectious diseases could display three different temporal patterns:

1. *Endemic*: relatively small fluctuations in monthly case counts, and only slow increase or decrease over the course of years (Ross listed leprosy and tuberculosis in this category)
2. *Outbreak*: constantly present but flaring up in epidemic outbreaks at frequent intervals (measles, malaria, dysentery)
3. *Epidemic*: Intense outbreaks followed by disappearance (plague, cholera)

Ross (1916, p. 205) asked "To what are these differences due? Why, indeed, should epidemics occur at all, and why should not all infectious diseases belong to the first group and remain at an almost flat rate?"

Ross's second motivation was to explain the characteristic shape of the *epidemic curve* for diseases in the third class. The epidemic curve is the time course of disease *incidence*, the number of new cases per unit time. Figures 6.1 and 6.2 show a few examples. The characteristic features are a symmetric or nearly symmetric rise and fall, with the outbreak terminating before all individuals susceptible to the disease have become infected. Because susceptibles still remain in the population when outbreaks terminate, it was argued by some at the time that outbreaks terminate because the pathogen loses infectivity; others hypothesized that the uninfected individuals must have been less susceptible to the disease.

Ross's (1916) model was a partial success, allowing him to show that the shape of epidemic curves could be explained without either of these hypotheses. His other goal, to explain different patterns of disease dynamics, was tackled a decade later by Kermack and McKendrick (1927). Current models are largely based on Kermack and McKendrick's modified versions of Ross's models, so we will consider those here. The models (SIR models) are formulated at the level of the available data: the numbers of individuals reported to contract the disease. Individuals are classified as being either *S*usceptible to the disease, *I*nfected by it, or *R*ecovered or *R*emoved. *R*-stage individuals are neither infectious nor infectable: either dead, or having immunity (permanent or temporary) against the disease.

The first of Kermack and McKendrick's basic models described a disease outbreak in a closed population of constant size:

$$\begin{aligned} dS/dt &= -\beta SI \\ dI/dt &= \beta SI - \gamma I \\ dR/dt &= \gamma I. \end{aligned} \tag{6.1}$$

Initial conditions are $S(0) = S_0 \approx N$, $I(0) = N - S_0 \approx 0$, $R(0) = 0$ where N is the total population size. Since $dS/dt + dI/dt + dR/dt = 0$ the total population size

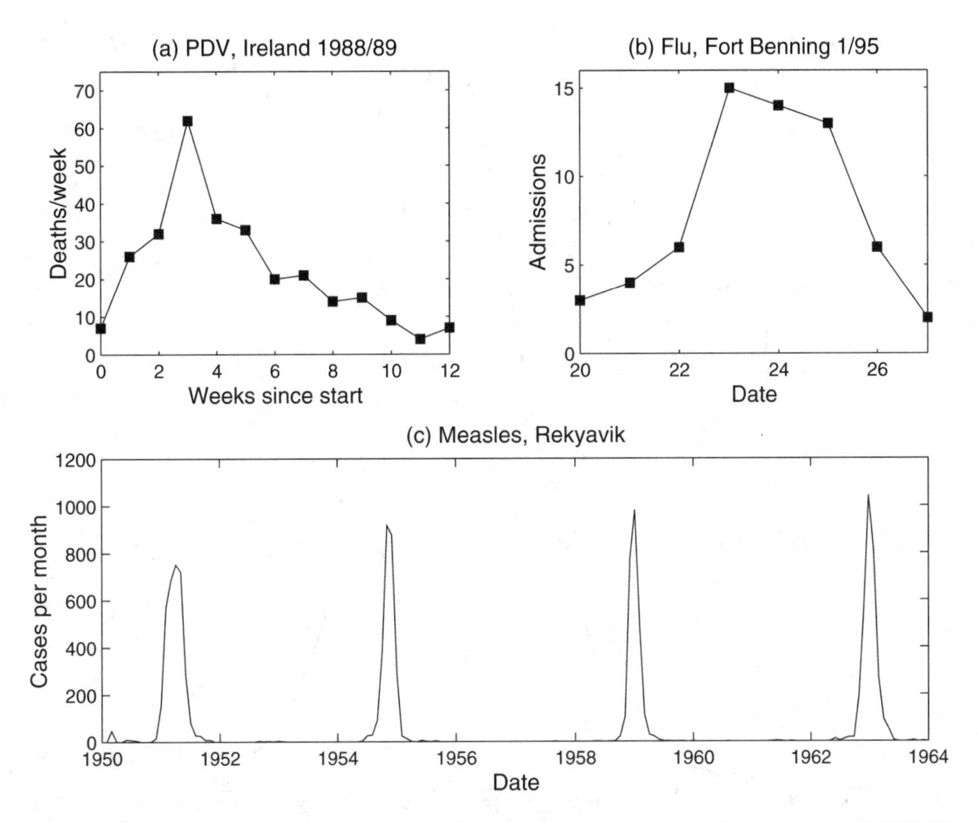

Figure 6.1 Examples of epidemic curves. (a) Phocine distemper virus in Northern Ireland 1988/89 (data from Figure 4 of Hall et al. 1992, provided by John Harwood). (b) An outbreak of influenza in Fort Benning, Georgia in 1995 (data from Davidson 1995). (c) Recurrent outbreaks of measles in Rekyavik, Iceland (data provided by Andrew Cliff, Department of Geography, University of Cambridge).

remains constant at N. The population is closed in the sense that no new susceptibles are added by births or immigration, and so long as R individuals are counted the population size is constant. Thus the assumption of constant population size is really that the only changes in population size are disease-induced deaths. The fraction of infected individuals, $I(t)/N$, is called the *prevalence* of the disease.

The first equation in model (6.1) is disease transmission resulting from contact between susceptibles and infectives. Ross (1916) justified this transmission rate as follows. Each infected individual transmits the pathogen to b individuals per unit time, but new cases arise only if the recipient individual is susceptible. Assuming a constant population of size N, the number of new cases per unit time is therefore $bI(S/N) = \beta SI$ where $\beta = b/N$. This form of transmission is called "mass action" (by analogy with the Law of Mass Action in chemical reactions) or "proportional mixing" (Anderson and May 1992). Mass action has been and still remains the

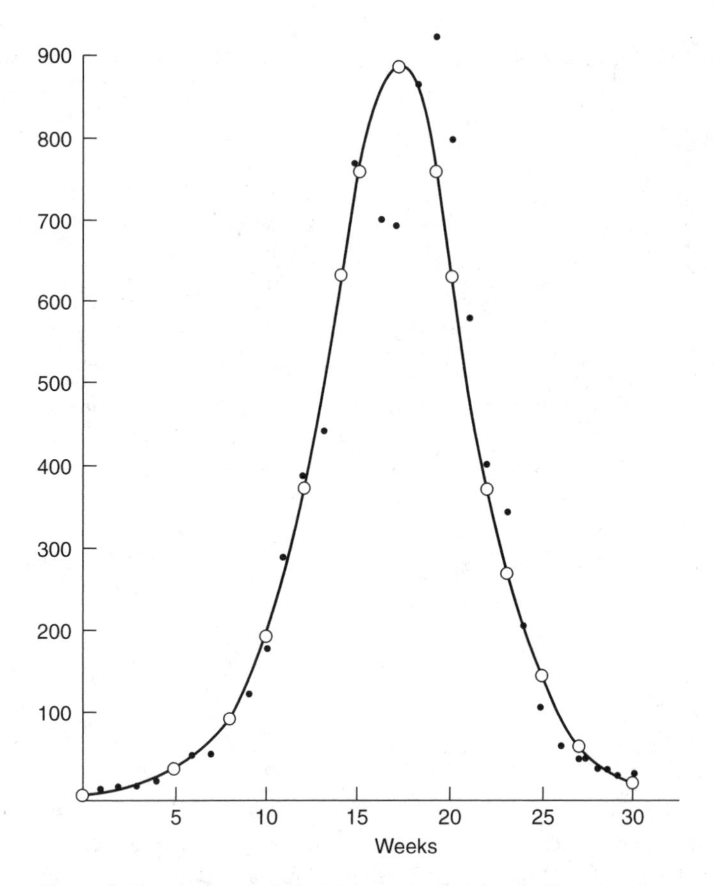

Figure 6.2 Deaths per week from plague in the island of Bombay from December 17, 1905 to July 21, 1906 (from Kermack and McKendrick 1927). The solid line is an approximate solution to their model for a disease with permanent removal—death or immunity—in the rat population on the island. It is compared with data on the human death toll on the assumption that "plague in man is a reflection of plague in rats."

most widely used transmission model; McCallum et al. (2002) review alternative models and empirical studies about the validity of the mass action model.

In the second equation, γ is the rate at which infected individuals recover from the disease (or die), at which point they transfer to the *R*ecovered class. The exit rate γ can be interpreted biologically as the inverse of the mean residence time in the compartment; this interpretation is very important for fitting these models to empirical data.

Whether *R*ecovered individuals are dead versus alive and immune is in one sense irrelevant for the future course of the epidemic, because in either case they have no impact on future infections. However, change in the number of living individuals invalidates Ross's (1916) derivation of the disease transmission rate.

If the living population is not constant, then to justify the βSI in [6.1] we have to assume that the rate of contacts per individual is proportional to the population size—if you double the number of people on the subway, then the kid with the runny nose infects twice as many people. If so, then the rate of new infections is $(bN)I(S/N) = \beta SI$ with $b = \beta$.

6.2 Rescaling the Model

What is the resulting shape of the epidemic curve? At first sight, it appears that we would need to see how the shape and behavior of solutions depend on three parameters: β, γ, and N. However, by rescaling the model (as in Chapter 4) we can reduce to a single parameter.

The benefit of rescaling is that the model becomes simpler just by changing the units of measurement for time and state variables. Usually the most effective rescalings are ones that render all variables in the rescaled model *dimensionless*. For example, S, I, R, and N are all "population size," measured in units like individuals/km^2 or individuals/m^2. The numerical values of these variables depend on the choice of units. However, if we look at the ratios $X = S/N, Y = I/N, Z = R/N$, their values will be the same regardless of the units used for population size. X, Y, Z are called *dimensionless* variables because their numerical values do not depend on the units of measurement.

The dynamic equations for our rescaled variables are easily derived:

$$\begin{aligned} dX/dt = (1/N)dS/dt &= -\beta SI/N \\ &= -(\beta N)(S/N)(I/N) \\ &= -bXY, \end{aligned}$$

and similarly

$$\begin{aligned} dY/dt &= bXY - \gamma Y \\ dZ/dt &= \gamma Y. \end{aligned}$$

This gets us down to two parameters: γ and the new composite parameter $b = \beta N$. We can get rid of one more parameter by defining a rescaled time variable $\tau = \gamma t$. Recall that the mean duration of infection is $1/\gamma$, so a unit increase in τ corresponds to real elapsed time equal to the mean duration of infection. We then have

$$\begin{aligned} dX/d\tau = dX/(\gamma dt) &= (1/\gamma)dX/dt \\ &= -(\beta N/\gamma)X. \end{aligned}$$

The step from $dX/d\tau$ to $(1/\gamma)dX/dt$ follows from the chain rule, $(dX/d\tau) \times (d\tau/dt) = dX/dt$, but the heuristic calculation in the last equation gets the right answer.

The conclusion is that $dx/d\tau$ depends on the single parameter combination $R_0 = \beta N/\gamma$. Doing the same with the other state variables we get the rescaled model

$$\begin{aligned} dX/d\tau &= -R_0XY \\ dY/d\tau &= R_0XY - Y \\ dZ/d\tau &= Y \end{aligned} \qquad [6.2]$$

with initial conditions $X(0) = X_0 \approx 1, Y(0) = Y_0 \approx 0, Z(0) = 0$.

An immediate prediction from this model is a threshold condition for an epidemic to occur. At time 0, $dY/d\tau = Y(R_0X_0 - 1) \approx Y(R_0 - 1)$, for a disease introduced at low incidence into the populations. Consequently, the disease prevalence increases if and only if $R_0 > 1$. Since $X(\tau)$ can only decrease over time, if Y is not increasing at time 0 it can never increase later, so the disease must die out.

The quantity

$$R_0 = \beta N/\gamma \qquad [6.3]$$

is called the "basic reproductive rate" of the disease, and can be interpreted as the expected number of new infections produced by a single infected individual introduced into a population of N susceptibles: βN infections per unit time, multiplied by the expected time $1/\gamma$ in the infectious stage. It therefore should be (and is) a very general property of epidemic models that a disease can be maintained in a population only if its R_0 (defined appropriately for the model) is greater than 1. Measures that reduced R_0 below 1 would then eradicate the disease, such as quarantine of infectives to reduce β or vaccination to reduce the number of susceptibles.

We can also show that epidemics in the model terminate before all susceptibles have become infected—thus achieving one of Ross's goals—and determine how many susceptibles remain.

In model [6.2] any individual who contracts the disease winds up eventually in Z. Since Z can only increase over time but can never go above 1, it must approach some limiting value $Z_\infty = \lim_{t\to\infty} Z(t)$, which is therefore the fraction of all individuals who contract the disease before it dies out. Z_∞ can be found by deriving a one-dimensional differential equation for $Z(t)$. By the chain rule $dX/d\tau = (dX/dZ)dZ/d\tau$ so

$$dX/dZ = (dX/d\tau)/(dZ/d\tau) = -R_0X;$$

hence $X(Z) = X(0)e^{-R_0Z}$. Using this expression for X and $Y = 1 - X - Z$, the third line of [6.2] becomes

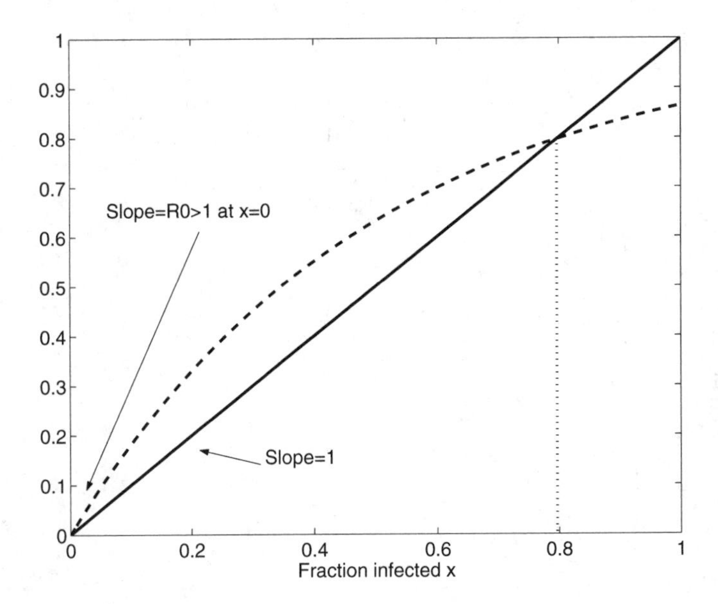

Figure 6.3 Graphical illustration that equation [6.5] has a unique solution between $x = 0$ and $x = 1$. The solution gives the approximate fraction of the population that contracted the disease over the course of an epidemic (Z_∞) in the Kermack-McKendrick SIR model, when the epidemic starts with a small number of infectives and the rest of the population susceptible.

$$dZ/d\tau = 1 - X(0)e^{-R_0 Z} - Z. \qquad [6.4]$$

As $Z(t) \to Z_\infty$, $dZ/d\tau$ decreases to 0, marking the end of the outbreak. When that occurs, since $X(0) \approx 1$ we must have (approximately)

$$Z_\infty = 1 - e^{-R_0 Z_\infty}.$$

Thus Z_∞ is the positive solution of the equation

$$x = 1 - e^{-R_0 x}. \qquad [6.5]$$

We see graphically that [6.5] has a unique solution between 0 and 1 so long as $R_0 > 1$ (see Figure 6.3), representing the fraction of the population that contract the disease before the outbreak collapses. The line $y = x$ has slope 1 and increases without limit. The curve $y = 1 - e^{-R_0 x}$ has slope R_0 at $x = 0$ but saturates to a limiting value of 1 as x increases. Thus the curves must intersect at some point Z_∞ between 0 and 1.

The relationship between Z_∞ and R_0 can be obtained by solving [6.5] for the inverse function, giving

$$R_0 = -\frac{1}{Z_\infty}\ln(1 - Z_\infty).$$

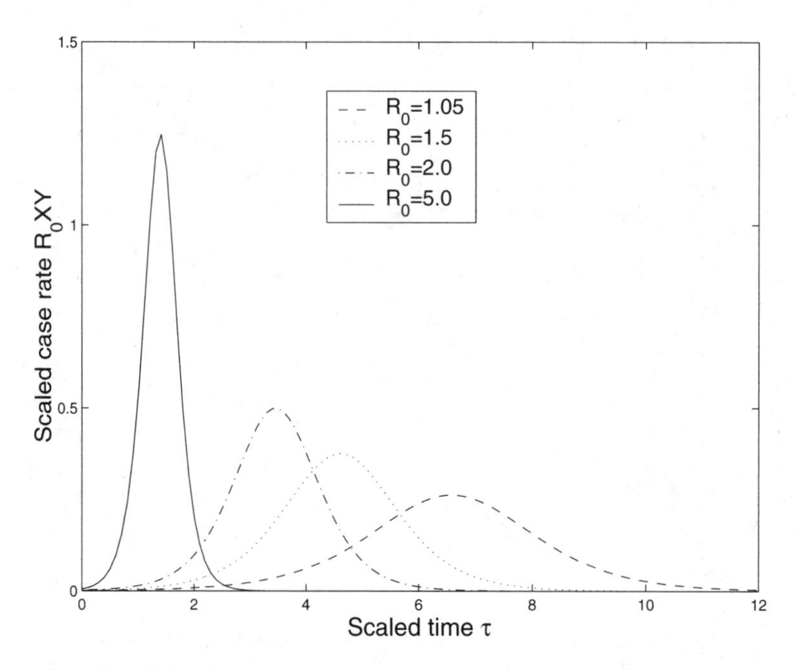

Figure 6.4 Epidemic curves (number of new cases per unit time) calculated from numerical solution of the Kermack-McKendrick model for a disease without recovery in a closed population of constant size. The plot shows $R_0X(\tau)Y(\tau)$ as a function of scaled time τ.

There is also an interesting approximation to [6.5] for R_0 near 1. In that case x is small, and we can use a two-term Taylor expansion $x \approx 1-(1-R_0x+(R_0x)^2/2)$ to obtain

$$Z_\infty \approx 2(R_0-1)/R_0{}^2 \approx 2(R_0-1) \text{ when } R_0 \approx 1.$$

This provides a possible explanation for occasional large outbreaks of a disease that is usually rare: small changes in infectivity as a result of conditions becoming more favorable for disease transmission can produce an epidemic affecting a significant fraction of the population. The approximation is actually an overestimate, but not by much. For example, the exact solution for $R_0 = 1.2$ is for 31% of the population to be infected before the epidemic burns out.

Finally, numerical solution can be used to find the shape of the epidemic curve predicted by the model [6.2], as a function of the single parameter combination R_0, for initial conditions $X(0) \approx 1$, $Y(0) \approx 0$, $Z(0) = 0$. The epidemic curve is defined as the rate at which new cases appear, that is, $R_0X(\tau)Y(\tau)$. Figure 6.4 shows that the model does indeed produce reasonable-looking epidemic curves. Higher values of R_0 naturally lead to a shorter and more intense epidemic, in addition to a higher final infected fraction Z_∞. Recall that τ measures elapsed time in units of the mean duration of infection, so the solutions show that if $R_0 = 5$ the epidemic only lasts about twice as long as the duration of the infection, while if $R_0 = 1.05$ it last for over ten times the duration of infection.

Exercise 6.1. An isolated village in Iceland experiences an outbreak of influenza in which 812 of the 1100 residents contract the infection. Estimate R_0 assuming that the outbreak started with a single case contracted from outside the village, with all others susceptible at the start of the outbreak.

Exercise 6.2. Find Z_∞ for [6.2] when $R_0 = 10, 20$, and 100.

6.3 Endemic Diseases and Oscillations

We turn now to Ross's second goal, understanding differences in dynamic patterns of incidence among endemic diseases. Consider, for example, two childhood diseases in New York City (measles and chickenpox) prior to the availability of vaccine (Figure 6.5). Both diseases show a pronounced annual cycle, most likely reflecting the higher transmission among children when schools are in session. However, statistical analysis confirms the presence of roughly two-year and three-year periodicities in measles, while the only significant periodicity in chickenpox is the annual cycle. What accounts for this difference?

In order for a disease to persist indefinitely there must be a supply of fresh susceptibles, either through recovery without immunity or through births. The simplest example is an SIS model with constant population size:

$$\begin{aligned} dS/dt &= -\beta SI + \gamma I \\ dI/dt &= \beta SI - \gamma I. \end{aligned} \qquad [6.6]$$

The acronym "SIS" indicates that when infected individuals recover they return to the susceptible class: there is no immunity conferred by infection. Gonorrhea, which we consider below, is a disease of this type.

This model is simple enough to solve. First, we can rescale it in the same way as the SIR model, getting

$$\begin{aligned} dX/d\tau &= -R_0 XY + Y \\ dY/d\tau &= R_0 XY - Y. \end{aligned} \qquad [6.7]$$

Second, since the $X + Y = 1$ we can replace X by $1 - Y$ in dI/dt to obtain $dY/d\tau = R_0 Y(1 - Y) - Y$. Then with a bit of algebra this re-arranges to

$$dY/d\tau = rY(1 - Y/K) \qquad [6.8]$$

where $r = R_0 - 1$, $K = (R_0 - 1)/R_0$. This is the well-known *logistic equation.* If $R_0 < 1$ the disease dies out. For $R_0 > 1$ the qualitative behavior of solutions is easy to determine by graphing $dY/d\tau$ as a function of Y: a parabola with its peak

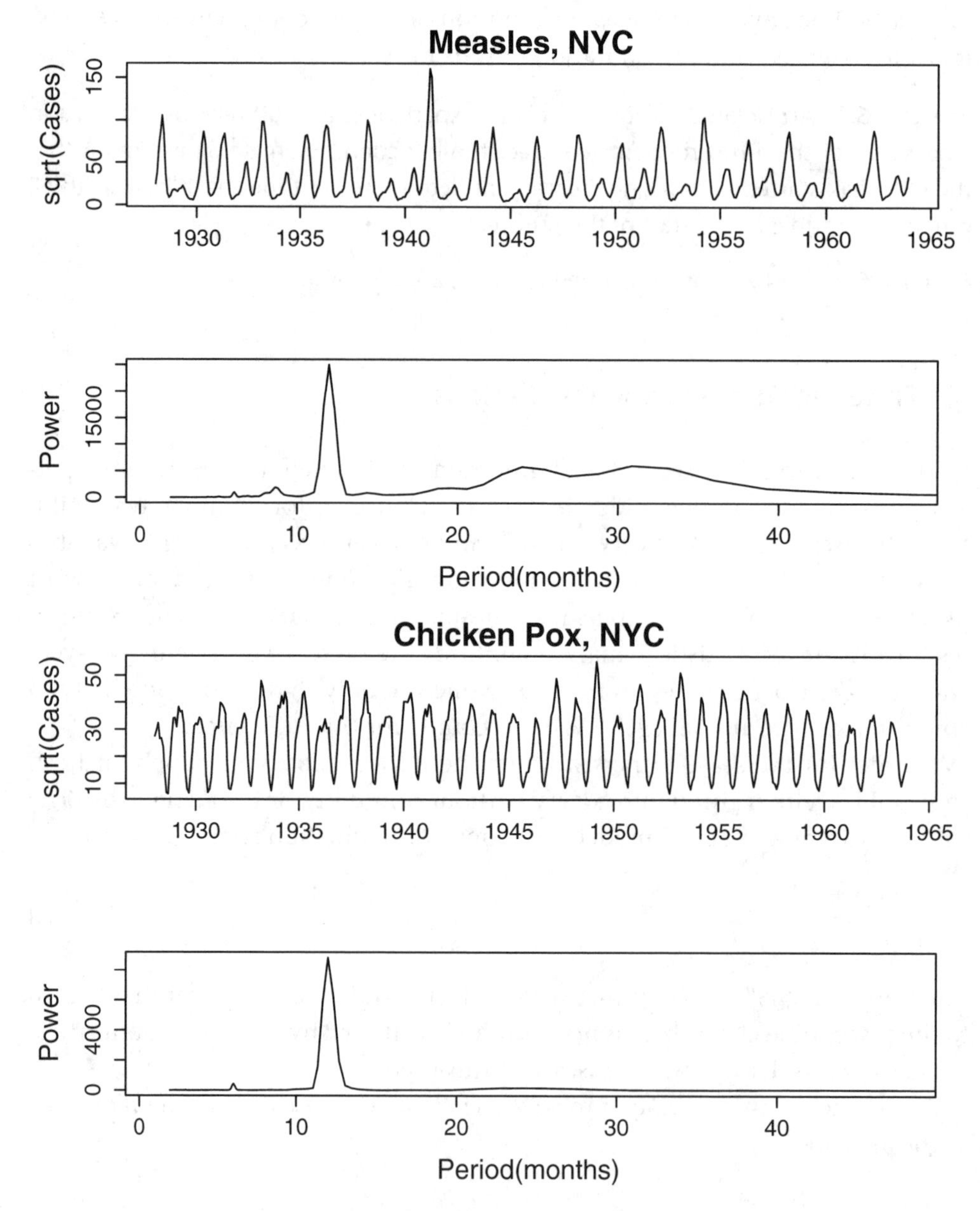

Figure 6.5 Monthly case report totals for measles and chickenpox in New York City prior to vaccination, plotted on square-root scale, and power spectra of the case report time series. The *power spectrum* of a time series represents the relative importance of different oscillation frequencies in the data. The spectra shown here confirm that chickenpox is dominated by a simple annual cycle, while measles shows a mix of annual, two-year, and three-year cycles.

at $Y = K/2$. From this we see that $Y(t) \to K$ from any initial value $Y(0) > 0$. We can also infer the qualitative shape of solutions. When $x(t)$ is below $K/2$, as x increases, dx/dt also increases: the second derivative is positive, hence $x(t)$ is concave up. Between $K/2$ and K, further increases in x leads to a decrease in dx/dt so $x(t)$ is concave down.

Because it reduces to a one-variable model, [6.7] always has a monotonic approach to steady state and cannot account for persistent oscillations in endemic diseases. Oscillations require a model in which the population passes through more disease states. The simplest example is an SIR model in which constant population size is maintained by a balance between births and deaths:

$$\begin{aligned} dS/dt &= \mu N - \beta SI - \mu S \\ dI/dt &= \beta SI - (\gamma + \mu)I \\ dR/dt &= \gamma I - \mu R. \end{aligned} \qquad [6.9]$$

Because population size is constant we only need the first two equations in (6.9). Following Anderson and May (1992) we rescale the variables to $X = S/N$ and $L = \beta I$. X is the fraction of susceptible individuals in the population, and L is the "force of infection"—the probability per unit time of becoming infected, for a susceptible individual. Because we are interested in the period of oscillations, we do not rescale time. The rescaled model is then

$$\begin{aligned} dX/dt &= \mu(1 - X) - LX \\ dL/dt &= (\gamma + \mu)L(R_0 X - 1) \end{aligned} \qquad [6.10]$$

where $R_0 = \beta N/(\gamma + \mu)$. R_0 has the same meaning as before: the mean number of new cases produced by a single newly infected individual added to a population of N susceptibles.

Exercise 6.3. What is the qualitative behavior of solutions to [6.8] starting from $x_0 > K$?

6.3.1 Analysis of the SIR Model with Births

The dynamics predicted by the SIR model with births are derived from linear stability analysis of steady states $(\bar{X}, \bar{L})$. There is always a disease-free steady state $(\bar{X}_0, \bar{L}_0) = (1, 0)$. An endemic steady state ($\bar{L}_1 > 0$) requires $R_0\bar{X} = 1$ (from setting $dL/dt = 0$); hence

$$\bar{X}_1 = 1/R_0. \qquad [6.11]$$

Then setting $dX/dt = 0$ we find

$$\bar{L}_1 = \mu(1 - \bar{X}_1)/\bar{X}_1 = \mu(R_0 - 1). \qquad [6.12]$$

Since $X < 1$ holds at all times (why?), the endemic steady state is biologically meaningful only when $R_0 > 1$, as expected from the meaning of R_0.

The endemic susceptible fraction $\bar{X}_1 = 1/R_0$ is a very general prediction, and will be true whenever an infective's rate of disease transmission is proportional to the susceptible fraction. By definition, a single newly infected individual dropped into a population with $X = 1$ (all susceptible) directly produces R_0 new infections, on average. At an endemic steady state, each newly infected individual must be exactly replacing itself, that is, producing 1 new infection rather than R_0—hence the susceptible fraction must be $1/R_0$.

The Jacobian matrix for [6.10] is

$$J(X,Y) = \begin{bmatrix} -\mu - L & -X \\ (\gamma+\mu)LR_0 & (\gamma+\mu)(R_0X-1) \end{bmatrix}. \qquad [6.13]$$

For the disease-free steady state we have

$$J(1,0) = \begin{bmatrix} -\mu & -1 \\ 0 & (\gamma+\mu)(R_0-1) \end{bmatrix} \qquad [6.14]$$

with eigenvalues $-\mu$ and $(\gamma+\mu)(R_0-1)$ (using the fact that the eigenvalues of a triangular matrix are the diagonal entries in the matrix). The disease-free steady state is therefore stable if $R_0 < 1$ and an unstable saddle if $R_0 > 1$. The stable eigenvector in the latter case is $(1,0)$—the X axis, which is also the stable manifold. If the population initially consists entirely of susceptible and recovered individuals, it converges to the disease-free steady state as recovered individuals die and are replaced by susceptibles.

The Jacobian for the endemic steady state is

$$J(\bar{X}_1,\bar{Y}_1) = \begin{bmatrix} -\mu R_0 & -R_0^{-1} \\ (\gamma+\mu)\mu(R_0-1)R_0 & 0 \end{bmatrix}. \qquad [6.15]$$

Since $R_0 > 1$ is necessary for this steady state to exist, the determinant is positive and the trace is negative, implying that the steady state is stable whenever there is an endemic steady state. To see if it is a spiral or a node, we compute the eigenvalues using the formula for a 2×2 matrix from Chapter 2:

$$\lambda = \frac{T \pm \sqrt{T^2 - 4D}}{2}$$

where $T = \text{trace}(J)$ and $D = \det(J)$. Applying this to [6.15] we get eigenvalues

$$\lambda_{1,2} = -\frac{1}{2}\mu R_0 \pm \frac{1}{2}\sqrt{\mu^2 {R_0}^2 - 4(\gamma+\mu)\mu(R_0-1)}. \qquad [6.16]$$

This expression can be understood using two approximations. First, if R_0 is just slightly above 1, the $(R_0 - 1)$ term within the square root will be dominated by $\mu^2 R_0^2$ so both eigenvalues are real, implying that the steady state is a node. However, endemic diseases typically have R_0 well above 1; for that situation recall that μ is the mortality rate and γ is the rate of recovery from the disease. To put it another way: $\mu = 1/(\text{mean lifetime})$ and $\gamma = 1/(\text{mean duration of the disease})$ so we typically have $\mu \ll \gamma$. Consequently, in the square root in [6.16] the terms involving μ^2 are dominated by the term involving μ, namely, $-4\gamma\mu(R_0-1) < 0$. Dropping the μ^2 terms we obtain the approximate eigenvalues

$$\lambda_{1,2} \approx -\frac{1}{2}\mu R_0 \pm i\sqrt{\gamma\mu(R_0 - 1)}. \qquad [6.17]$$

These are complex conjugates, implying that the endemic steady state becomes a spiral when R_0 is large, and the approach to steady state will be oscillatory.

The eigenvalues also give us the (approximate) period of the decaying oscillations. With complex conjugate eigenvalues $\lambda = a \pm ib$, the solutions to the linearized equations are a linear combination of $e^{at}\cos(bt)$ and $e^{at}\sin(bt)$, thus having period $T = 2\pi/b$. For [6.17] we therefore have the period

$$T \approx 2\pi/\sqrt{\gamma\mu(R_0 - 1)} = 2\pi\frac{1}{\sqrt{\gamma}}\frac{1}{\sqrt{\mu(R_0 - 1)}}. \qquad [6.18]$$

$1/\gamma$ is the mean duration of the disease. $\mu(R_0 - 1)$ is the force of infection at the endemic steady state, the probability per unit time of moving from susceptible to infectious. Its inverse is therefore the mean time an individual spends in the susceptible class, which is the mean age at infection in this model since all individuals are born as susceptibles. Calling these δ and α, respectively, we therefore have

$$T \approx 2\pi\sqrt{\alpha\delta}. \qquad [6.19]$$

For comparisons with data we need to estimate the model parameters. Some are easy to come by. N is the total population size; for example, for New York in the prevaccination era we could take 5 million as a representative value. $1/\gamma$ is the mean disease duration. For measles this is estimated at 12 to 14 days, so with time measured in years we could estimate $\gamma \approx 365/13 \approx 28$.

Estimating β is harder. The transmission rate involves the frequency of contacts, and the fraction of contacts that actually lead to disease transmission. Neither of these is easy to observe or estimate directly. In addition, different kinds of contacts (social, in school, within the family, on public transportation, etc.) each occur at different rates and with a different chance per contact of disease being transmitted.

An alternative approach is *calibration*, which means adjusting parameters to make model solutions correspond to the data as well as possible. With calibration, we use the model and data on state variables to infer the value of model parameters. *This assumes that the model is valid*, a dangerous assumption because we are unlikely to have independent tests of the model's validity. Nonetheless calibration is very widely used because it is rare to have a complex biological model in which all parameters can be estimated from direct data on process rates.

Here, we can use calibration to estimate β as follows. Assuming a disease at steady state with constant force of infection, we derived above that $\mu(R_0 - 1) = 1/\alpha$ where α is the mean age at infection. Rearranging, we get

$$\hat{R}_0 = 1 + \frac{1}{\alpha\mu} = 1 + \frac{\text{mean lifespan}}{\text{mean age at infection}} \qquad [6.20]$$

and then the resulting estimate $\hat{\beta} = \hat{R}_0(\gamma + \mu)/N$. The mean age at infection can be inferred from age-specific case reports, assuming a roughly even age distribution (e.g., Grenfell and Anderson 1985); for measles in England and Wales 1948–1968, this gave $\alpha \approx 5$ years.

Anderson and May (1992) suggested that the oscillations about the endemic steady state in the SIR model with births could account for the observed dynamics in measles and other endemic childhood diseases. Although cycles in the model are damped, Anderson and May (1992) argued that a number of mechanisms would continually perturb the system, leading to sustained oscillations at periods similar to [6.18]. Two suggested mechanisms were finite-population effects—the "coin-tossing" nature of disease transmission, especially when the number of infecteds is low—and seasonal variation in transmission rate due to school vacations. This hypothesis was tested by comparing predicted periods against periodicities observed in the data for a number of diseases (Anderson and May 1992, Table 6.1), and in many cases the fit is good. For example, for measles in developed countries they estimated mean age at infection of 4–5 years, and disease duration of 12 days, giving approximate period $2\pi\sqrt{4.5(12/365)} = 2.4$ years. This compares well to the observed two- and three-year periodicities (Figure 6.5). However, for chickenpox they estimate mean age at infection of 6–8 years, and disease duration 18–23 days, giving predicted periods of 3.4–4.5 years, for which there is no evidence in the data. So the damped-cycles hypothesis is only part of the story.

There is an enormous and still growing literature about the processes underlying dynamic patterns in endemic diseases. For childhood diseases it appears that multiple factors are involved, including those raised by Anderson and May (1992): seasonal variation in transmission rates and demographic stochasticity. Neither of these alone is sufficient. Models without seasonal variation in transmission cannot reproduce the clear annual periodicity observed in virtually all childhood diseases (Schaffer et al. 1990; see Figure 6.6). Deterministic models with seasonal forcing require unrealistically high levels of seasonal variation in order to mimic, through deterministic chaos, the complex multiannual patterns seen in measles (Ellner et al. 1995). But models incorporating both finite-population effects and plausible levels of seasonal variation have been able to account for the main features of the pre-vaccination oscillations in measles (Ellner et al. 1998; Finkenstad and Grenfell 2000; Grenfell et al. 2002), and for effects of vaccination on spatiotemporal patterns in measles and pertussis (Rohani et al. 1999). In addition, variation in birth rates and population size may contribute to changes over longer time scales, such as the transition in the New York measles data from more complex dynamics to a regular biennial oscillation (Ellner et al. 1998; Earn et al. 2000).

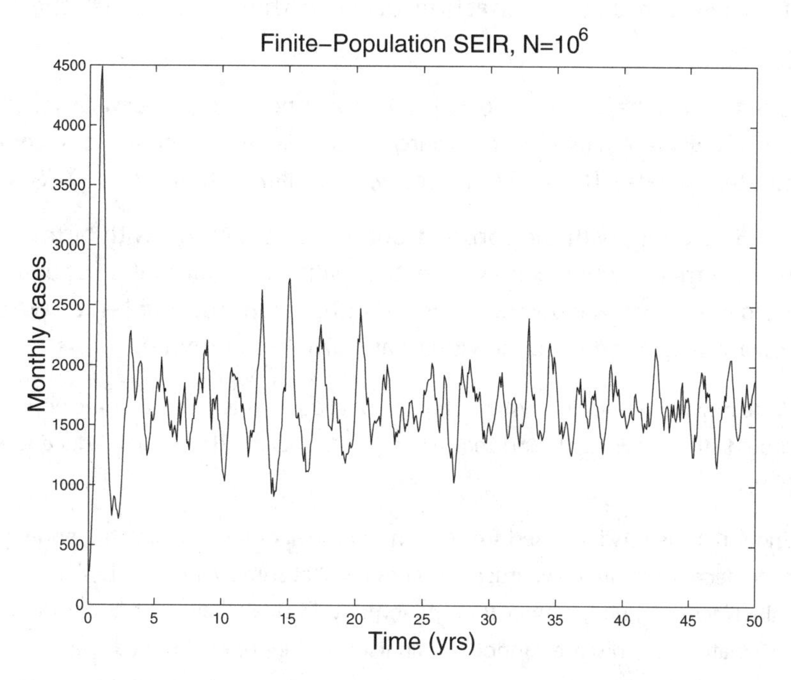

Figure 6.6 Output from a finite-population SEIR model with parameters appropriate for pre-vaccination dynamics of measles in a city of 1 million, but without any seasonal variation in the transmission rate.

6.3.2 Summing Up

We can now return to the biological questions posed at the start of this chapter: what accounts for differing patterns of dynamics in infectious diseases? We have seen that differences in epidemic dynamics emerge from biological differences in the interactions between different pathogens and their host that are reflected in the basic structure of the appropriate model:

- A highly infectious disease with permanent removal or immunity ($S \to I \to R$ model without births, [6.1]), leads to a classic epidemic curve and terminates before all susceptibles are infective.
- A disease where immunity following infection is temporary ($S \rightleftarrows I$ model, [6.6]) leads to a stable endemic state.
- A less infectious disease with permanent removal or immunity ($S \to I \to R$ model with birth and death, [6.9]) can lead to an endemic state with oscillations.

That is not to say that every endemic disease is adequately described by one of these simple models. The essential message is that qualitative properties at the level of individual hosts and pathogens create qualitative differences at the whole-population level, and the connection between these is made by the dynamic models.

Exercise 6.4. What happens in [6.16] if R_0 becomes really, really large? Is this a realistic possibility—that is, just how large must R_0 be to change our conclusions about the steady state? [Hint: if R_0 is "really, really large" then $(R_0 - 1)/R_0 \approx 1$.]

Exercise 6.5. Starting with the constant-population SIR model with births, equation [6.9], suppose that newborns are vaccinated, with the result that a fraction $p \leq 1$ of all newborn individuals are born as removed rather than susceptible (i.e., we consider here a disease where removed individuals are alive but immune).

(a) Write down the resulting system of differential equations. Note that the population size should still be constant, and that when $p = 0$ your model should reduce to equation [6.9].

(b) Show that as p is increased from 0, the number of infectives at the endemic steady state decreases until it eventually reaches $\bar{I} = 0$ at some value $p < 1$. This is sometimes called *herd immunity*: even though some individuals have not been immunized by vaccination, the disease cannot sustain itself in the population as a whole.

Exercise 6.6. Starting again from [6.9]:

(a) Modify the model to include *vertical transmission*, meaning that offspring of an infected parent have probability v of being born as infected rather than susceptible.

(b) Use linear stability analysis of the trivial steady state ($S = N, I = R = 0$) to study how vertical transmission affects the conditions for persistence of the disease.

Exercise 6.7. Starting once again from [6.9]:

(a) Add a latent phase (E) of individuals who have been infected but are not yet infective—they carry the disease but do not transmit it to others—and write out the resulting system of differential equations. This is called the SEIR model.

(b) Derive the expression for R_0 in this model.

(c) Write a script file to solve this model numerically, for a disease with $R_0 = 15$ (as estimated for measles) in a population of 3,000,000 and with the latent and infectious stages each lasting one week. Do numerical experiments on this model, with seasonal variation in contact rate, to test the claim that the latent proportion $E(t)/(E(t) + I(t))$ remains roughly constant. Describe your

- *methods:* describe the simulations that you conducted;
- *results:* give a verbal summary of the results;
- *conclusions:* Was the claim valid?

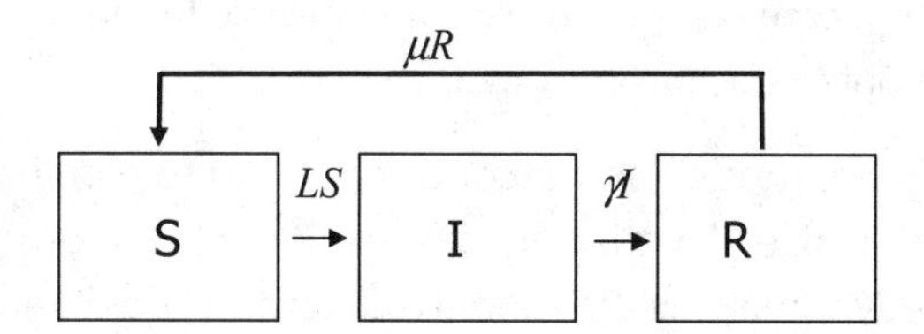

Figure 6.7 Compartment diagram for discrete-event SIR model.

Include a few well-chosen and well-designed graphs to support your claims. [Note: seasonal variation in contact rate is often modeled by $\beta(t) = \bar{\beta}(1 + \phi \cos(2\pi t))$ with time measured in years. Values of ϕ between 0.1 and 0.3 are considered credible for measles, depending on who you ask. Routines for numerically solving differential equations may have trouble for larger values of ϕ, especially if you start with $S(0) \approx N$ so that there is a massive initial outbreak.]

Exercise 6.8. This exercise involves constructing a discrete-event SIR model; an introduction to discrete-event models and how to simulate them is given in the online Computer Lab materials. Figure 6.7 shows the compartment diagram for an SIR model in a population of constant size N. Here L is the force of infection given by $\beta(t)(I + I_0)$, $\beta(t) = \bar{\beta}(1 + \phi \cos(2\pi t))$, where I_0 represents an external pool of infectives (cousins in Connecticut, etc.) whose presence keeps the disease from ever dying out completely. Constant population size is maintained by having each death (from R) balanced by birth of a new susceptible. The corresponding transition matrix for an individual with time step $dt = 1$ day is

$$A = \begin{bmatrix} 1-L & 0 & \mu \\ L & 1-\gamma & 0 \\ 0 & \gamma & 1-\mu \end{bmatrix}. \qquad [6.21]$$

(a) Write a script file to simulate a discrete-event version of the model for 50 years with time step $dt = 1$ day, and produce a plot of the daily number of new cases, using parameter values (for time measured in days) $N = 1,000,000$, $\mu = 0.015/365$, $R_0 = 16$, $\gamma = 1/12$, $I_0 = 1$, $\phi = 0.01$.

(b) Can your model explain the qualitative difference between measles and chickenpox dynamics prior to vaccination—that is, the presence of persistent multiannual periodicities in measles but not in chickenpox? Assume $R_0 = 16$ for measles, 10 for chickenpox, and the disease durations are 12 days for measles, 21 days for chickenpox. For purposes of this exercise accept the current view that $\phi < 0.25$, with smaller values being more plausible. Nothing is really known about I_0. Turn in a write-up explaining how you obtained your answer to this question, with well-chosen graphs or tables to support your arguments. Some case-report data are available from this book's web page, but remember that only a fraction of cases are reported.

Exercise 6.9. Modify the model (as a second script file) to eliminate the randomness due to finite population size. That is, any Binomial(N, p) random variable is replaced by its expected value Np, making the model a deterministic difference equation with a time-step of 1 day. Can this model explain the difference between measles and chickenpox?

6.4 Gonorrhea Dynamics and Control

In the 1970s Herbert Hethcote and James Yorke used epidemic models to study the control of gonorrhea in the United States. Their work illustrates how qualitative insights derived from simple models can have important practical implications. Bringing models and data into contact also led to an important conceptual advance that now plays an important role in HIV/AIDS research and public health policy, and also has been applied to the spread of computer "viruses." This section is largely based on their monograph (Hethcote and Yorke 1984).

The bacterium causing gonorrhea, *Neisseria gonorrhoeae*, lives only on mucus membranes and dies within seconds outside the human body. In one sexual exposure an infected woman has a 20–30% chance of transmitting the disease, while an infected man has a 50–70% chance. Gonorrhea incidence in the United States tripled between 1965 and 1975, and by the early 1980s there were roughly 1 million cases reported per year, implying roughly 2 million actual cases based on estimated reporting rates. Gonorrhea is a public health concern mainly because of its consequences in women: it is a major cause of pelvic inflammatory disease, infertility, and ectopic pregnancy, and facilitates the transmission of HIV. No vaccine against gonorrhea is available.

6.4.1 A Simple Model and a Paradox

A simple model for gonorrhea can be based on three basic properties. First, infection does not confer immunity. Second, the latent period can be omitted because it is very short (1–2 days) compared to the average duration of infection (about 1 month). Third, because there is only weak seasonal variation in the case reports, it is not necessary to include seasonal variation in the contact rate. The simplest model is therefore an SIS model with constant population size, which reduces to a single equation for the number of infectives,

$$dI/dt = \beta I(N - I) - \gamma I. \qquad [6.22]$$

As noted above, this is a logistic model whose solutions converge to the steady state $I = N(1 - 1/R_0)$ whenever $R_0 > 1$.

The paradox comes from the fact that, as usual, the endemic fraction of susceptibles in model [6.22] is $1/R_0$. The endemic fraction of susceptible therefore provides an estimate of R_0. Hethcote and Yorke (1984) reasoned as follows:

> We estimate that the actual yearly incidence of gonorrhea in the United States is 2.0 million and that the population at risk is approximately 20 million. If the average duration of infection is one month, then the number of cases at any given time is 166,667 which is less [than] 1% of the active population.

So $1 < R_0 < 1.01$, which portrays gonorrhea as a disease on the brink of extinction. That does not square well with its long-term persistence and its threefold increase in incidence within a decade. In addition, a rough estimate of the actual value of R_0 was obtained from the effects of a screening program in 1973–1975 (Yorke, Hethcote, and Nold 1978). It was estimated that the program decreased the average infectious period (and thus R_0) by 10%, and resulted in a 20% reduction in the rate of new case reports. The fraction infectious at any given time was therefore reduced by a fraction $(0.9)(0.8) = 0.72$. Thus, $1 - 1/(0.9R_0) \approx 0.72(1 - 1/R_0)$, giving $R_0 \approx 1.4$. But if R_0 is this large and most of the population is susceptible, the disease should be very rapidly increasing—tripling within a year rather than a decade.

6.4.2 The Core Group

How can we reconcile this contradiction? In order for the disease to be at steady state or slowly growing, the fraction of susceptibles among individuals contacted by an infected individual must be close to $1/R_0$. If we accept that $R_0 \approx 1.4$, then $1/R_0 \approx 0.7$ so the disease incidence must be about 30% *among individuals contacted by an infective.*

Since disease incidence in the general population is much lower than 30%, mixing between susceptibles and infectives cannot be random. Instead, Hethcote and Yorke concluded, there must be a *core group* of individuals, mostly transmitting the disease to each other, in which the disease is at much higher incidence than in the general population.

The importance of the core group had an immediate impact on programs for gonorrhea control (e.g., St. John and Curran 1978, WHO 1978, quoted by Hethcote and Yorke 1984). At that time, the main control measure in the United States was mass screening of women at public health clinics in order to identify asymptomatic carriers, who were considered to be the main reservoir for the disease. The presence of a core group implied that control programs should target the core group rather than the general population. The question, then, is how to do that most effectively.

The simplest model that could be used to examine control strategies is an SIS model that distinguishes between core (group = 1) and noncore (group = 2) subpopulations:

$$\boxed{S_1} \rightleftarrows \boxed{I_1}$$
$$\boxed{S_2} \rightleftarrows \boxed{I_2}. \qquad [6.23]$$

This model ignores a good bit of reality—for example, it does not distinguish men from women, or symptomatic from asymptomatic infectives—and it is simplistic to posit a sharp division between core and noncore. However, it is the natural first step.

Assuming constant size for the core and noncore subpopulations, [6.23] reduces to a pair of differential equations for the number of infecteds in each group:

$$dI_1/dt = (\lambda_{11}I_1 + \lambda_{12}I_2)X_1 - \gamma I_1$$
$$dI_2/dt = (\lambda_{21}I_1 + \lambda_{22}I_2)X_2 - \gamma I_2. \qquad [6.24]$$

Here λ_{ij} is the number of effective (pathogen-transmitting) contacts per unit time of a group-j individual with persons in group i, and $X_i = S_i/N_i$ is the fraction of susceptible individuals in group i. This is Ross's mass action model, where each infected individual has a constant number of contacts, and the rate of new infections is limited by the fraction of contactees already infected.

To use the model we have to specify its parameters, including the as-always unobservable contact rate parameters λ_{ij}. To reduce the number of parameters, Hethcote and Yorke made the so-called *proportionate mixing* assumption, that frequencies of contact between individuals are proportional to their *activity levels* a_i, defined as the average number of effective contacts per unit time for an individual in group i. The fractional activity level of group i is then $b_i = a_iN_i/A$, where $A = a_1N_1 + a_2N_2$. Note that the b's only depend on the relative population sizes and activity levels:

$$b_i = a_iN_i \Big/ \sum_j a_jN_j = 1 \Big/ \sum_j (a_j/a_i)(N_j/N_i). \qquad [6.25]$$

The proportionate mixing assumption is that each individual (of either group) has encounters with core versus noncore individuals in proportions b_1:b_2.

Proportionate mixing simplifies things considerably—we then have

$$dY_i/dt = a_i\left(\sum_j b_jY_j\right)X_i - \gamma Y_i. \qquad [6.26]$$

That is, the rate at which group i susceptibles contract the disease is given by the product of their activity level a_i—their rate of effective contacts—and the fraction of contacts who are infected, $\sum_j b_jY_j$.

The model is thus specified by the activity levels a_i, the relative sizes of the two groups, and the mean disease duration. Some idea of the core group's size and parameters can also be determined, if the core is identified as individuals having repeated infections in a relatively short time period. For example, one study cited by Hethcote and Yorke (1984) found that 6.7% of 7347 patients at venereal disease clinics fit this description, and were responsible for over 22% of the cases seen at the clinic.

The importance of the core group results from the fact that the product $b_1 Y_1$ appears in the transmission rate for all groups. If the core is especially active ($a_1 \gg a_2$) and consequently has a much higher disease prevalence ($Y_1 \gg Y_2$), the core group can be the primary source of new cases even if they are a small fraction of the population.

Exercise 6.10. This exercise illustrates the potential importance of a small but active core group. Suppose $N_1 = N_2/50$ and $a_1 = 10a_2$, with the overall contact rate such that the steady-state prevalence in the noncore population is 3%.

(a) By solving [6.26] for the endemic steady state, show that steady-state prevalences $\bar{Y}_i$ scale with activity level according to

$$\frac{\bar{Y}_i}{1 - \bar{Y}_i} = a_i \bar{Y}/\gamma$$

where $\bar{Y} = \sum_j b_j \bar{Y}_j$ is the average fraction of infected contacts.

(b) Use the result of (a) to show that $\bar{Y}_1 = 10\bar{Y}_2/(1 + 9\bar{Y}_2)$.

(c) Show that the fractional activity levels are $b_1 = 1/6, b_2 = 5/6$, and hence 5/6 of all contacts are with noncore individuals.

(d) Contacts with infecteds in core versus noncore occur in the ratio $(b_1\bar{Y}_1)$:$(b_2\bar{Y}_2)$. Combine the results of (c) and (d) to show that this ratio is > 1; hence most infections are contracted from a core group member.

6.4.3 Implications for Control

The two-group model can now be used to evaluate alternative control measures by adding each to the model, and comparing their effects on disease incidence. For example, a strategy of randomly screening individuals in the at-risk population, and treating those infected, would be modeled as

$$dY_i/dt = a_i \left(\sum_j b_j Y_j \right) X_i - \gamma Y_i - g Y_i \qquad [6.27]$$

where g is the screening rate. Hethcote and Yorke (1984) developed similar models for other strategies:

- *Rescreening:* Treated individuals are rescreened a short period after recovery from the disease, and retreated if necessary.

- *Contact tracing:* When a case is treated, try to identify the persons to whom the patient may have given the disease (potential infectees) or from whom they may have gotten it (potential infectors), and treat any who are infected.

Which of these works best? To find out, Hethcote and Yorke (1984) went through the exercise of estimating parameters for these models insofar as possible, and considering a range of possibilities for unknown parameters. They then did the same for a far more complex model with twelve state variables: susceptible, symptomatic infected, and asymptomatic infected among men and women in the core and noncore groups. But those efforts were hardly necessary, because the results can be inferred from how successful each strategy is at finding core group members:

- General random screening finds core versus noncore individuals in proportion to the group size $N_1{:}N_2$, and so does a bad job of treating the core.
- Rescreening finds core versus noncore individuals in proportion to their disease incidence, $I_1{:}I_2 = N_1 Y_1{:}N_2 Y_2$.
- Potential infectee tracing finds core versus noncore individuals in proportion to their activity levels, $b_1{:}b_2 = N_1 a_1{:}N_2 a_2$.
- Potential infector tracing finds core versus noncore individuals in proportion to the rate at which they transmit the disease to susceptibles, $b_1 Y_1{:}b_2 Y_2$.

(The difference between infectee and infector tracing is that potential infectees may or may not have been infected, whereas Hethcote and Yorke assumed that an infected individual will know who gave them the disease.) Because $Y_1 > Y_2$ and $a_1 > a_2$ we see that rescreening and infectee tracing will both outperform general screening, and (by similar comparisons) infector tracing will outperform both of these. Hethcote and Yorke's (1984) simulations confirmed these conclusions: infector tracing was found to be the most effective by far, for all parameter sets considered.

The concept of a core group and the importance of targeting the core for treatment was a major factor in the 1975 revision of U.S. control measures for gonorrhea. The new programs emphasized contact tracing and rescreening of people identified as core group members on the basis of frequent reinfection. The result was an immediate and long-lasting reversal of the increase in gonorrhea incidence (Figure 6.8), indicating that the new measures were substantially more effective.

Later models have relaxed the assumption of a sharp distinction between core and noncore, in two ways. The first is to allow multiple groups. Anderson and May (1992, Chapter 11) review models for HIV with an arbitrary number of groups, where group-j individuals constitute a fraction $P(j)$ of the population and have effective contacts at rate $a_j = aj$. They derived the important formula

$$R_0 = aT(E[j] + \mathrm{Var}(j)/E[j]). \qquad [6.28]$$

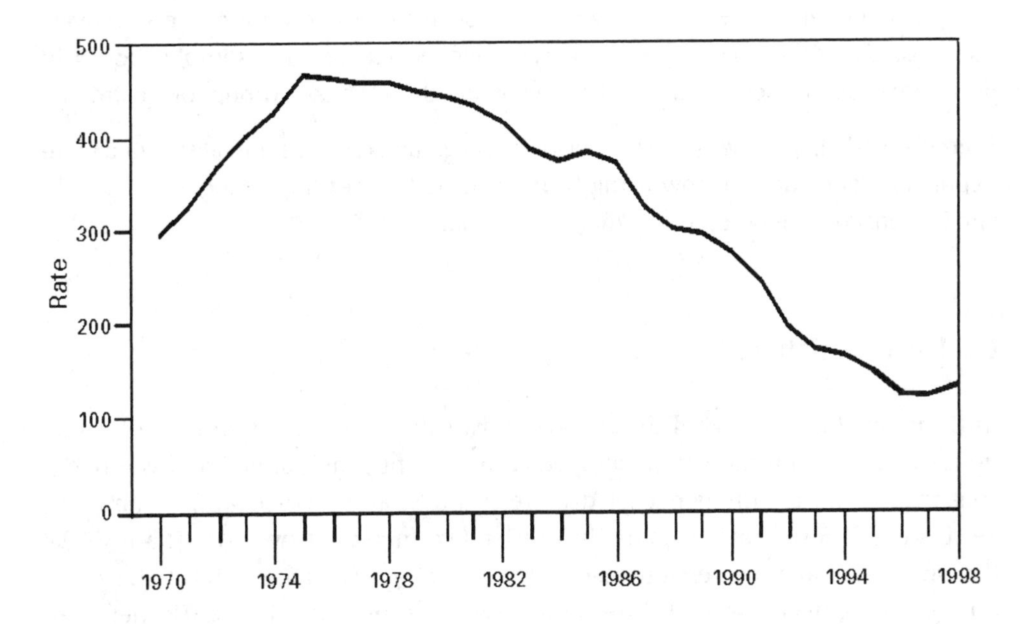

Figure 6.8 Reported gonorrhea rate (cases per 100,000 population) in the United States (from CDC 2000)

Here T is the mean duration of infection, and $E[j]$ and $\mathrm{Var}(j)$ are the mean and variance of the frequency distribution of contact rate: $E[j] = \sum_j jP(j)$, $\mathrm{Var}(j) = \sum_j P(j)\,(j - E[j])^2$. As always the disease persists if $R_0 > 1$, so [6.28] implies that individuals in the upper tail of the activity level distribution have a disproportionate effect on disease persistence.

The second generalization is modeling epidemics on social networks, where each individual is explicitly linked to a finite number of other individuals, and an infected individual has a constant probability per unit time of infecting individuals to whom it is linked. This is a very active research area now; Newman (2003) and Moreno et al. (2002) are good introductions. As in multiple-group models, persistence of the disease can be strongly influenced by the fact that individuals differ greatly in their rate of contacts with others in the population.

Exercise 6.11. We said above that rescreening finds core and noncore individuals in proportion to disease incidence in each group. Why is this? Similarly, give a verbal justification for the expressions given above for the proportions of core versus noncore individuals found by the infector tracing and infectee tracing strategies.

Exercise 6.12. What happens if gonorrhea-infected individuals can't always identify correctly the person who infected them? Propose a model for this situation and discuss its implications.

Exercise 6.13. Consider a power-law distribution for the activity rate in the multigroup HIV model, $P(j) = Cj^{-(2+\gamma)}$, which has been suggested by some recent investigations. Show that for $0 < \gamma \leq 1$, R_0 is infinite. What would this imply about HIV persistence? How does it relate to the idea of targeting the core group for treatment?

Exercise 6.14. Explain why the model in the previous exercise is unrealistic for a finite population, and suggest how it might be modified. Does this change the model's implications for disease control strategies, and if so, how?

6.5 Drug Resistance

Antimicrobial and antiviral drugs have drastically reduced the impact of infectious diseases on humans in developed countries, but their effectiveness is being challenged by the emergence of drug-resistant strains. For example, between 1991 and 1996 the rate of penicillin resistance increased by more than 300%; the rate of cefotaxime resistance in *Streptococcus pneumoniae*, one of the main causes of ear infections in children, increased by more than 1000% (Butler et al. 1996). Within an individual receiving drug treatment, the drug creates an environment where drug-resistant strains are at an advantage. As a result, if treatment fails—for example, if the patient does not comply with the treatment regime or does not complete it—that individual may then be carrying and transmitting drug-resistant strains of the disease. The U.S. Centers for Disease Control and Prevention has identified antibiotic resistance as a significant public health problem and has initiated a National Campaign for Appropriate Antibiotic Use aimed at reducing the spread of antibiotic resistance.

Two current concerns are multidrug resistant (MDR) tuberculosis, and antiretroviral resistant HIV. Tuberculosis (TB) is a major global public health burden, with over 9 million cases per year and $\sim$ 25% mortality; untreated or drug-resistant cases would have $\sim$ 50% mortality rate (Dye et al. 2002). TB strains are classified as MDR if they are resistant to the two main first-line drugs used to treat TB, isoniazid and rifampicin. In 2000 only 3.2% of all new TB cases were estimated to be MDR, but MDR prevalence of 10–14% has been estimated for specific regions in Eastern Europe, Asia, and Africa (Dye et al. 2002). The incidence of TB is currently increasing by about 2%/year, and there is concern that this might reflect the emergence of MDR strains that could lead to a significant global rise in TB prevalence.

Combination antiretroviral (ARV) therapies for HIV, involving simultaneous treatment with three or more different drugs, are currently the most effective available treatment. In use since 1996, these have substantially decreased the death rate from AIDS (Blower et al. 2001). However, strains resistant to the three-drug "cocktail" have emerged and have been sexually transmitted. Blower et al.

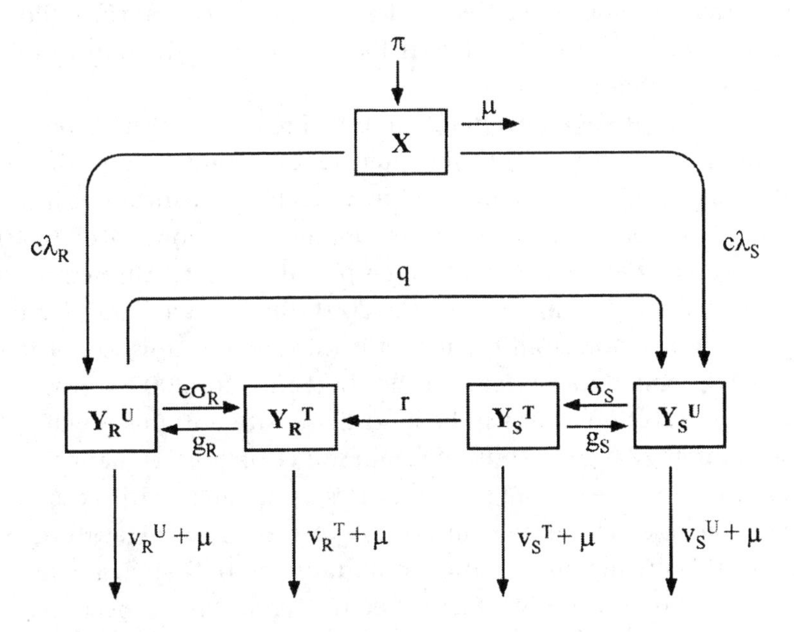

Figure 6.9 Compartment diagram from Blower et al. (2000) for their model for HIV transmission dynamics in the presence of antiretrovivral therapy, with both resistant and non-resistant strains circulating in the population.

(2000, 2001) used a relatively simple model to evaluate the magnitude of the threat posed by ARV-resistant HIV strains, and to evaluate possible responses.

The model (Figure 6.9) describes the transmission dynamics of HIV in the presence of antiretroviral therapy, with resistant and nonresistant strains being transmitted. In structure it is an SIR model with two complications: distinguishing between treated and untreated cases, and between resistant and nonresistant strains of the disease. Dropping the recovered population as usual, the state variables are the numbers of susceptible individuals (X), and four classes of infected individuals (Y) with S and R indicating drug-sensitive versus drug-resistant strains, and T and U indicating treated versus untreated individuals. The λ's are the force of infection, calculated from the number of infected individuals in each category and the infectiousness (β) of each type of infection. Parameters σ and g represent the rates of individuals entering and leaving treatment. Untreated drug-resistant infections revert to drug sensitive at rate q, while treated drug-sensitive infections acquire drug resistance at rate r.

Data-based estimates for several model parameters were available for the San Francisco gay community, while others parameters were less certain and probability distributions were used to represent the relative likelihood of different possible values. For example Blower et al. (2001) allowed the rate r at which ARV drug resistance develops in treated cases to range between 10% and 60% per year. Because little is known about ARV resistant strains, and in particular about their transmissibility relative to drug-sensitive strains, they allowed the relative transmissibility to vary between 1% and 100% of the transmissibility of the drug-sensitive strain.

Given this wide range of uncertainty, model predictions were generated for a large number of random draws of parameter values according to the distributions representing parameter uncertainty, and then studied statistically as if they were the results of an experiment—which in a sense they are. Blower et al. (2001) were able to compare model predictions with empirical estimates through 1999; data published later allowed comparison out to 2001 (Blower et al. 2003, Figure 6.10).

A surprising prediction from the model is that the transmission of resistance is low, and will remain low at least in the short run: for 2005, it was predicted that most new infections will still be by drug-sensitive strains (median 84.4%, interquartile range 72–94%). Thus, the main source of drug-resistant cases is conversion of drug-sensitive to drug-resistant cases: individuals with a drug-resistant infection are at risk themselves, but do not pose a major threat to the general population. This prediction has some important practical applications. First, it says that combination ARV will remain effective on most new infections, and can continue to be used on newly diagnosed cases. Second, efforts to limit the spread

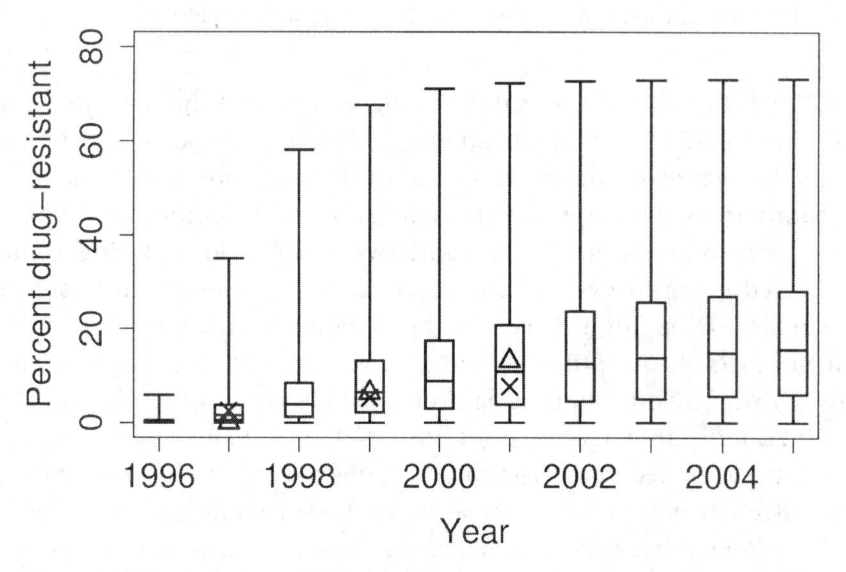

Figure 6.10 Fraction of new HIV infections that are resistant to combination ARV treatment: theoretical predictions versus empirical data for San Francisco (from Blower et al. 2003). Model simulations were run over the time period 1996–2005, with initial conditions corresponding to estimated values for 1996. Boxes enclose the interquartile range (25th to 75th percentiles) of model outcomes and bars show outlier cutoffs; the bars inside the boxes are the median values. Triangles show resistance to non-nucleoside reverse transcriptase inhibitor, and crosses show resistance to protease inhibitor, in a study of 243 newly infected individuals in San Francisco from 1996 to 2001 who had no previous exposure to ARV drugs.

of drug-resistant strains should focus on minimizing the rate of conversion from sensitive to resistant cases—by delaying treatment as long as possible, and trying to enforce strict compliance with the treatment program. Finally, the lack of resources to monitor patient compliance in developing countries implies that drug resistance is likely to be more of a problem than in developed countries.

6.6 Within-Host Dynamics of HIV

Models for infectious diseases at the population level have recently been adapted to model the proliferation of viral diseases within a single host (Nowak and May 2000). In this section we present some relatively simple models that were developed to analyze clinical studies of HIV. Results from these models provided the first inkling of the massive battle waged by the immune system against HIV during the chronic period of infection, and had a major impact on the treatment of HIV infection.

The primary targets of HIV-1 are CD4-positive T-lymphocytes. Infection begins when a viral particle (virion) encounters an activated T-cell, and the viral envelope binds with the CD4 receptor on the cell membrane. The cell membrane and viral envelope fuse, and the viral core enters the cell. The host cell's genetic machinery is commandeered and it begins making multiple copies of the viral RNA. New virions form within the cell and then bud off, carrying along some of the host membrane as a new viral envelope.

Disease progression after HIV infection, if untreated, has three phases. An initial acute phase is marked by high viral loads and flu-like symptoms. The second phase is largely asymptomatic: viral loads fall to a quasi-steady-state and remain there for a period of a few months to a decade. During this time T-cells slowly decline. Finally, there is immune system failure followed by death from opportunistic infections (Figure 6.11).

Because viral and T-cell levels change very slowly, the asymptomatic stage was assumed to be a period when the virus was relatively inactive and nothing much was happening. The development of antiretroviral drugs made it possible to test this assumption. Ho et al. (1995) and Wei et al. (1995) treated chronically infected patients with then newly developed drugs (reverse transcriptase and protease inhibitors) which prevent the virus from infecting additional cells. The surprising result was an extremely rapid exponential decay in viral load (Figure 6.12).

The simplest model to explain these findings posits that viral production is totally shut down by drug treatment. Then

$$dV/dt = -cV \qquad [6.29]$$

where V is the viral load and c the virion clearance rate. The value of c is then the slope of $log\ V$ versus t, and was estimated by fitting a straight line to the

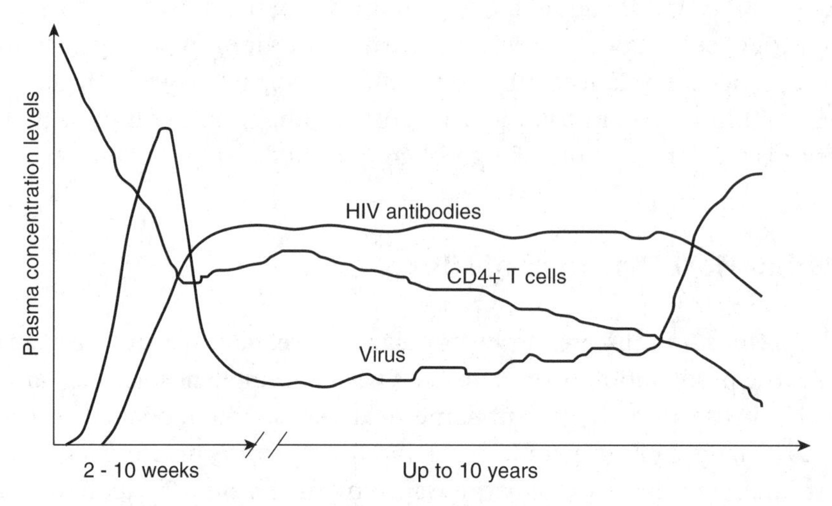

Figure 6.11 Schematic depiction of the typical course of HIV infection in an adult. The early peak in viral load corresponds to the primary infection or "acute" phase. Also shown are T-cell dynamics (from Perelson and Nelson 1999).

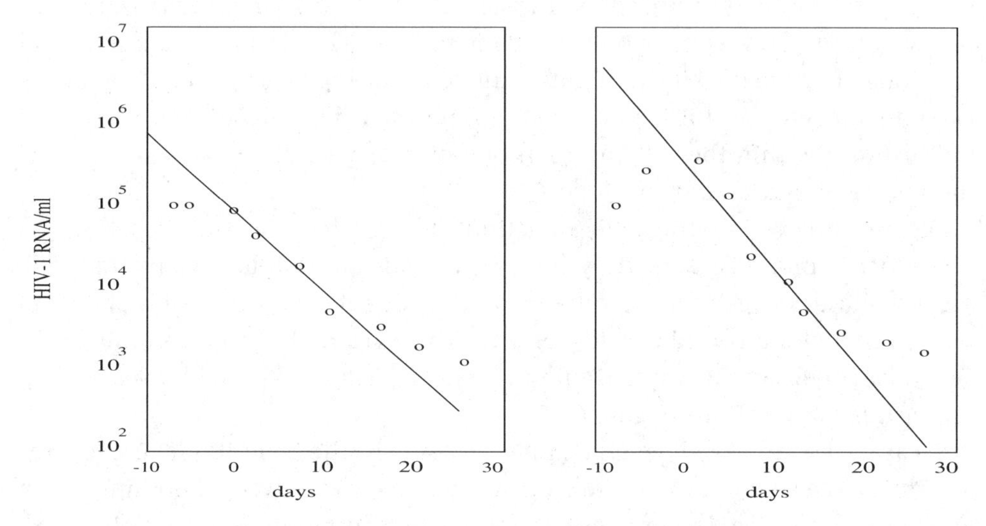

Figure 6.12 Decay of plasma viral load in two patients following treatment with a protease inhibitor. Treatment is initiated at $t = 0$ (from Perelson and Nelson 1999, using data from Ho et al. 1995).

exponential decay phase of the data (Figure 6.12). The viral half-life is then $t_{1/2} = log\,2/c$. Ho et al. (1995) estimated $t_{1/2} = 2.1 \pm 0.4$ days based on 20 patients, and Wei et al. (1995) estimated $t_{1/2} = 1.8 \pm 0.9$ days based on 22 patients. These estimates revealed that the asymptomatic stage is actually very dynamic, with rapid virus production balanced by rapid clearance.

However, the simple model [6.29] confounds two processes: the clearance of existing free virions, and the clearance of infected cells that are producing new virus. Estimating both clearance rates required a model in which both processes are explicitly represented (Ho et al. 1995, see also Perelson and Nelson 1999; Nowak and May 2001):

$$dT/dt = \lambda - dT - kVT$$
$$dT^*/dt = kVT - \delta T^* \qquad [6.30]$$
$$dV/dt = N\delta T^* - cV$$

where T are uninfected T-cells, T^* are productively infected cells, and V represents free virus. This is essentially an SI model for the spread of infection in a population of T-cells, with infection occurring through contacts between susceptible T-cells and free virions. T-cells are "born" at rate λ, have per capita mortality rate d, and are infected at rate kVT—a "mass action" model for contacts between susceptible cells and virus. Infected cells die or are cleared at a rate δ. Virus V is generated at a rate $N\delta$ per infected T-cell T^* and cleared at a per capita rate c. N is called the "burst size" and represents the total number of free viral particles produced by an infected cell over its lifetime. Since the mean lifetime of an infected cell is $1/\delta$, the burst size N corresponds to an instantaneous virion production rate of $N\delta$.

Perelson et al. (1996) applied this model to clinical data. Five patients received a protease inhibitor, and their HIV-1 RNA concentrations were measured frequently over the next 7 days. Protease inhibitors cause infected cells to produce noninfectious virions. In the presence of a protease inhibitor (assumed for simplicity to be 100% effective), the model becomes

$$dT/dt = \lambda - dT - kV_I T$$
$$dT^*/dt = kV_I T - \delta T^*$$
$$dV_I/dt = -cV_I \qquad [6.31]$$
$$dV_{NI}/dt = N\delta T^* - cV_{NI}.$$

Here V_I is infectious virus, and V_{NI} is noninfectious virus. Both are cleared at rate c.

Given the short duration of the experiment and assuming a relatively large pool of uninfected cells T, it was assumed that T remained at its steady state value T_0 for the duration of the study. The model then reduces to

$$dT^*/dt = kV_I T_0 - \delta T^*$$
$$dV_I/dt = -cV_I \qquad [6.32]$$
$$dV_{NI}/dt = N\delta T^* - cV_{NI}.$$

This can be solved by matrix methods, or sequentially as follows. Prior to treatment all virus is infectious, that is, $V_I(0) = V_0$, therefore

$$V_I(t) = V_0 \exp(-ct).$$

Assuming that the patient was at steady state prior to treatment, the second line of [6.31] implies that

$$kV_0T_0 = \delta T_0^*. \quad [6.33]$$

Substituting these into the equation for T^* yields

$$dT^*/dt = -\delta T^* + \delta T_0^* \exp(-ct) \quad [6.34]$$

with initial condition $T^*(0) = T_0^*$. Since this equation is linear it can be solved explicitly (as outlined in exercises below), to obtain

$$T^*(t) = \frac{T_0^*}{(c-\delta)} \cdot [c\exp(-\delta t) - \delta\exp(-ct)].$$

The equation for noninfectious virus can be solved similarly, after substituting in the expression for T^* and imposing the initial condition $V_{NI}(0) = 0$. The resulting expression for the total free virus $V(t) = V_I + V_{NI}$ is then

$$V(t) = V_0\exp(-ct) + \frac{cV_0}{(c-\delta)} \cdot \left\{ \frac{c}{(c-\delta)} \cdot [\exp(-\delta t) - \exp(-ct)] - \delta t\exp(-ct) \right\}. \quad [6.35]$$

Values of c and δ were estimated for each patient by fitting this equation to the measurements of plasma viral load (Figure 6.13). The clearance rate of free virus c was found to be relatively rapid, with the half-life estimated to be 0.24 ± 0.06 days. Thus, half the virion population is replaced every 6 hours. Death of productively infected cells T^* was found to occur more slowly, with an estimated half-life of 1.55 ± 0.57 days.

During the approximate steady state prior to treatment, virus production must balance the clearance, cV. Using their estimate of c and measured steady-state viral loads V_0, Perelson et al. estimated that 10.3×10^9 free virions were produced per day prior to drug treatment. This was an order of magnitude higher than the original estimates by Ho et al. (1995) and Wei et al. (1995).

The implication for treatment was that HIV has the potential to evolve very rapidly in response to selection imposed by the immune system, or by drug treatment. Combining turnover rate estimates with estimates of genome size and mutation rate, Coffin (1995) concluded that all possible point mutations in HIV arise thousands of times each day in a chronically infected adult. This provided a simple mechanistic explanation for the rapid development of resistance to single-drug treatments. The current practice of administering three-drug cocktails—which would require simultaneous mutations at three different sites to confer drug resistance—arose directly from these conclusions. As noted in the last section, multidrug resistance is developing much more slowly.

Exercise 6.15. Here is one way to find the solution to [6.34]. General theory for first-order linear differential equations tells us to expect a solution of the form $T^*(t) = A\exp(-ct) + B\exp(-\delta t)$ for some constants A and B. Find the solution by

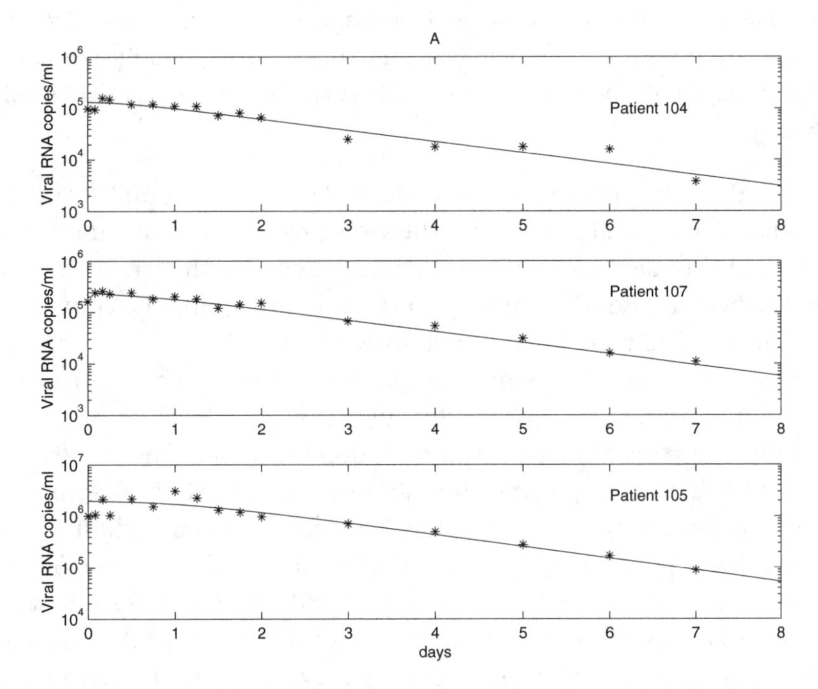

Figure 6.13 Viral load data (symbols) versus model predictions (solid line) during first phase of viral decay after onset of treatment ($t = 0$) (from Perelson and Nelson 1999, after Perelson et al. 1996).

substituting this trial form into [6.34] and finding A, B to satisfy the initial conditions and the differential equation.

Exercise 6.16. Here is another. Equation [6.34] can be rearranged as

$$dT^*/dt + \delta T^* = \delta T_0^* \exp(-ct).$$

(a) Let $x(t) = \exp(\delta t)T^*(t)$ and use the chain rule to show that $\exp(-\delta t)dx/dt = dT^*/dt + \delta T^*$.

(b) Using the result of part (a) show that $x(t)$ satisfies a differential equation of the form $dx/dt = f(t)$, which implies that $x(t) = x(0) + \int_0^t f(s)ds$.

(c) The rest is algebra: plug in $x(0) = T^*(0)$, use [6.33] and simplify, to derive [6.35].

6.7 Conclusions

Infectious disease models provide some of the best examples of the practical value of simple dynamic models. Simple models, deliberately stripped down to bare essentials so that they could be fully understood, led to *qualitative* insights, such as

- endemic steady-states of diseases with permanent immunity will be spirals,
- contact tracing is the best way to target the core group of gonorrhea carriers,
- rapid turnover of HIV-1 allows it to rapidly evolve resistance to any single-drug therapy.

In principle these insights could also be gleaned from more complex and realistic models, that would presumably make the same predictions. But complex models can be as hard to understand as the real-world system that they represent. Simpler models are then an essential starting point for disentangling the complexities of complex models and complex real-world systems.

Infectious diseases also provide examples of situations where complex models are essential because the quantitative details matter—such as predicting just how rapidly drug-resistant HIV will spread, deciding what control measures will be sufficient to halt the spread of foot-and-mouth disease (as described in the Preface), or determining which individuals should be given highest priority for limited vaccine supplies. Our next two chapters are devoted to some of the complexities that we have so far ducked. The next adds *space*, where the variation in state variables across location as well as across time must be taken into accout. After that we look more closely at models that try to represent the complexity of biological systems by modeling at the level of individual agents, and therefore can only be studied by computational methods.

6.8 References

Anderson, R. A., and R. M. May. 1992. *Infectious Diseases of Humans: Dynamics and Control.* Oxford University Press, Oxford, U.K.

Blower, S. M., H. B. Gershengorn, and R. M. Grant. 2000. A tale of two futures: HIV and antiretroviral therapy in San Francisco. Science 287: 650–654.

Blower, S. M., A. N. Aschenbach, H. B. Gershengorn, and J. O. Kahn. 2001. Predicting the unpredictable: Transmission of drug-resistant HIV. Nature Medicine 7: 1016–1020.

Blower, S. M., A. N. Aschenbach, and J. O. Kahn. 2003. Predicting the transmission of drug-resistant HIV: Comparing theory with data. The Lancet Infectious Diseases 3: 10–11.

Butler, J. C., et al. 1996. The continued emergence of drug-resistant Streptococcus pneumoniae in the United States: An update from the CDC's Pneumococcal Sentinel Surveillance System. Journal of Infectious Diseases 174: 986–993.

CDC (Centers for Disease Control and Prevention). 2000. Gonorrhea—United States, 1998. Morbidity and Mortality Weekly Report 29: 538–542.

J. M. Coffin. 1995. HIV population dynamics in vivo: Implications for genetic variation, pathogenesis, and therapy. Science 267: 483–489.

Campbell, C. L., and L. V. Madden. 1990. *Introduction to Plant Disease Epidemiology.* John Wiley & Sons, New York.

Davidson, D. R. 1995. Influenza Outbreak in Basic Training, Fort Benning, GA. Medical Surveillance Monthly Report 1:7. U.S. Department of Defense, U.S. Army Center for Health Promotion and Preventive Medicine (Provisional), Aberdeen Proving Ground, MD 21010-5422.

Dye, C., B. G. Williams, M. A. Espinal, and M. C. Raviglione. 2002. Erasing the world's slow stain: Strategies to beat multidrug-resistant tuberculosis. Science 295: 2042–2046.

Earn, D.J.D., P. Rohani, B. M. Bolker, and B. T. Grenfell. 2000. A simple model for complex dynamical transitions in epidemics. Science 287: 667–670.

Ellner, S. P., B. A. Bailey, G. V. Bobashev, A. R. Gallant, B. T. Grenfell, and D. W. Nychka. 1998. Noise and nonlinearity in measles epidemics: Combining mechanistic and statistical approaches to population modeling. American Naturalist 151: 425–440.

Finkenstadt, B. F., and B. T. Grenfell. 2000. Time series modelling of childhood diseases: A dynamical systems approach Journal of the Royal Statistical Society Series C—Applied Statistics 49: 187–205.

Grenfell, B. T., and R. M. Anderson. 1985. The estimation of age related rates of infection from case notifications and serological data. Journal of Hygiene 95: 419–436.

Grenfell, B. T., O. N. Bjornstad, and B. F. Finkenstadt. 2002. Dynamics of measles epidemics: Scaling noise, determinism, and predictability with the TSIR model. Ecological Monographs 72: 185–202.

Hall, A., P. Pomeroy, and J. Harwood. 1992. The descriptive epizootiology of phocine distemper in the UK during 1988/89. Science of the Total Environment 115: 31–44.

Hethcote, H. W., and James A. Yorke. 1984. *Gonorrhea Transmission Dynamics and Control.* Lecture Notes in Biomathematics Vol. 56. Springer-Verlag, New York.[1]

Ho, D. D., A. U. Neumann, A. S. Perelson, W. Chen, J. M. Leonard, and M. Markowitz. 1995. Rapid turnover of plasma virions and CD4 lymphocytes in HIV-1 infection. Nature 373: 123–126.

Hudson, P. J., A. Rizzoli, B. T. Grenfell, H. Heesterbeek, and A. P. Dobson. 2002. *The Ecology of Wildlife Diseases*. Oxford University Press, Oxford.

Kermack, W. O., and W. G. McKendrick. 1927. A contribution to the mathematical theory of epidemics. Proceedings of the Royal Society of London, Series A 115: 700–721.

Kermack, W. O., and W. G. McKendrick. 1932. Contributions to the mathematical theory of epidemics. II. The problem of endemicity. Proceedings of the Royal Society of London, Series A 138: 55–83.

McCallum, H., N. Barlow, and J. Hone. 2002. How should pathogen transmission be modelled? Trends in Ecology and Evolution 16: 295–300.

Moreno, Y., R. Pastor-Satorves, and A. Vespignani. 2002. Epidemic outbreaks in complex heterogeneous networks. European Physics Journal B 26: 521–529.

Newman, M. E. J. 2003. The structure and function of complex networks. SIAM Review 45: 167–256.

Nowak, M. A., and R. M. May. 2000. *Virus Dynamics: Mathematical Principles of Immunology and Virology.* Oxford University Press, Oxford, U.K.

Perelson, A. S., A. U. Neumann, M. Markowitz, J. M. Leonard, and D. D. Ho. 1996. HIV-1 dynamics in vivo: Virion clearance rate, infected cell life-span, and viral generation time. Science 271: 1582–1586.

Perelson, A. S., and P. W. Nelson. 1999. Mathematical analysis of HIV-1 dynamics in vivo. SIAM Review 41: 3–44.

Rohani, P., D. J. D. Earn, and B. T. Grenfell. 1999. Opposite patterns of synchrony in sympatric disease metapopulations. Science 286: 968–971.

Ross, R. 1916. An application of the theory of probabilities to the study of *a priori* pathometry. Part I. Proceedings of the Royal Society of London, Series A 92: 204–230.

Schaffer, W. M., L. M. Olsen, G. L. Truty, and S. L. Fulmer. 1990. The case for chaos in childhood epidemics. Pages 138–166 in S. Krasner (ed.), *The Ubiquity of Chaos.* American Association for the Advancement of Science, Washington, DC.

Wei, X. P., S. K. Ghosh, M. E. Taylor, V. A. Johnson, E. A. Emini, P. Deutsch, J. D. Lifson, S. Bonhoeffer, M. A. Nowak, B. H. Hahn, M. S. Saag, and G. M. Shaw. 1995. Viral dynamics in human-immunodeficiency-virus type-1 infection. Nature 373: 117–122.

York. J. A., H. W. Hethcote, and A. Nold. 1978. Dynamics and control of the transmission of gonorrhea. Sexually Transmitted Diseases 5: 51–56.

[1]This book is out of print but has been placed in the public domain by the authors and is available at http://www.math.uiowa.edu/ftp/hethcote/lnb56.pdf

7 Spatial Patterns in Biology

Spatial pattern is a fascinating aspect of biological systems, from the shape and size of organisms to the geographic distribution of species. Indeed, "morphogenesis," the generation of pattern, has long been regarded as one of *the* central problems of biology, and one that requires theory and models to address. The book *On Growth and Form* by D'Arcy Thompson, published in 1917, continues to be widely read and admired—a remarkable achievement in a discipline that has undergone rapid change. The conviction that morphogenesis is based upon mechanisms whose principles we can discover remains strong. The possible mechanisms can be classified in terms of the physical and/or chemical forces involved, or they can be classified in terms of the types of models that are used to reproduce observed patterns. Moreover, it is evident that different mechanisms operate in different systems. Rather than attempt to survey the different possibilities, we shall examine a single mechanism, called a reaction-diffusion system, and study two examples in which reaction-diffusion mechanisms have been shown to yield observed spatial patterns.

Reaction-diffusion models were proposed as an explanation of morphogenesis by Alan Turing in his last paper. Turing was a remarkable individual who was at the center of British efforts to break German codes during World War II. This work has been dramatized in novels, film, and plays (Hodges 2000). Turing also invented the *Turing machine*, now one of the fundamental concepts in computer science. In the work that we discuss here, Turing hypothesized that the combination of molecular diffusion and chemical reaction of substances he called *morphogens* could lead to instabilities in the homogeneous distribution of these substances (Turing 1952). While Turing describes three biological examples of morphogenesis (the tentacles of hydra, gastrulation, and the whorl patterns of plants such as woodruff), his work did not include analysis of data from these systems.

Many attempts have been made subsequently to demonstrate that Turing mechanisms are responsible for the creation of patterns in different systems.

Some of these have been successful, but there are natural patterns in which other forces also play a critical role in the pattern formation. To cite one example, Odell and Oster (1980) investigated the role of traction and mechanical forces in the process of gastrulation. Even in the cases in which reaction and diffusion do provide the primary basis for pattern formation, identification of the morphogens and their reactions with each other has required years of effort by many investigators. Nonetheless, certain patterns displayed by reaction-diffusion systems are very robust. These patterns can be studied in relatively simple reaction systems and the insight gained from them can be used to guide experimental research that seeks to elucidate the details of the biological mechanisms of different pattern-forming systems.

7.1 Reaction-Diffusion Models

Diffusion is the process by which the random motion of molecules due to thermal fluctuations leads to changes in the concentration of chemical species in space and time. We will talk later about animal species, and why diffusion models are often good approximations to their movements. There are two approaches to the derivation of the equations that describe diffusion. The first (historically) starts at the "macroscopic" level, treating chemical concentrations as smoothly varying functions of space and time, and making assumptions about the net behavior of large numbers of molecules in solution. A more fundamental and general approach is based on a "microscopic"-level analysis of individual molecules.

To introduce the individual approach, we begin with a *lattice model*. Consider the integer-valued points on the real line, and a molecule jumping randomly among these points. During each time step from t to $t+1$, the molecule randomly takes a step to one of the two neighboring points, moving right or left with equal probability. This model is an example of a one-dimensional *random walk*. It is a caricature of the fact that in liquids and gases, molecules are constantly in motion, changing direction as they collide with one another. This *Brownian motion* was discovered in the nineteenth century by Robert Brown who observed the motion of pollen grains floating in water with a microscope.

The outcome of this random-walk diffusion model is summarized by the probability distribution of the particle's location. Let X_n denote the particle's location after n time units, starting from $X_0 = 0$. The choice of random steps to the right and left can be generated by tossing a fair coin n times, assigning a right step to H and a left step to T. Let R and L be the total number of steps to the right and left. If at the end of n steps the particle is at location k, we then have

$$\begin{aligned} R + L &= n \\ R - L &= k \end{aligned} \qquad [7.1]$$

and hence $2R = n + k \Rightarrow R = (n+k)/2$ [note that n and k must either be both even, or both odd, since R is an integer: after an even (odd) number of steps the particle is at an even (odd) location]. The probability that $X_n = k$ is therefore given by the binomial formula for the probability of getting $(n+k)/2$ H's in n tosses of a fair coin,

$$Pr\{X_n = k\} = \binom{n}{(n+k)/2} \left(\frac{1}{2}\right)^n. \qquad [7.2]$$

This holds for $|k| \leq n$ with k and n either both even or both odd, otherwise $Pr\{X_n = k\} = 0$.

We study approximations of [7.2] for large k and n via the central limit theorem. The central limit theorem is a generalization of the DeMoivre-LaPlace limit theorem for the binomial distribution that applies to more general models than the simple random walk. For our randomly walking particle, let z_j be the change in location on the jth step, so that

$$X_n = z_1 + z_2 + \cdots + z_n.$$

Each of the z's is independent of the others, and they all have the same probability distribution of possible values, ± 1 with equal probability 1/2. Consequently they all have the same mean $E[z] = 0$, and variance $\text{Var}(z) = E[z^2] - E[z]^2 = 1 - 0 = 1$. The central limit theorem states that for a sum X_n of n independent and identically distributed random variables with common mean μ and variance σ^2, the distribution of $(X_n - n\mu)/\sqrt{n}\sigma$ converges to a Gaussian (normal) distribution with mean 0 and variance 1. Informally, we express this by saying that the distribution of X_n is approximately Gaussian with mean $n\mu$ and variance $n\sigma^2$.

For our random walk model, we therefore conclude that the particle's location after n steps has an approximately Gaussian distribution, with mean and variance

$$E[X_n] = 0, \qquad \text{Var}[X_n] = n. \qquad [7.3a]$$

The central limit theorem also applies to more general situations where exact binomial calculations are no longer easy. For example, suppose that at each time-step jumps to the left or right of some fixed size $w > 0$ each occur with probability $p \leq 1/2$, and the particle stays put with probability $1 - 2p$. In this case z_j has three possible values $(-w, 0, w)$. We still have $E[z_j] = 0$, but now $\text{Var}(z_j) = 2pw^2$; hence X_n is approximately Gaussian distributed with mean and variance

$$E[X_n] = 0, \qquad \text{Var}[X_n] = 2pw^2 n. \qquad [7.3b]$$

Exercise 7.1. Suppose that at each time-step the particle can move to the right with probability p and to the left with probability q where $p + q \leq 1$. Assuming steps are of fixed size w, generalize [7.3b] (which applies if $p = q$) to this situation (and note that if $p \neq q$ the walk is biased: $E[z_j] \neq 0$).

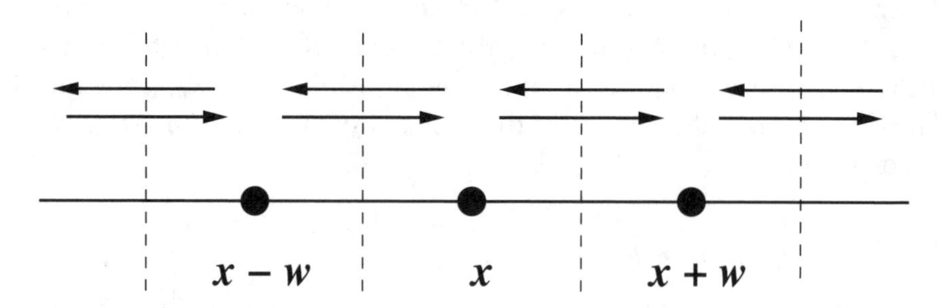

Figure 7.1 Depiction of particles undergoing a random walk on the line with general step-size w.

The random walk model has the advantage of simplicity, but space is not a lattice, so we want continuous analogs for this analysis of random walks. To get a continuous model we will scale down the increments of time and distance to 0 in such a way that the distribution of particle locations approaches a limit. Specifically, we suppose that a move of length w to the left or right can occur every h units of time, with $p \leq 1/2$ being the probability of a move in each direction, see Figure 7.1. The Gaussian approximation [7.3b] tell us how w and h should be related to each other, in order to reach a limit: the variance in a particle's location at time t has to remain the same. The number of steps up to time t is $n = t/h$ so the variance is

$$2pw^2 n = 2pw^2/h \times t = 2Dt \tag{7.4}$$

where we have defined $D = pw^2/h$. Thus, we rescale so that D remains constant (or at least approaches a limiting value as $w, h \to 0$).

We also consider the overall behavior of a large number of particles, obeying the same random walk model but moving independently of each other. Let $C(x,t)w$ be the number of particles in the interval centered on x (between the dashed vertical lines surrounding x in Figure 7.1); thus $C(x,t)$ is the "concentration at x" at time t, the number of particles per unit area. As a result of the moves that occur between t and $t+h$ we then have

$$C(x,t+h)w = C(x,t)w - 2pC(x,t)w + p[C(x-w,t)w + C(x+w,t)w]. \tag{7.5}$$

That is, the number in the interval at time $t+h$ is the number there at time t, minus those who leave, plus those who move in. The expressions in [7.5] for the numbers of moving particles are really the expected number (total number of particles $\times$ the probability of a move); by ignoring deviations from the expected number we are tacitly assuming that there are "many" particles in each compartment, even when C is small.

Now we do Taylor series expansions of C in space and time,

$$\text{(a)}\quad C(x,t+h) - C(x,t) = \frac{\partial C}{\partial t}(x,t)h + O(h^2)$$

$$\text{(b)}\quad C(x+w,t) - C(x,t) = \frac{\partial C}{\partial x}(x,t)w + \frac{\partial^2 C}{\partial x^2}(x,t)\frac{w^2}{2} + O(w^3) \qquad [7.6]$$

$$\text{(c)}\quad C(x-w,t) - C(x,t) = \frac{\partial C}{\partial x}(x,t)(-w) + \frac{\partial^2 C}{\partial x^2}(x,t)\frac{(-w)^2}{2} + O(w^3).$$

Equation [7.5] says that $(a) = p \times [(b) + (c)]$. Expressing this in terms of the right-hand sides of [7.6], we get that

$$\frac{\partial C}{\partial t}(x,t)h + O(h^2) = pw^2 \frac{\partial^2 C}{\partial x^2}(x,t) + O(w^3).$$

Rearranging, and recalling that $pw^2/h = D$, we get

$$\frac{\partial C}{\partial t}(x,t) + O(h) = D\frac{\partial^2 C}{\partial x^2}(x,t) + O(w^3/h).$$

Finally we let $h, w \to 0$ with $D = pw^2/h$ constant. Since $w^3/h = Dw/p$, the $O(w^3/h)$ term goes to 0, and we get the *diffusion equation*

$$\frac{\partial C}{\partial t}(x,t) = D\frac{\partial^2 C}{\partial x^2}(x,t). \qquad [7.7]$$

This equation is a continuous description of the law of diffusion for particles undergoing Brownian motion on the line. To relate this back to the random walk model, consider the situation where there is a single particle (so the integral of C over space is equal to 1 at all times), starting at location 0. A direct calculation shows that

$$p(x,t) = \left(\frac{1}{4\pi Dt}\right)^{1/2} \exp\left(\frac{-x^2}{4Dt}\right) \qquad [7.8]$$

is a solution to [7.7] with these properties. This is the formula for a Gaussian distribution with mean zero and variance $2Dt$, which is exactly the Gaussian approximation for the underlying random walk model.

In the plane, the corresponding equation for the concentrations $C(x,y,t)$ of Brownian motion particles is

$$\frac{\partial C}{\partial t} = D\left(\frac{\partial^2 C}{\partial x^2} + \frac{\partial^2 C}{\partial y^2}\right). \qquad [7.9]$$

The simple form of this equation results from the fact that motion in the vertical direction is independent of the motion in the horizontal direction. Similar to our one-dimensional equation, this can be derived by dividing the plane up into squares of side w, writing the balance equation for the numbers of particles moving into and out of each box, and scaling space and time in a way that preserves a Gaussian approximation for the distribution of individual particles.

To reach these simple equations we have made the heroic assumptions that particles take steps of fixed size, at fixed time intervals. Nonetheless, there is considerable experimental evidence supporting these equations as a quantitatively accurate description of Brownian diffusion. One reason for this is that our assumptions can actually be relaxed considerably—for example, by allowing particles to have a probability distribution of step lengths, that might correspond to the free path of a Brownian particle between one collision and the next, or the flight distance of an insect moving from one flower to another in a field. In that case, the simple diffusion equation emerges as the leading-order terms in an expansion involving higher-order derivatives in space (e.g., Goel and Richter-Dyn 1974), so long as conditions are homogeneous in space. A second reason is the central limit theorem, which yields a Gaussian approximation depending only on the mean and variance of the displacements in each unit of time, and how displacements in successive time intervals are correlated with each other. Over a macroscopic time span long enough that many steps occur, any "microscopic" model that gets these ingredients right, necessarily leads to the same approximate Gaussian distribution for the long-term distribution, and hence to the same partial differential equation for the concentration profile of many such particles. Thus, the long-term spread of some animal populations is described well by a simple two-dimensional diffusion model, even though the underlying assumptions about individual movement steps are rarely satisfied (Turchin 1998). However, the diffusion model fails when animals exhibit distinct short-range and long-range movement behaviors, so that the macroscopic displacements on the time scale of interest may result in part from a small number of large moves, leading to non-Gaussian patterns of spread. A third, related reason is that the diffusion equation can be derived from macroscopic-level properties. In particular the same balance equation [7.5] results from *Fick's Law of Diffusion*, which asserts that the rate of material flux across a surface (e.g., the dashed vertical lines in Figure 7.1) is proportional to the concentration gradient across the surface (see, e.g., Berg 1983 or Edelstein-Keshet 1988 for a derivation of the diffusion equation from Fick's Law). So any microscopic model which implies Fick's Law will again yield the same diffusion equation.

Boundary conditions are an important ingredient in obtaining a well-posed initial value problem for the diffusion equation. If our particles are random-walking within a bounded interval I on the line, or a bounded spatial area A in the plane, the diffusion equation does not specify what happens when particles reach the boundary. Additional boundary conditions must be imposed for there to be a unique solution to the diffusion equation. The appropriate boundary conditions reflect properties of the biological system being modeled. Two common boundary conditions are *no flux*, and *constant concentration* (in time). No-flux boundary conditions correspond to a situation where particles cannot move across the boundary. The equations describing this are that the directional derivative of the concentration, in the direction perpendicular to the boundary, is identically 0. In the case of an interval $I = [a, b]$ on the line, no-flux boundary conditions are simply

$$\frac{\partial C}{\partial x}(x, t) \equiv 0 \text{ at } x = a \text{ and } x = b.$$

Constant boundary conditions arise when the concentration at the boundary is fixed by external conditions. For example, the region of interest may be immersed in a large reservoir of fluid which is well mixed and maintains a constant concentration. Particles can move across the boundary, but the external reservoir is so large that there is no significant change in its concentration of particles due to this movement. For example, the ion fluxes of membrane currents can be small enough that they make little change in intracellular or extracellular concentrations of the ions.

In the absence of spatial variation in concentrations, systems of equations for chemical reactions have the form

$$\dot{c}_i = f_i(c_1, \ldots, c_k).$$

Here c_i is the concentration of species i, there are k species altogether and the f_i give the rate of change of c_i due to chemical reactions. A reaction-diffusion model assumes that chemical concentrations may vary in space, and that the changes in chemical concentrations at a given location come from both chemical reactions and diffusion. In two space dimensions the model is

$$\frac{\partial c_i}{\partial t} = f_i(c_1, \ldots, c_k) + D_i \left(\frac{\partial^2 c_i}{\partial x^2} + \frac{\partial^2 c}{\partial y^2} \right) \qquad [7.10]$$

where each c_i is a function of (x, y, t).

7.2 The Turing Mechanism

Diffusion is a force that acts to homogenize concentrations. An initial distribution of concentrations will evolve to a spatially uniform state if diffusion is the only force acting on a substance and the boundary conditions are compatible with constant concentration in the interior of a domain. Therefore, it is surprising that instability of a reaction-diffusion system can give rise to spatial pattern for initial concentrations that are close to a stable equilibrium of the reaction system alone. Intuition says that diffusion should make things spatially smoother and more stable. The discovery that diffusion can destabilize an otherwise stable equilibrium was made by Alan Turing in 1952. We give one version of his argument here.

The systems we will consider are reaction-diffusion equations for a pair of chemical species on the unit circle. The circle is used as a domain because it is one dimensional, compact, and has no boundary. We further assume that the kinetic equations for the chemical reactions have a stable steady state and have been linearized about that state. Letting $u_1(x,t)$ and $u_2(x,t)$ represent the *departures* of the chemical concentrations from their steady-state values, we obtain equations of the form

$$\begin{aligned}\frac{\partial u_1}{\partial t} &= a_{11}u_1 + a_{12}u_2 + D_1\left(\frac{\partial^2 u_1}{\partial x^2}\right)\\ \frac{\partial u_2}{\partial t} &= a_{21}u_1 + a_{22}u_2 + D_2\left(\frac{\partial^2 u_2}{\partial x^2}\right)\end{aligned} \qquad [7.11]$$

for the linearized system. Here x is the angular coordinate along the circle and the u_i satisfy periodic boundary conditions $u_i(x+2\pi,t) = u_i(x,t)$ since u_i is a function on the circle.

Now $u_1(x,t) = u_2(x,t) \equiv 0$ is a solution of this system that represents the homogeneous steady-state concentrations. We want to investigate the stability of this steady state, looking for instabilities of the homogeneous steady state that give rise to spatial patterns. This is more difficult than studying the stability of equilibria for ordinary differential equations because the state of the system at each time is a pair of functions $(u_1(x), u_2(x))$ describing the spatial distribution of concentrations. We can think of the set of possible concentration functions $(u_1(x), u_2(x))$ as forming an infinite dimensional phase space space and the partial differential equations [7.11] as defining a vector field on this phase space. Theorems about existence and uniqueness of solutions are more complicated in this infinite-dimensional context, but we plunge ahead anyway.

The equations are linear, so we seek solutions constructed from the eigenvalues and eigenvectors of the right-hand side of the equations. Since the states of the system are functions, the eigenvectors of this problem are called eigenfunctions. To find eigenfunctions, we utilize the fact that

$$\frac{d^2 \sin(nx)}{dx^2} = -n^2 \sin(nx).$$

Of course, a similar formula holds for $\cos(nx)$. Substituting

$$(w_1(t)\sin(nx), w_2(t)\sin(nx)) = (u_1(x,t), u_2(x,t))$$

into the reaction diffusion equation gives

$$\frac{dw_i}{dt}\sin(nx) = (a_{i1}w_1 + a_{i2}w_2 - n^2 D_i w_i(t))\sin(nx)$$

for $i = 1, 2$. We conclude that the functions whose spatial dependence is $\sin(nx)$ comprise a two-dimensional space that is *invariant*: solutions with initial conditions in this space stay in the space. Dividing the last equations by $\sin(nx)$ removes the x dependence and leaves us with the pair of linear differential equations

$$\frac{dw_1}{dt} = a_{11}w_1 + a_{12}w_2 - n^2D_1w_1$$
$$\frac{dw_2}{dt} = a_{21}w_1 + a_{22}w_2 - n^2D_2w_2 \qquad [7.12]$$

for the coefficients $w_1(t), w_2(t)$ of spatial patterns whose spatial dependence is $\sin(nx)$. If the origin is a stable equilibrium of this system, then initial conditions will decay to the zero solution. On the other hand, if the origin is an unstable equilibrium then nonzero initial conditions of the form $(w_1(0)\sin(nx), w_2(0)\sin(nx))$ yield solutions of [7.11] that grow in amplitude.

The matrix of equation [7.12] is

$$A(n) = \begin{pmatrix} a_{11} - n^2D_1 & a_{12} \\ a_{21} & a_{22} - n^2D_2 \end{pmatrix}. \qquad [7.13]$$

Note that $n = 0$ gives the matrix for homogeneous perturbations to the zero solution of [7.11] and corresponds to the stability of the steady state for the reaction equations with no diffusion.

Spatial patterns will arise if the spatially homogeneous steady state is stable against spatially homogeneous perturbations ($n = 0$), but is unstable against some spatially inhomogeneous perturbations ($n > 0$). Our goal, therefore, is to analyze [7.13] to determine when this situation can occur.

There will be stability against homogeneous perturbations if all eigenvalues of $A(0)$ have negative real part. That is, we must have

$$a_{11}a_{22} - a_{12}a_{21} > 0 \text{ (positive determinant) and } a_{11} + a_{22} < 0 \text{ (negative trace).}$$

We next seek a value of n with $A(n)$ having an eigenvalue with positive real part. The trace of $A(n)$ is $a_{11} + a_{22} - n^2(D_1 + D_2)$ which decreases with n. So the trace of $A(n)$ is negative for all n. Thus the only way to create a matrix with positive eigenvalue is to have the determinant become negative:

$$a_{11}a_{22} - a_{12}a_{21} - n^2(D_1a_{22} + D_2a_{11}) + n^4D_1D_2 < 0.$$

If $D_1 = D_2$, this too is impossible because $a_{11} + a_{22} < 0$ and the determinant increases with n. However, Turing observed that with diffusion constants D_1 and D_2 that differ, the determinant may become negative for positive values of n. For example, if we set $a_{11} = 1$, $a_{12} = -6$, $a_{21} = 1$, $a_{22} = -4$, $D_1 = 0.5$, $D_2 = 10$, and $n = 1$, then

$$A(1) = \begin{pmatrix} 0.5 & -6 \\ 1 & -14 \end{pmatrix},$$

which has determinant -1 but $a_{11}a_{22} - a_{12}a_{21} = 2 > 0$ and $a_{11} + a_{22} = -3 < 0$. It is easily seen that the signs of the reaction coefficients a_{11} and a_{22} must be opposite in any example of the Turing mechanism. This example demonstrates that diffusion can destabilize a homogeneous stable steady state for a system of chemical reactions.

These observations on the conditions required for the Turing mechanism are the starting point for a class of reaction-diffusion models based upon local activation and long-range inhibition, introduced by Gierer and Meinhardt in 1972. The "activator" u_1 is a substance for which deviations from its equilibrium state will be amplified in the absence of diffusion or coupling to the inhibitor, that is, $a_{11} > 0$. This is usually referred to as *autocatalysis* or *positive feedback*. The inhibitor returns to equilibrium when perturbed, and the requirement that $a_{11} + a_{22} < 0$ implies that it is "stronger" than the activator; that is, the exponential rate at which it returns to equilibrium is faster than the activator grows. In order for $D_1 a_{22} + D_2 a_{11} > 0$, we must have $D_2 > D_1$: the rate of diffusion of the inhibitor must be larger than the rate of diffusion of the activator. Finally, to have stability of the homogeneous steady state, we required $a_{11}a_{22} - a_{12}a_{21} > 0$. Since $a_{11}a_{22} < 0$, this implies also that $a_{12}a_{21} < 0$. For this we assume that the activator stimulates production of the inhibitor ($a_{21} > 0$) and that the inhibitor limits the production of the activator ($a_{12} < 0$). The combination of fast diffusion of the inhibitor together with its effect on limiting production of activator is termed *lateral inhibition*. A large region of activator will stimulate production of inhibitor that diffuses and reduces production of inhibitor. In the appropriate circumstances, this leads to a nonuniform distribution of activator with local peaks that are small enough to be in balance with the inhibitor whose production they stimulate nearby. The Gierer-Meinhardt model and its followers are nonlinear models that satisfy these conditions, with nonlinearities that prevent unbounded growth of the concentrations of activator and inhibitor.

7.3 Pattern Selection: Steady Patterns

The Turing argument does not help much to understand *which* spatial patterns will emerge when diffusion together with chemical reactions destabilizes a homogeneous steady state of concentrations. Leopards have spots, zebras have stripes and there is a profusion of intricate patterns on butterfly wings. Figures 7.2 and 7.3 show examples of these patterns. We can hypothesize that the leopard got its spots from a reaction-diffusion mechanism. More specifically, we want to know which patterns are produced by which reaction-diffusion systems. This information is helpful in investigations of the biological mechanisms underlying animal coat patterns and the differentiation of tissues that takes place during development of an organism. From a mathematical perspective, our entry point is to regard the reaction-diffusion system [7.10] as a dynamical system and seek to extend the concepts for analyzing dynamical systems to *spatially extended* sys-

Figure 7.2 An example of animal coat patterns: the eastern chipmunk, *Tamias striatus*. (Photograph ©Marie Read, used with permission.)

tems in which spatial pattern and time varying dynamics can interact with one another. Since we have not discussed them, we focus on the patterns themselves first.

Equilibrium solutions of the reaction-diffusion system [7.10] satisfy the equations

$$D_i \left(\frac{\partial^2 c_i}{\partial x^2} + \frac{\partial^2 c_i}{\partial y^2} \right) = f_i(c_1, \ldots, c_k) \qquad [7.14]$$

together with the boundary conditions that were imposed for [7.10]. This is still a system of partial differential equations which may have many solutions. We ask how many solutions there are and how they depend upon the D_i, f_i, and the boundary conditions. Once the solutions to [7.14] have been determined, we can examine their stability as solutions of the system [7.10] which includes time dependence. Similar questions arise in many physical problems, and there is an extensive body of mathematics that has been created to help us solve systems of partial differential equations like [7.14]. We draw upon this mathematics to help us get started.

Figure 7.3 An example of insect color patterns: two beetles from the genus *Anoplophora* with different patterns. Left: *Anoplophora medenbachii* (Ritsema), female, from Java, Indonesia. Right: *Anoplophora mamaua* (Schultze), male, from Mindoro Island, Phillipines. (Photographs by Kent Loeffler, Department of Plant Pathology, Cornell University. From W. Lingafelter and E. R. Hoebeke. 2002. Revision of *Anoplophora* (Coleoptera: Cerambycidae). Entomological Society of Washington, Washington, D.C. Used with permission.)

The simplest version of the *boundary value* problems that we seek to solve is a single, linear equation in one dimension with zero boundary conditions:

$$-D\frac{d^2u}{dx^2} + au = 0$$
$$u(0) = U(L) = 0. \qquad [7.15]$$

$u(x) \equiv 0$, the function that is identically zero, is always a solution of this equation. We are interested in finding additional solutions that would represent a spatial pattern. If $a > 0$, then the solutions of the equation $Dd^2u/dx^2 + au = 0$ are trigonometric functions of the form $u(x) = c_1 \cos(\sqrt{a/D}x) + c_2 \sin(\sqrt{a/D}x)$. To see whether this u satisfies the boundary conditions of [7.15], we evaluate $u(0) = c_1$ and $u(L) = c_1 \cos(\sqrt{a/D}L) + c_2 \sin(\sqrt{a/D}L)$. Thus the boundary conditions are satisfied if $c_1 = 0$ and $c_2 \sin(\sqrt{a/D}L) = 0$. If $\sqrt{a/D}L$ is an integer multiple of π, $\sin(\sqrt{a/D}L) = 0$ and the boundary condition at L is satisfied for all values of c_2. If $aL^2 = (n\pi)^2$, then the nonzero solutions of the equation have $n - 1$ zeros in the interior of the interval $[0, L]$.

Exercise 7.2. Show that if $a < 0$, then the solutions of the equation $Dd^2u/dx^2 + au = 0$ are sums of exponential functions, and the only solutions of the boundary value problem [7.15] is the zero function.

To interpret the implications of this equation, we can regard either a or L as a parameter that varies. For example, if we are interested in a developing organism that is growing and there is diffusion reaction process governed by equation [7.15], then when the organism reaches a critical length $L = n\pi/\sqrt{a/D}$, this process could produce a "segmented" spatial pattern with n segments. However, the model [7.15] seems deficient in two respects: (1) the pattern appears only at the critical lengths $L = n\pi/\sqrt{a/D}$ and then disappears again at longer lengths, and (2) the amplitude of the pattern is not determined.

To "improve" model [7.15], we consider a slight change that makes it nonlinear:

$$D\frac{d^2u}{dx^2} + au - u^3 = 0$$
$$u(0) = u(L) = 0. \qquad [7.16]$$

The term u^3 added to the differential equation is nonlinear. It serves to limit the growth of u due to the reaction terms of the system. Despite the nonlinearity, we can still analyze properties of the system [7.16] by viewing the differential equation as a dynamical system. To do so, we introduce another dependent variable v that represents du/dx and consider the vector field

$$\frac{du}{dx} = v$$
$$\frac{dv}{dx} = \frac{-au + u^3}{D}. \qquad [7.17]$$

Solutions $u(x)$ of the equation $Dd^2u/dx^2 + au - u^3 = 0$ correspond exactly to solutions $(u(x), v(x))$ of [7.17] with $v = du/dx$. The variable x plays the role of the "time" variable in the dynamical system [7.17]. For $a > 0$, the vector field [7.17] has three equilibrium points at $u = -\sqrt{a}, 0, \sqrt{a}$ and $v = 0$. The eigenvalues of the equilibrium at the origin are purely imaginary: $\pm\sqrt{a/D}i$. The key to understanding the phase portrait of system [7.17] is that the function

$$E(u, v) = \frac{1}{2}v^2 + \frac{a}{2D}u^2 - \frac{1}{4D}u^4$$

is constant on trajectories.[1] The level sets of E surrounding the origin are closed curves, so there is a family of trajectories that are periodic orbits. See Figure 7.4. To solve the boundary value problem [7.16], we want to pick out trajectories that

[1]To prove this, use the chain rule to differentiate E with respect to x:

$$\frac{dE}{dx} = \frac{\partial E}{\partial u}\frac{du}{dx} + \frac{\partial E}{\partial v}\frac{dv}{dx} = \left(\frac{a}{D}u - \frac{1}{D}u^3\right)v + v\left(\frac{-au + u^3}{D}\right) = 0.$$

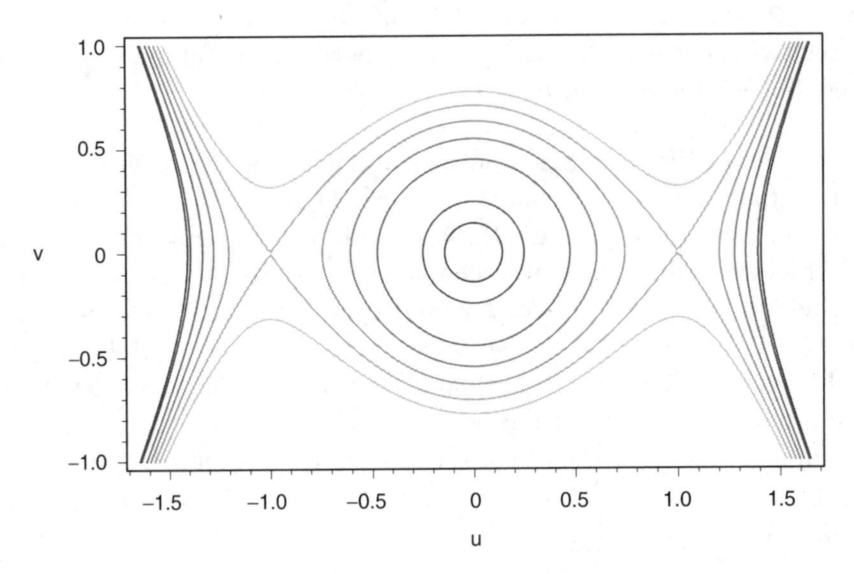

Figure 7.4 Trajectories of the vector field [7.17] are level curves of the function $E(u, v) = (1/2)v^2 + (a/2D)u^2 - (1/4D)u^4$.

have $u = 0$ at times 0 and L. Due to the symmetry of the equation, this will happen if there is an integer n so that the periodic orbit has period $2L/n$. A calculation with *elliptic functions* implies that the periods of the periodic orbits grow as their amplitude grows. The range of values for the periods is from $2\pi/\sqrt{a/D}$, the value determined by the eigenvalues at the origin, to ∞, the value obtained as the periodic orbits approach the saddle equilibria. This implies that each time $L = n\pi/\sqrt{a/D}$, a new solution will emerge at the origin and grow in amplitude. Thus, the number of solutions increases as L increases. Our analysis does not address the more difficult question of which of these solutions (if any) is stable as an equilibrium of the reaction-diffusion system [7.10], but it does limit the possible steady patterns.

Exercise 7.3. This exercise uses calculus to show that the period E of the solutions to [7.17] increases with amplitude. The periodic solutions lie along the curves $(1/2)v^2 + (a/2D)u^2 - (1/4D)u^4 = c$. A more convenient parametrization is to write

$$2c = \frac{ab^2}{D} - \frac{b^4}{2D}$$

where b is the value of $|u|$ at its intersection with the u-axis. Solving for v along this curve,

$$v = \pm\sqrt{\frac{a}{D}(b^2 - u^2) - \frac{1}{2D}(b^4 - u^4)}.$$

Substituting this expression into $du/dx = v$, we obtain

$$\frac{du}{dx} = \pm\sqrt{\frac{a}{D}(b^2 - u^2) - \frac{1}{2D}(b^4 - u^4)}.$$

After a half-period $E/2$ beginning at the point $(u, v) = (-b, 0)$, the solution reaches the point $(u, v) = (b, 0)$. The fundamental theorem of calculus then gives

$$2b = \int_0^{E/2} \frac{du}{dx} dx$$

as an integral formula for determining the value of $E(b)$.

- Apply a change of variables $x = bw$ to "normalize" the domain of u to $[-1, 1]$ in the formula for E.
- Differentiate the normalized integral with respect to b to conclude that the period E increases with b.

We turn next to the steady spatial patterns produced by solving system [7.14] when diffusion is taking place in a two-dimensional region. The Turing mechanism postulates that the geometry of these solutions determines the spatial patterns observed in developing organisms. This hypothesis can be studied in the context of patterns on the coats of animals like zebras, giraffes and leopards and on the wings of butterflies. The principal biological task in verifying the hypothesis is to identify the diffusible morphogens. This quest has been pursued for decades, but has been given a large boost during the past twenty years by genomic techniques.

When the amplitude of the solutions to equation [7.16] is small, these solutions are almost sinusoidal. This is no accident. Linearization of the equation around the trivial solution leads back to system [7.15]. With two space dimensions the corresponding problem is

$$D\left(\frac{\partial^2 u}{\partial x^2} + \frac{\partial^2 u}{\partial y^2}\right) + au = 0 \qquad [7.18]$$

with boundary conditions $u = 0$ along a curve that bounds a region R in the plane. As in the system [7.15], there will be specific values of a for which there are nonzero solutions. The Turing mechanism postulates that the geometry of these solutions determines the spatial patterns observed in developing organisms. This prompts us to ask how the solutions of this eigenvalue problem depend upon the shape of the region R. We resort to numerical methods to solve [7.18] on arbitrary domains, but there do not appear to be any simple answers as to how the solutions depend upon the boundary as was the case for the one-dimensional problem.

The Turing hypothesis is very attractive, but is it correct? Decades of research have given rise to a large body of information without producing a simple yes or no answer. As always, biology is complicated. The organism that has been and continues to be studied the most intensively is the fruit fly *Drosophila*. The development of the segmentation pattern on the abdomen with its bristles and the development of the wing of *Drosophila* have been studied intensively using genetic techniques (Gurdon and Bourillot 2001). One of the key concepts in

Turing's hypothesis is the presence of morphogens whose concentration gradients control the development of specific structures in the organism. Morphogens for *Drosophila* development have been identified conclusively using genetic techniques, along with signaling pathways for their genetic expression and pathways that are "activated" by the morphogens. Identification of these morphogens took a very long time because they are proteins that are active at extremely low concentrations of 10^{-11} to 10^{-9} M. Spatial patterns of morphogen concentration are difficult to measure, but appear correlated with the structures they induce as predicted by Turing's hypothesis. The morphogens bind to receptors on the surfaces of cells and lead to gene expression through a signaling cascade triggered by the surface binding events. Thus, the cellular response to morphogen concentration is complex with many steps that can interact, each potentially affecting the outcome of the developmental fate of a cell.

The second aspect of the Turing hypothesis is that the mechanism for the transport of morphogens is diffusion. Here the evidence suggests that diffusion is hardly the only transport mechanism. Proteins are large molecules that diffuse slowly, and at the extremely low concentrations of the morphogens, the force of diffusion is small. Moreover, the rates of diffusion are affected by other factors than the morphogen concentration. For example, the Hedgehog protein is a morphogen in *Drosophila* that is modified after translation by cleavage and then covalent linking with cholesterol. These variant forms of Hedgehog have very different diffusion rates in the extracellular matrix of the *Drosophila* embryo (Strigini and Cohen 1999). In addition to diffusion, modes of active transport and "relay" mechanisms that involve repeated secretion and internalization of morphogens from cell to cell seem to be present in some systems. Thus, Turing mechanisms for pattern formation appear more complicated than the simplest reaction-diffusion models in that the morphogens act by triggering whole signaling pathways, and diffusion may be augmented by more "active" transport processes. Nonetheless, one can ask whether observed developmental patterns match solutions of reaction-diffusion equations. In some cases, the answer is yes; for example, in butterfly eyespots (Monteiro et al. 2001). Dynamical models will be a useful tool in unraveling additional details about morphogenesis in developing organisms.

7.4 Moving Patterns: Chemical Waves and Heartbeats

Thus far we have considered equilibrium solutions of reaction-diffusion systems. These are by no means the only solutions, or the only ones that are important in biological systems. For example, propagation of action potentials along nerves is a reaction-diffusion process that is essential to the function of the nervous system. The complex anatomy of biological systems gives rise to diffusion in one-dimensional fibers, two-dimensional sheets, or three-dimensional volumes. Visualizing time-dependent phenomena in three-dimensional tissue is a challenge, so most of the time-dependent spatial patterns that have been studied are two-dimensional patterns on surfaces. The example we discuss here is electrical stimulation of the heartbeat, a vitally important part of our lives. Failure of a coordinated heartbeat leads to death within a matter of minutes. Our hearts are thick enough that this is a three-dimensional process, but we treat the problem as two dimensional as do most studies of spatial patterns of electrical activity in the heart. This section gives a glimpse of the mathematics used to study time-dependent solutions of reaction-diffusion systems, pointing to references that go farther. Throughout the section we work with reaction-diffusion systems in two space dimensions.

The simplest types of time-dependent solution to reaction-diffusion systems are called *traveling waves*. These are solutions in which the time dependence takes the form of a fixed spatial pattern translating in time. If one introduces a coordinate system that translates with the pattern, then the solution of the reaction-diffusion system appears steady in this coordinate system. Analytically, planar traveling waves solve ordinary differential equations that are derived from the reaction-diffusion system in the following way. The orientation of the traveling wave is given by a *wave vector* (k_x, k_y) that specifies the normal direction to the lines on which the solution is constant. We denote the *wave speed* by s. The special form of the traveling wave solution is then expressed by stating that the concentrations $c_i(x, y, t)$ are functions of the scalar variable $\xi = k_x x + k_y y - st$. More formally, there are functions $\bar{c}_i$ such that $c_i(x, y, t) = \bar{c}_i(k_x x + k_y y - st) = \bar{c}_i(\xi)$ is a solution of the equation [7.10]. If we substitute $\bar{c}_i$ into [7.10], then we obtain a system of ordinary differential equations with independent variable ξ. Using the chain rule, we find

$$\frac{\partial c_i}{\partial t} = \frac{d\xi}{dt}\frac{d\bar{c}_i}{d\xi} = -s\frac{d\bar{c}_i}{d\xi}; \quad \frac{\partial c_i}{\partial x} = k_x\frac{d\bar{c}_i}{d\xi}; \quad \frac{\partial c_i}{\partial y} = k_y\frac{d\bar{c}_i}{d\xi}.$$

Therefore

$$f_i(c_1, \ldots, c_k) + D_i(k_x^2 + k_y^2)\left(\frac{d^2\bar{c}_i}{d\xi^2} + s\frac{d\bar{c}_i}{d\xi}\right) = 0.$$

This is a second-order ordinary differential equation that is converted into a vector field by introducing additional dependent variables for $d\bar{c}_i/d\xi$.

The geometry of traveling waves is determined by their dependence upon the single independent variable ξ. In (x, y, t) "space-time," the level surfaces of the ξ are planes whose normal vector is $(k_x, k_y, -s)$. A train of water waves is a good example to help us visualize the behavior of traveling waves. The crests and troughs of the wave are perpendicular to the vector (k_x, k_y) at all times, and they move at a speed s in the direction of this vector. Although traveling waves are observed in many phenomena, the boundary conditions of physical models are seldom compatible with traveling-wave solutions. Typically, the waves will be deformed near the domain boundary to "match" the constraints imposed by the boundary conditions with the propensity of the medium to support traveling waves.

Exercise 7.4. Consider the linear system of reaction diffusion equations

$$\frac{\partial c_1}{dt} = d\left(\frac{\partial^2 c_1}{\partial x^2} + \frac{\partial^2 c_1}{\partial y^2}\right) + ac_1 - bc_2$$

$$\frac{\partial c_2}{dt} = d\left(\frac{\partial^2 c_2}{\partial x^2} + \frac{\partial^2 c_2}{\partial y^2}\right) + bc_1 + ac_2.$$

When does this system have traveling wave solutions of the form

$$c_1(x, t) = \cos(k_x x - st)$$

$$c_2(x, t) = \sin(k_x x - st)?$$

When these solutions exist, find a formula for traveling-wave solutions whose wave vector is parallel to (k_x, k_y).

Planar traveling waves have a space-time symmetry: the wave profiles at different times are translations of each other. We exploit this symmetry to reduce the partial differential equation to an ordinary differential equation when computing the shape of a traveling wave. There are additional mathematical reasons to expect that stable time-dependent solutions of reaction-diffusion systems might have symmetries, and we can search for solutions that have different types of symmetries than traveling waves. Such patterns were first observed in experiments with thin layers of a chemically reacting fluid. Two striking types of solutions with rotational symmetries, *target patterns* and *spiral waves*, are produced by the oscillating *Belousov-Zhabotinsky* (BZ) chemical reaction. Oscillating chemical reactions have been known for about a century (e.g., Bray 1921), but they have been seen as a chemical curiosity at best for much of that period. The laws of thermodynamics tell us that a closed system of reactions eventually comes to equilibrium, just as friction eventually brings a mechanical system that is not acted upon by external forces to rest. Nonetheless, reactions can oscillate for a long time and reactors which receive a steady inflow of reactants can sustain oscillations indefinitely. Epstein and Showalter (1996) survey different types of spatio-temporal phenomena in varied chemical oscillators. The system that has been studied most extensively is the BZ reaction. It was discovered by Belousov in 1951, but his colleagues were skeptical of his work and widespread awareness of his work came only much later [see the account by Winfree (1987)]. The BZ reaction itself involves oxidation of malonic or citric acid by bromate ions, catalyzed by cerium. Addition of an indicator dye to the medium shows the different phases of the reaction as vivid color changes. When the reaction takes place in

Figure 7.5 Spiral patterns of the BZ reaction (Reprinted with permission from A. T. Winfree, S. Caudle, G. Chen, P. McGuire, and Z. Szilagyi, Chaos 6: 617 (1996). Copyright 1996, American Institute of Physics).

a thin layer, it is capable of displaying striking spatial patterns like those shown in Figure 7.5. The pattern shown in this figure separates into domains, in each of which a spiral pattern rotates around a "core" at its center, typically with a period on the order of a minute. Target patterns are time-dependent structures in which circular waves propagate radially from a center.

The spatial patterns produced by the BZ system have been studied extensively in the laboratory. There has been a lively interaction between experiment, theory, and simulation surrounding this fascinating system (Field and Burger 1985). Some of the research has sought to construct good dynamical models for the BZ reaction. Producing models that agree quantitatively with the observations has proved to be difficult, despite the fact that the molecules participating in the reaction are small. Another aspect of the research has been to investigate carefully the spatial patterns formed by the BZ reaction and to study their stability. This research draws heavily upon the roles of both topology and symmetry in dynamical systems. Spiral waves have the property that they look steady in a *rotating* coordinate system that rotates with the angular velocity of the waves. However, unlike traveling waves, one cannot reduce the equations describing spiral waves to ordinary differential equations. Thus, research on spiral waves has depended heavily upon numerical simulations of partial differential equations, especially simpler model systems exhibiting spiral waves than those describing the detailed kinetics of the BZ reaction or the electrical excitability of the heart. Dwight Barkley has written a freely distributed computer code, `ez-spiral`, for the reaction-diffusion system

$$\begin{aligned} \frac{\partial u}{\partial t} &= \frac{\partial^2 u}{\partial x^2} + \frac{\partial^2 u}{\partial y^2} + \epsilon^{-1} u(1-u)\left(u - \frac{v+b}{a}\right) \\ \frac{\partial v}{\partial t} &= D_v\left(\frac{\partial^2 v}{\partial x^2} + \frac{\partial^2 v}{\partial y^2}\right) + u - v \end{aligned} \qquad [7.19]$$

which serves as such a model system.

Exercise 7.5. Download the program `ezspiral` and experiment with the patterns that it produces. Alternatively, implement your own program of this type. Implementation requires that the spatial derivatives of the reaction-diffusion system be discretized. The simplest discretizations are given by the finite-difference formulas

$$\frac{\partial u}{\partial t}(x,y,t) \approx \frac{u(x,y,t+h) - u(x,y,t)}{h}$$

$$\frac{\partial v}{\partial t}(x,y,t) \approx \frac{v(x,y,t+h) - v(x,y,t)}{h}$$

$$\begin{aligned} \frac{\partial^2 u}{\partial x^2}(x,y,t) &\approx \frac{(\partial u/\partial x)(x,y,t) - (\partial u/\partial x)(x-k,y)}{h} \\ &\approx \frac{u(x+k,y,t) - u(x,y,t) - (u(x,y,t) - u(x-k,y,t))}{k^2} \end{aligned}$$

$$\begin{aligned} \frac{\partial^2 u}{\partial y^2}(x,y,t) &\approx \frac{(\partial u/\partial y)(x,y,t) - (\partial u/\partial y)(x,y-k,t)}{k} \\ &\approx \frac{u(x,y+k,t) - u(x,y,t) - (u(x,y,t) - u(x,y-k,t))}{k^2}. \end{aligned}$$

Here h is the *time-step* of the method and k is the spacing of points in a regular rectangular mesh used to represent the spatial domain. Simulation of the model requires boundary conditions as well, and these can be difficult to implement for domains with irregular boundaries. `Ezspiral` uses periodic or no-flux boundary conditions in a rectangular domain. The periodic boundary conditions are easiest to implement and correspond to solve the system on a two-dimensional torus that has no boundary. If we ignore arteries and veins, the surface of the heart itself can be regarded as having the topology of a two-dimensional sphere without boundary, but it is harder to discretize the spatial derivatives of the reaction-diffusion equation on a sphere. No flux boundary conditions can be implemented simply and efficiently with a trick. The flux of u flowing between two adjacent sites at (x, y) and $(x - k, y)$ is given by $u(x, y) - u(x - k, y)$. To implement no-flux boundary conditions, we extend the computational domain by adding another row of points to each edge of the boundary. We then assign values to u and v on these extra rows to be equal to those on the adjacent site inside the domain. This allows us to use the formulas above to compute the spatial derivatives at grid points in the domain along the boundary, and the values that we compute implement the no-flux boundary conditions.

The insights derived from studies of the spatial patterns produced by the BZ reaction have been applied to biological systems, notably by Winfree's theory of ventricular fibrillation in the heart (Winfree 1987). Strong contractions of the ventricles provide the main force for pumping the blood through the lungs and body. Contractions of the heart muscle are triggered by action potentials that propagate by a reaction-diffusion process in a coordinated traveling wave. The action potentials are initiated at the sinoatrial node on the right atrium, spread across the atria to the atrioventricular node, and then spread through the ventricles along the system of Purkinje fibers. Myocardial tissue does not oscillate without stimulation, but fires action potentials in response to a small stimulation. Each spreading wave of electrical excitation is analogous to a grass fire: a fire front propagates through the grass as one burning patch of grass ignites its neighbors. However, the action potentials in the heart are clearly different from grass fires in that the tissue recovers its ability to fire action potentials following a brief refractory period after an action potential. The pumping action of the heart depends on the spatial coherence of the action potential waves.

Heart malfunction can be fatal within minutes or hours. *Ventricular fibrillation* is an immediately life-threatening cardiac arrhythmia in which the spatial pattern of electrical action potentials in the heart becomes more complex and disordered, prompting the heart muscle to quiver. Blood flow largely stops, leading to rapid death. Regulations in the United States now require that many public institutions such as schools and airports have electrical defibrillators available for use. These devices shock the heart in an attempt to synchronize action potentials and restore the normal spatial patterns of action potentials. Many scientists believe that ventricular fibrillation is preceded by the formation of spiral waves of action potentials in the heart. Witkowski et al. (1998) published a striking visualization of spiral waves in the heart. The frequency of the spiral waves is typically much higher than that of the normal heartbeat, a condition called *tachycardia*. Once spiral waves or ventricular fibrillation are established, the heart needs to be given a substantial shock in order to restore normal excitation and contraction. Winfree's 1987 book, *When Time Breaks Down*, is a lively, nontechnical introduction to these theories about ventricular fibrillation.

Propagation of cardiac action potentials can be modeled by systems of equations that couple diffusion of ions to Hodgkin-Huxley-type models for the membrane potential of cardiac cells. These equations have precisely the same form as reaction-diffusion systems, with gating equations for channels and an equation for the membrane potential replacing the chemical reactions in the system. Since the equations have the same form, the same methods can be used in their math-

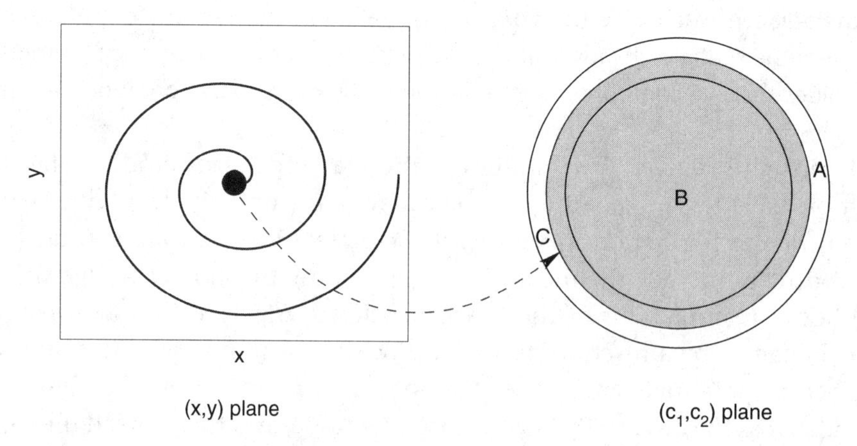

Figure 7.6 Map of spiral wave in the (x, y) plane into the (c_1, c_2) concentration plane. The dark core of the spiral in the (x, y) plane is mapped into the shaded region in the (c_1, c_2) plane, stretching over the disk B. The boundary of the core is mapped onto the curve C bounding the shaded region, making one turn around B. All of the points outside the core are mapped into the annulus A.

ematical analysis. To the extent that spatial patterns for these systems do not depend upon the details of their "kinetic" terms, we can gain insight into the spatial patterns of cardiac arrhythmias by studying reaction-diffusion systems. Since heart tissue is a difficult medium to work with, we draw inspiration from studies of the BZ reaction to guide investigations of the electrical activity of the heart.

Winfree (1987) describes topological principles related to the spatial patterns formed by time-dependent solutions of reaction-diffusion equations, especially spiral waves. We shall discuss these principles in the setting of systems in which there are two chemical concentrations. At any time, the state of the system gives a map of the "spatial" (x, y) plane into the (c_1, c_2) "concentration" plane. Figure 7.6 illustrates this map. Assume there is a spiral wave pattern in which there is a "core" disk outside of which the concentrations always lie in an annulus A contained in the concentration plane.[2] The complement of an annulus has two components, one bounded and one unbounded. We denote the bounded component of the complement by B. Consider the concentration map on the boundary of the core disk. The image will be a curve C in the annulus A that returns to its starting point. The number of times that C winds around B is a *topological* property of C that does not change if C is continuously deformed. Consequently, if the solution evolves continuously as a spiral wave, the image of the concentration map outside the spiral core will continue to surround B the same number of times. Thus, spirals have a propensity to persist: very large per-

[2] An annulus is a region that can be continuously deformed into a region of the form $0 < a < x^2 + y^2 < b$.

turbations are required to destroy them unless they are spontaneously unstable. Typically, instability of the spiral manifests itself in the emergence of more disordered, complex patterns that are described as *spiral turbulence*. Spiral turbulence is widely regarded as a good model for ventricular fibrillation. These properties help explain why defibrillation is required to "restart" a normal sinus rhythm in a heart that is in ventricular fibrillation. They also help explain why a sudden shock like a sharp blow to the chest can sometimes trigger a normal heart to go into a state of fibrillation and cause sudden cardiac death.

Retaining the terminology from the previous paragraph, topology also implies that if C winds around B a nonzero number of times, then the image of the core disk must contain all of B. The concentrations in the core of the spiral must be different from those found in the arms of the spiral. This precludes the possibility of developing models for the spiral core in terms of a single angular "phase variable" of concentration. Models with a single phase variable are frequently used to study the synchronization of coupled oscillators (Strogatz 2003), but we cannot model the core of spiral waves in this way. Similar arguments also explain why spirals in the BZ system are typically observed as counter-rotating pairs. The experimental methods that have been used to producing spirals in the BZ system begin with a disturbance in the interior of a homogeneous pattern. Following the disturbance, the net change of phase around the region of disturbance is zero. As spirals develop near the disturbance, the sum of the angular changes produced by their arms must still sum to zero. The simplest pattern that does this is a pair of counter-rotating spirals.

The occurrence of spiral waves was surprising when they were first observed, but even more surprising were observations that show that the spiral cores are capable of coherent motions along flowerlike patterns that resemble those produced by a spirograph! See Figure 7.7. The shape of these meanders has been explained by Barkley as a result of the symmetries of the reaction-diffusion equations (Barkley 1994). In addition to these meandering spirals, in other circumstances the spirals become more unstable and degenerate into disordered patterns that have been described as spatio-temporal turbulence. Using *Fitzhugh-Nagumo* equations equivalent to those used in the `ez spiral` model [7.19], Winfree (1991) produced a diagram of spatio-temporal bifurcations of spirals in a reaction-diffusion system. He hypothesized that disordered patterns evolving from spiral waves are the root cause of ventricular fibrillation in the heart.

As with the Turing mechanism for morphogenesis, we are faced with evaluating the validity of a theory based upon simple dynamical models. The spiral-wave theory of ventricular fibrillation proposes a mechanism for one of the leading causes of death: there are 300,000 people who die of sudden cardiac arrest in the United States yearly. The theory suggests that sudden cardiac death often results from the inability of the heart to pump blood when spiral waves become unstable. Clearly, we want not only a conceptual understanding of ventricular fibrillation

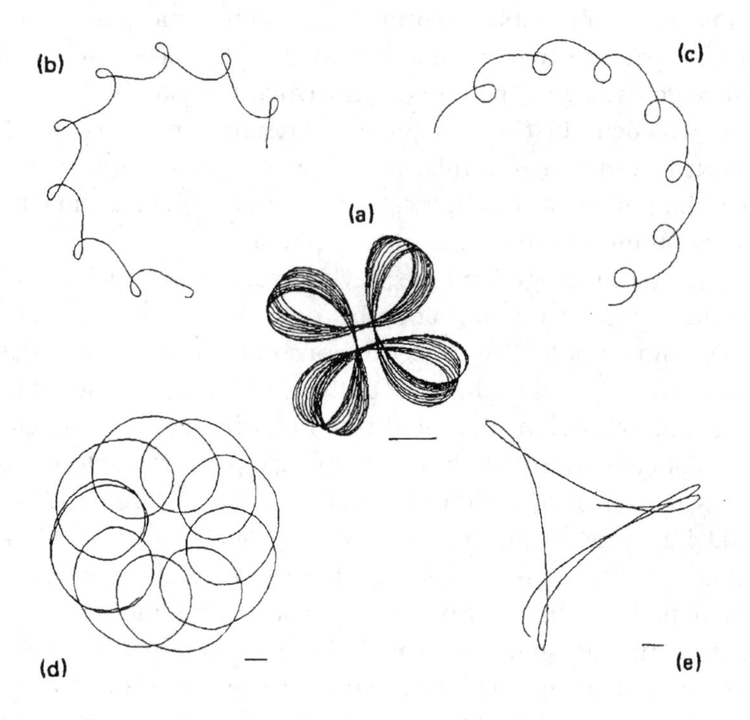

Figure 7.7 Meandering paths of spiral cores in simulations of a reaction-diffusion system (from Winfree 1991).

but also the means of preventing its occurrence. From a clinical perspective, research on ventricular fibrillation focuses on the breakdown of normal action potential propagation in the sinus rhythm. Two primary causes for block of action potentials are (1) damage to heart tissue that prevents normal conduction of action potentials and (2) changes in cellular properties that lead to abnormally long action potential durations and inadequate time to recover from a refractory period for a normal rhythm. For example, Fox et al. (2002) observed that period-doubling bifurcations in the chemical kinetics seem to lead to instability of wave propagation. Spiral waves and *scroll waves* (Winfree and Strogatz 1983; Keener 1989), their three-dimensional counterparts, lead to ventral tachycardia, degrading the ability of the heart tissue to recover from its refractory period to keep up with the pacing from the wave core. Damaged tissue can give rise to dead regions around which spiral-like waves of action potentials can propagate. Winfree placed emphasis on spiral waves and their topological properties as fundamental aspects of cardiac arrhythmias. Whether or not these topological properties are as important in sudden cardiac death as proposed by Winfree, his theories have had an important influence in focusing attention on action potential propagation as a reaction-diffusion process that is central to cardiac function and disease. Computational models of this process will undoubtedly remain an important tool in research on the heart.

7.5 References

Barkley, D. 1994. Euclidean symmetry and the dynamics of rotating spiral waves. Physical Review Letters 72: 164–167.

Barkley, D. ezspiral_3_1.tar.gz, http://www.maths.warwick.ac.uk/ barkley/ez_software.html

Berg, H. C. 1983. *Random Walks in Biology*. Princeton University Press, Princeton, NJ.

Bray, W.C. 1921. A periodic reaction in homogeneous solution and its relation to catalysis. Journal of the American Chemical Society 43: 1262–1267.

Edelstein-Keshet, L. 1988. *Mathematical Models in Biology*. Random House, New York.

Epstein, I. R., and K. Showalter. 1996. Nonlinear chemical dynamics: Oscillations, patterns and chaos, Journal of Physical Chemistry 100: 13132–13147.

Field, R. J., and Burger, M. (eds.). 1985. *Oscillations and Traveling Waves in Chemical Systems*. Wiley, New York.

Fox, J. J., R. F. Gilmour, Jr., and E. Bodenschatz. 2002. Conduction block in one dimensional heart fibers. Physical Review Letters 89: 198101–4.

Gierer, A., and H. Meinhardt. 1972. A theory of biological pattern formation. Kybernetik 12: 30–39.

Goel, N., and N. Richter-Dyn. 1974. *Stochastic Models in Biology*. Springer, New York.

Gurdon, J., and P.-Y. Bourillot. 2001. Morphogen gradient interpretation. Nature 413: 797–803.

Hodges, A. 2000. *Alan Turing: The Enigma* Walker & Co., London.

Keener, J. P. 1989. Knotted scroll wave filaments in excitable media. Physica D 34: 378–390.

Monteiro, A., V. French, G. Smit, P. Brakefield, and J. Metz. 2001. Butterfly eyespot patterns: Evidence for specification by a morphogen diffusion gradient. Acta Biotheoretica 49: 77–88.

Murray, J. D. 1993. *Mathematical Biology*. Springer, New York.

Odell, G., and G. Oster. 1980. The mechanical basis of morphogenesis III: A continuum model of cortical contraction in amphibian eggs. Journal of Mathematical Biology 65–89.

Odell, G., G. Oster, B. Burnside, and P. Alberch. 1980. A mechanical model for epithelial morphogenesis. Journal of Mathematical Biology 9: 291–295.

Odell, G., G. Oster, P. Alberch, and B. Burnside. 1981. The mechanical basis of morphogenesis I: Epithelial folding and invagination. Developmental Biology 85: 446–462.

Strigini, M., and S. Cohen. 1999. Formation of morphogen gradients in the *Drosophila* wing. Seminars in Cell and Developmental Biology. 10: 335–344.

Strogatz, S. 2003. *Sync: The Emerging Science of Spontaneous Order*. Hyperion, New York.

Turchin, P. 1998. *Quantitative Analysis of Movement: Measuring and Modeling Population Redistribution in Animals and Plants*. Sinauer Associates, Sunderland MA.

Turing, A. M. 1952. The chemical basis of morphogenesis. Philosophical Transactions of the Royal Society of London, Series B 237: 37–72.

Winfree, A. T. 1987. *When Time Breaks Down*. Princeton University Press, Princeton, NJ.

Winfree, A. T. 1991. Varieties of spiral wave behavior: An experimentalist's approach to the theory of excitable media. Chaos 1: 303–334.

Winfree, A. T., S. Caudie, G. Chen, P. McGuire, and Z. Szilagy. 1996. Quantitative optical tomography of chemical waves and their organizing centers. Chaos 6: 617–626.

Winfree, A. T., and S. H. Strogatz. 1983. Singular filaments organize chemical waves in three dimensions: 3. Knotted waves. Physica D 9: 333–345.

Witkowski, F. X., L. J. Leon, P. A. Penkoske, W. R. Giles, W. L. Ditto, and A. T. Winfree. 1998. Spatiotemporal evolution of ventricular fibrillation. Nature 392: 78–82.

8 Agent-Based and Other Computational Models for Complex Systems

So far we have stuck to our general theme: simple models, using tried-and-true approaches that have a a mature mathematical theory to help us understand how the models behave. An understanding of simple models and hands-on experience with them are essential prerequisites for working effectively with complicated models.

The purpose of this chapter is to round out the picture by examining some models whose complexity puts them beyond the reach of mathematical analysis. These models are based on simulating each individual unit in a system, generically called an "agent." In Chapter 4 we briefly considered systems biology—large systems of differential equations for the interactions among numerous genes, signal transduction pathways, and other within-cell processes. The only thing special about those models is their size, dozens to hundreds of state variables. Agent-based models, in contrast, represent a totally different modeling paradigm. Each agent is represented explicitly, and the model consists of context- and state-dependent rules for agents' possible actions and the consequences that result from them. The term "agents" brings to mind individual organisms, but agent-based models have also been developed for molecules, genes, cells, and tissue segments. The goal of agent-based models is to explain or predict the dynamics of "macroscopic" properties from the rules that operate at the "microscopic" level of system components interacting with their local environment and other agents.

We have already seen one simple agent-based model: discrete-event simulations of an epidemic in a finite population (Chapter 6). In that model each agent is characterized by a single state variable—disease state—and has a simple decision rule for state transitions. The appeal of agent-based models is that it is not much harder to program situations where agents are characterized by many state variables and have more elaborate decision trees for choosing actions. Based on a review of agent-based models for human systems, Bonabeau (2002) identified some properties that favor agent-based modeling:

1. Agents have discontinuous "either-or" responses to other agents' behaviors.
2. Agents adapt, learn, or otherwise modify their behavior based on past experience.
3. Agents are heterogeneous, either inherently or due to learning and adaptation.
4. Interactions are localized—for example, with spatial neighbors or a small network of social contacts—but the pattern of interactions is dynamic.

Peck (2004) offers a more concise summary:

> When an analytic model can be used, it should be used. They are simpler, clearer and usually can be fit to data to make their interpretation much easier. However, some processes are inescapably complex. This is the domain of the complex simulation model.

It is important to realize that "complex" in this quote does not just mean "complicated." It refers to phenomena that occur with a large number of interacting components or processes, but not with a small collection of the same type of entities. In this sense, discrete-event epidemic models (Chapter 6) are not complex simulation models: in that case the new phenomena arise when there are few enough agents that coin-tossing stochasticity becomes important, while large numbers of agents are well described by a differential equation model. In either case, phenomena beyond the present reach of mathematical analysis can nonetheless be studied by simulation.

The argument against elaborate computational models, discussed in Chapter 4, is that they replace a complex biological system that we don't understand with a computer model that we don't understand either. So after presenting examples of agent-based models, we consider two approaches that facilitate the process of understanding a complex model's properties and predictions.

1. A computer model can be studied "experimentally" in much less time than the real system, with no practical obstacles to changing parameters or modifying system properties. When done *successfully*, this puts computational models on an equal footing with simpler models whose properties can be studied analytically. But success requires really exploring the full range of possibilities, which may be hard for large models with many parameters and assumptions. We describe methods for identifying which parameters and assumptions have the largest effects on the behavior of a model, so that attention can focus on the most important ones.
2. A complementary approach is to derive an approximate model which is easier to understand. Although this is often done in an *ad hoc* way, there are also some systematic approaches that can be tried.

Exercise 8.1. Based on the considerations discussed above, suggest a biological question that could profitably be addressed by an agent-based model, and explain why an agent-based approach is reasonable. Be original: do a literature search to make sure that your proposed study has not already been published, and describe how you conducted your search. Don't just do just a web search—use Science Citation Index or some other appropriate bibliographic resource for the area of biology.

8.1 Individual-Based Models in Ecology

In ecology, agent-based models are known as "individual-based" models (DeAngelis and Gross 1992). The agents are single organisms, each characterized by a set of "*i*-state variables" specifying attributes that can differ between individuals and change over time. The total number of state variables in the model is then the product of the number of individuals, and the number of *i*-state variables per individual.

8.1.1 Size-Dependent Predation

Individual-based models need not be complicated, because it may be very easy to describe assumptions at the individual level. An example is the model by Rice et al. (1993) for survival of larval fish in the presence of size-dependent predation. The model was developed to explore how individual variation in growth rate might affect the fraction of individuals surviving the larval stage and becoming a juvenile. The model has a daily time step and follows a cohort of larvae over 60 days, with individual size (length) as the only *i*-state variable.

- Fish i has initial length 12 mm and growth rate $g_i = (G + z_i)$ mm/day, where $G = 0.2$, 0.4, or 0.6, and z_i is chosen from a normal distribution with mean 0, and standard deviation $\sigma = 0.04$, 0.08, or 0.16.
- Each day, each fish has a 20% chance of encountering a predator.
- On encounter with a predator, the chance of a fish being eaten depends on the size ratio: $P(\text{capture}) = -0.33 + 0.15\times$ (predator length/prey length), truncated onto the interval [0, 1]. This equation was based on experiments for 90 mm alewife preying on larval bloater. It assumes that predators are mature fish, all roughly the same size, and not changing in size over the time period being modeled.
- At the end of each day, the size of each surviving fish is increased by its growth rate.

The assumption of constant growth rates set at birth is unrealistic, so Rice et al. (1993) considered a second model with dynamic random growth rates:

$$g_i(t+1) = g_i(t) + d_i(t),$$

where $d_i(t)$ was drawn each day from a Gaussian distribution with zero mean and standard deviation $\sigma_d = 0.08$. In this model there is no intrinsic tendency for growth rate to go up or down (the mean change is 0), but as time goes on individuals become more variable in growth rate.

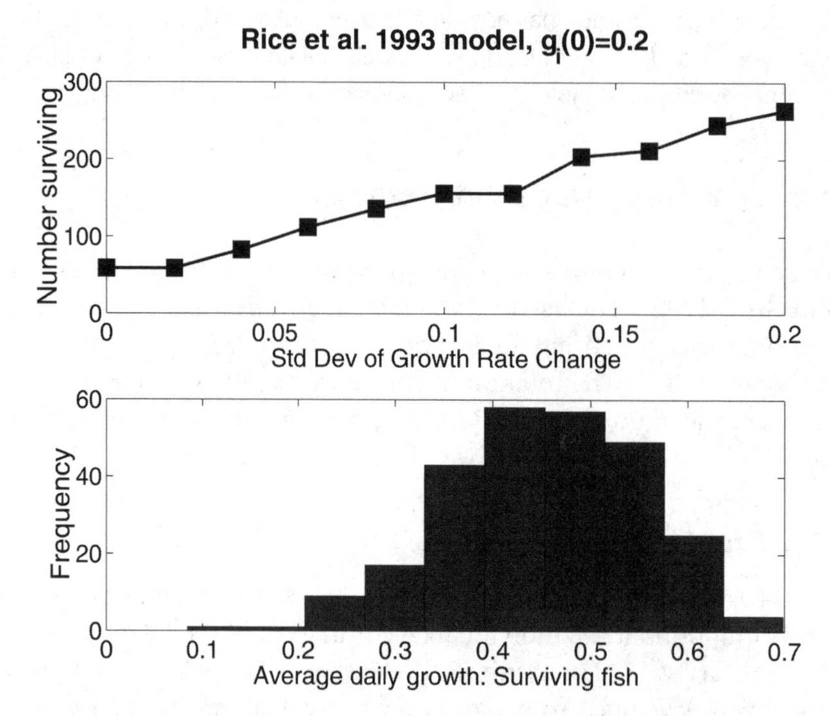

Figure 8.1 Results from simulations of the Rice et al. (1993) individual-based model for larval fish growth and survival in the presence of size-dependent predation. Individuals were started at 12 mm with $g_i(0) = 0.2$ mm/d, and were constrained to have growth rate between 0 and 0.8 mm/day, as in Rice et al. (1993).

The main result from this model was that increases in either the mean or the variance of growth rate increased the probability of survival. The effect of the mean is no surprise. The effect of the variance says that survivors are mostly the lucky few individuals who achieve above-average growth rates. In simulations of the second model, up to 90% of the surviving individuals had actual average growth rates above the initial mean growth rate, depending on model parameters. Growth rate variability could more than double the number of survivors, with most survivors having above-average growth rates (Figure 8.1).

One could use a matrix model based on size rather than an agent-based model. For example, the "stages" could be size-classes 0–0.2mm, 0.2–0.4mm, 0.4–0.6mm, and so on. However, the model would actually require a two-dimensional classification, by size and growth rate, because the probability of size-class transitions depends on growth rate. So a matrix model would need to have a lot of stages, and a large and complicated matrix to describe the odds of moving between categories. In contrast, coding up the individual-based model is a simple exercise (Table 8.1).

This model is *theoretical* even though it is *computational*. Larval growth is a highly simplified abstraction, predation risk is constant even though prey abun-

```
nfish=5000; ndays=50; %5000 fish for 50 days
sizes=zeros(ndays,nfish); growth=zeros(ndays,nfish);
alive=ones(1,nfish);
sigmag=0.08; % standard deviation of changes in growth

% initial sizes and growth rates
sizes(1,1:nfish)=0.4*ones(1,nfish);
growth(1,1:nfish)=0.4;

%iterate the model
for jday=1:(ndays-1);
    % Predation. For simplicity of coding, the dead can die again.
    meetpred=rand(1,nfish)<0.2; %does a predator find me?
    pdie=-0.33 + 0.15*(90./sizes(jday,:)); %if so....
    eaten=rand(1,nfish)<pdie;
    m=find(meetpred.*eaten>0);   % do I live or die?
    alive(m)=0;
    % find sizes and growth rates for tomorrow
    sizes(jday+1,:)=sizes(jday,:)+growth(jday,:);
    growth(jday+1,:)=growth(jday,:)+sigmag*randn(1,nfish);
    growth(find(growth<0))=0;
    growth(find(growth>0.8))=0.8;
end;

% plot growth rate of survivors
m=find(alive>0); growthrate=(sizes(ndays,m)-sizes(1,m))/ndays;
hist(growthrate);
```

Table 8.1 MATLAB code for the Rice et al. (1993) model with random changes in growth rate over time.

dance is changing rapidly and predators are changing in size, the causes of growth variation are left unspecified, and so on. In order to focus on the main theoretical question—interactions between variation in growth rate and size-dependent predation—everything else is represented simply or ignored.

8.1.2 Swarm

Enquist and Niklas (2001) used an individual-based model to explain the constancy of size-frequency distributions in forests. The empirical pattern is that the number of trees of trunk diameter x in a given area is proportional to x^{-2}. To explain this pattern they developed a simulation model in which trees compete with each other for space and light and follow size-dependent rules for allocating assimilated energy. Enquist and Niklas (2001) wrote their model in Swarm, an open-source toolkit for developing agent-based models. The Swarm Development Group (wiki.swarm.org) claimed 300–500 users as of 2002, and lists applications to cell biology, animal behavior, ecology, economics, sociology, and military strategy.

Enquist and Niklas's (2001) simulations begin by scattering a user-specified number of seeds across a finite area. These may be all of one species, or else are individually assigned a species-specific biomass, light requirement, and dispersal range. Each tree's total energy intake is partitioned into stem, leaf, and reproduction, and converted into biomass of each type according to "allometric" allocation rules. These rules specify the relationship between the total biomass of the individual and its parts, and were derived from a theoretical model for such scaling relationships in vascular plants. A tree's stem, leaf, and reproductive biomass determine its height, canopy diameter, and seed production, respectively. Some of the available reproductive biomass may be reserved for future use. Propagules are dispersed randomly over a circle whose radius depends on the height of the parent and the weight of the seed. Individuals growing under the canopies of other, larger individuals receive less light, and therefore grow more slowly, or may die, depending on their light requirements (differing among species). There is also ongoing mortality unrelated to light limitation.

Simulations of the model replicate the size-frequency spectrum of natural forest ecosystems (frequency $\sim$ size^{-2}). They were also reported to replicate another pattern in real forests, that the total biomass per area at steady state is roughly constant—independent of latitude, elevation, and the number of species present. Enquist and Niklas (2001) concluded that "the invariant properties identified for real plant communities emerge from the allometric rules that influence the behaviour of individual plants competing for space and limited resources."

8.1.3 Individual-Based Modeling of Extinction Risk

VORTEX (Lacy et al. 2003) is an agent-based model for population viability analysis (PVA)—quantitative analysis to estimate the risk of extinction for a population, group of populations, or species, over a finite time horizon (typically 10 to 100 years), or to compare possible management options for reducing the risk of extinction. VORTEX takes an individual-based approach to modeling the extinction process in structured populations, and incorporates genetic processes in addition to the demographic processes included in matrix models (Chapter 2).

There are two motivations for an agent-based approach. The first is that extinction necessarily occurs in a small population where the randomness of individual "coin tossing" is likely to be important. The second is that, as in Rice et al. (1993), individuals are characterized by so many state variables that a matrix model would be unwieldy. The first two are age and sex: a user specifies a maximum possible age, and the age at first breeding for females and males. Juvenile (i.e., nonbreeding) individuals have an age-specific survivorship. Survival of adults is independent of age until they reach the maximum, at which point death is certain. Adult females have an annual probability of breeding, which may be constant or vary randomly, and a user-specified probability distribution for the number of offspring when breeding occurs (i.e., a breeding female has j offspring with probability $a_j, j = 1, 2, \ldots, M$). Adult males have an annual probability of mating; the mate for each mating female is chosen at random from the pool of mating males for that year. If the user chooses a monogamous mating system, then each male is only allowed to mate with one female. The pairing up of mating individuals is made independently each year—the model does not allow long-term pair bonds.

The third state variable is genotype, in order to model inbreeding depression. As a population shrinks there will be a loss of genetic variability and a higher rate of matings between closely related individuals ("inbreeding"). "Inbreeding depression" refers to the fact that matings between close relatives tend to produce fewer viable offspring. This may result from deleterious recessive alleles, which are expressed more often in inbred individuals due to their greater chance of getting the same allele from both parents, or from a general advantage of heterozygosity. VORTEX accounts for both of these possible effects. The program tracks the kinship between all individuals, and offspring of closely related parents have a decreased chance of survival in their first year. Each individual in the starting population is assigned ten unique recessive lethal alleles—two each at five hypothetical loci—and the model tracks the genotype of each individual at these loci in subsequent generations, using random numbers to simulate the genotype of offspring based on those of their parents. Offspring having identical alleles at any of the five loci are killed immediately upon birth.

As additional complications, the birth and survival rates can be affected by population density, environmental variation, and catastrophes, in ways determined by user-specified parameters. The population may be subdivided into up to fifty distinct subpopulations, with the user specifying different parameter values for each subpopulation and the degree of migration between subpopulations. The flexibility of VORTEX has allowed it to be used for a wide range of species; a recent program manual (Miller and Lacy 2003) lists over 100 publications or reports using VORTEX; see Table 8.2. The price of flexibility is that a user has to specify lots of parameters (Table 8.3), or else accept default values that may not apply to their species.

The value of VORTEX hinges on its predictions being accurate enough for real-world decision making, even in the typical case where many parameter estimates are imprecise or simply lacking. Results on this count are mixed. For example,

Birds	Other vertebrates
Humbolt penguin	Asiatic lion
Whooping crane	Baird's tapir
Capricorn silvereye	Florida panther
Bearded vulture	Giant panda
Red-cockaded woodpecker	Chinese river dolphin
Kirtland's warbler	Tree kangaroo
Attwater's prairie chicken	Lion-tailed macaque
Missssissippi sandhill crane	Leadbeater's opossum
Javan hawk-eagle	Northern white rhinoceros
White-winged wood duck	Sumatran tiger
Hihi	Hawaiian monk seal

Table 8.2 Some of the animal species for which PVA has been carried out using Vortex, based on Appendix III of Miller and Lacy (2003)

Required
Age at first breeding α, males and females
Mating system
Maximum longevity
Mean % of adults breeding each year, males and females
Variance in % breeding each year, males and females
Maximum litter size M
Litter size distribution a_j, $j = 0, 1, \ldots, M$
Age-specific mortality p_x, $x = 0, 1, \ldots$, males and females
Magnitude of environmental variability, in survival and in fecundity
Correlation of environmental variability between survival and fecundity
% inbreeding depression due to lethals
Optional to create more complex models
Frequency of catastrophes
Effect of catastrophes on survival and on reproduction
Number of populations
Which sexes disperse?
Which ages disperse?
Survival during dispersal
% of individuals dispersing
Effects of population density on survival, fecundity, dispersal, etc.

Table 8.3 Partial list of the parameters involved in specifying a VORTEX model

Lindenmayer et al. (2000) tested VORTEX by using it to predict the abundance of three species of arboreal marsupials across 39 remnant patches of native *Eucalyptus* forest in southeastern Australia, embedded in a landscape dominated by introduced pine forest. Life-history parameters were available for all three species, but little was known about their dispersal, so Lindenmayer et al. (2000) considered five different plausible scenarios:

1. No migration
2. Migration to all patches equally likely
3. Migration only to the nearest patch
4. Migration to each other patch proportional to patch size divided by distance (or distance^2) from the patch of origin

Good agreement between observed and predicted abundances could be obtained, but only for one of the five dispersal scenarios (size/distance^2), specific values of the overall migration rate, and the mortality rate during dispersal, and if differences in patch quality were taken into account (patch quality was based on how well the available vegetation matched the dietary preference of the species). Lindenmayer and Lacy (2002) similarly found, for two small mammals in the same habitat, that actual patterns of patch occupancy and abundance were obtained in only a fraction of the plausible scenarios. In contrast, Penn et al. (2000) found very good agreement between observed population trends in two koala populations, and the trends predicted by VORTEX using independent data to estimate parameters.

Brook et al. (1999) compared extinction forecasts for whooping crane from Vortex and five other commonly used programs. The whooping crane has been reduced to a single wild flock of roughly 150 individuals, and about 130 captive individuals (Brook et al. 1999). The whooping crane has been monitored since it was recognized as endangered in 1939, giving 57 years of demographic data that could be used to estimate model parameters. The models differed as to which processes were included, with the following included in only some of the packages: catastrophes, inbreeding depression, breeding structure (whooping crane is monogamous), correlation between survival and recruitment variation, sex ratio and mate availability, and the ability to specify an arbitrary probability distribution for litter size. Each model was used to generate 500 replicate projections for 50 years starting from a population of 18 individuals.

All models were parameterized from exactly the same information but they produced significantly different predictions. Predicted mean size after 50 years ranged from 66 to 129, and the risk of extinction from 1% to 7.6%. Moreover, there were significant differences between two versions of VORTEX. In version 5 males are first paired with females, and it is then determined whether or not the female will produce a litter that year. This version predicted a mean final size of 35 individuals, and a 9.4% risk of extinction over 50 years. Starting from version 6 males are only paired with females who will produce a litter that year if they have a mate. This change doubled the mean final population size, and halved the predicted extinction risk. As a control for their comparisons Brook et al. (1999) developed a simplified crane life history that could be represented in all the programs. In these "standardized" simulations the models produced very similar results. Thus the different predictions were due to differences in basic assumptions rather than the simulation methods used in the programs.

Even VORTEX does not include everything. For example, infectious diseases play a significant role in regulating many animal populations (Hudson et al. 2002) and have been implicated in population declines of a number of threatened species including African wild dogs, Arctic foxes, mountain gorilla, and rainforest toads (Cleaveland et al. 2002). Diseases also appear to pose an increasing threat to marine species (e.g., Harvell et al. 1999; Ward and Lafferty 2004). Using an agent-based model Hess (1996) showed that increased opportunities for individual movement between remnant habitat patches—which reduces extinction risk in the absence of disease—could instead increase a small population's risk of extinction in the presence of disease.

Individual-based modeling remains controversial in ecology. Proponents argue that individual-based models can be based on assumptions that are derived directly from observations of individual behavior and its consequences, and can incorporate all important variables (e.g., Huston et al. 1988; deAngelis et al. 1998). This is counterbalanced by the need for detailed information on all important variables and the potential for seemingly harmless assumptions to have large consequences, such as the mating rules in Vortex.

Exercise 8.2. Download Vortex (currently at vortex9.org) and use it to replicate the koala PVA in Penn et al. (2000) using the model parameters listed in Table 3 of that paper. Explore how sensitive the predicted extinction risk is to the assumed frequency of catastrophic years (1 year in 20, in the published results).

8.2 Artificial Life

Artificial life refers to the construction of artifical entities, mechanical or computational, that share some properties of biological life. Here we consider only one aspect: the use of agent-based simulations as models of long-term evolution.

For many basic questions the history of life on Earth constitutes a single unreplicated data point. As a result there is considerable controversy about the role of chance versus necessity—if the process were started again from scratch, would the second run look like the first (Gould 1990)? For example,

- Would life necessarily emerge from the primordial soup?

- The universality of the genetic code argues for a single origin. Of the code's many features—using DNA to store information, use of base-pair triplets as the units to specify amino acids, and so on—which if any are essential for life to proliferate and diversify, and which are just happenstance?
- Would multicellular life necessarily emerge? If so, would it necessarily diversify into a suite of taxa similar to what we now have and a similar spectrum of ecological roles (primary producers, herbivores, predators, parasites, and so on)?

Chance has almost certainly played a significant role, for example the importance ascribed to meteor impacts in mass extinction events. Proponents of chance point to such events and argue that the history of life would be very different if the process were started from scratch a second time. One purpose of artificial life is to provide additional instances of evolution under similar sets of rules. We can then see how widely outcomes vary when a set of rules is run multiple times. And by comparing models we can try to identify which properties of the underlying rules determine the consistent features of the "organisms" and "ecosystems" that evolve.

8.2.1 Tierra

Tierra (Ray 1991, 1992) is the best-known example of long-term evolution in digital organisms. Tierra simulates competition between bit segments that strive to self-replicate within a computer's memory. Allowing these "organisms" to compete for a limiting resource (CPU time) and to reproduce with random mutations yields a digital analog of organic evolution.

A Tierran organism is a sequence of 0/1 bits occupying a linear block of memory. Each five-bit segment of the organism corresponds to a meaningful instruction, a structure analogous to the actual genetic code. Tierra is seeded with a few individuals who replicate and spread through memory. "Mutations"—random copying errors—occur during replication, creating genetic diversity. There are also "somatic" mutations (random $0 \leftrightarrow 1$ bit flips within existing organisms) that occur every time step. Once memory fills to some specified level, organisms begin to die and their space in memory is deallocated. Mutations conferring faster reproduction or lower mortality increase in frequency.

The limiting resource in Tierra is time. Organisms take turns getting a slice of the computer's CPU time allocated to the instructions in their "genome." Organisms that use fewer instructions to replicate propagate faster, so there is selection for having a short genome without superfluous instructions. Mortality is age dependent: individuals are "born" at the bottom of a queue that advances toward the "reaper" over time. Individuals are also bumped up or down in the queue according to whether or not they successfully execute certain specific instructions.

In Ray's initial experiments (Ray 1992), Tierra was seeded with a single self-replicating "ancestor" organism of 80 instructions, which was designed to be a minimal self-replicating algorithm with no other functionality. The ancestor examines itself to determine where in memory it begins and ends, indicated by specific bit strings. It then calculates its size, allocates a memory block for an offspring, and uses a copy procedure embedded in its own code to replicate its genome into the offspring memory block. It then registers the daughter as a new organism, puts it at the bottom of the CPU time and "reaper" queues, and starts trying to breed again.

According to Ray (1992) the first ancestor was created as a test case for debugging Tierra, and was expected to just spread and replicate. Instead, it initiated a complex evolutionary arms race.

- *Parasites* evolved that lacked the copy procedure and therefore were shorter and faster-replicating. Instead, they located and used the copy procedure from other nearby individuals. Their analog in real life would be viruses that exploit the replication machinery of the cells that they infect. Parasites usually emerged within the first few million instructions in a run.
- Organisms with *resistance to parasitism* then emerge, and may exclude parasites for long periods of time.
- Parasites could also be excluded by *hyperparasites*. When the parasites tries to use them for self-replication, the hyperparasite resets some of the parasite's memory registers so that the parasite replicates the hyperparasite instead of itself. These drive the parasite to extinction and become dominant in the population.
- Such conditions of high genetic uniformity favor the evolution of sociality. *Social hyperparasites* emerge, which are short and rapidly breeding but can only reproduce by exploiting nearby organisms that are genetically similar to themselves (e.g., their "tail" and part of a neighbor's "head" are needed for reproduction).
- This situation was then exploited by *cheaters*, who intercept the CPU pointer identifying the block of code that is being executed as it passes from one hyperparasite to another, and redirect it to their own genome.

This evolutionary sequence required about 1 billion instructions. In the process the creatures spontaneously developed sexual reproduction. "Sexual" organisms were producing offspring that were partial copies of themselves and partial copies of others in the population—a blending akin to sexual reproduction.

Exercise 8.3. Based on the results from Tierra, we might predict that if multicellular life has evolved on other planets, it will suffer from viruses—diseases that replicate themselves using the host's cellular replication mechanism. Eventually Star Fleet will come back with the data. What might you do now, on planet Earth, to challenge or support this prediction?

8.2.2 Microbes in Tierra

Yedid and Bell (2001) used Tierra to model short-term adaptation in microbes, motivated by two hypotheses for the mode of evolutionary adaptation in microbial cultures. The dominant paradigm is "periodic selection": most of the time the culture has a single dominant genotype, but occasionally a beneficial mutation of the dominant type arises and then replaces its ancestor. The genealogy of successive dominant genotypes is then predicted to be a direct line of descent. In a new culture representing a novel environment for the organism, many different mutations will be helpful and adaptation will be rapid. Once those have become fixed, the rate of beneficial mutations will dwindle and the rate of adaptation is expected to diminish.

Yedid and Bell (2001) argued that some experimental results were inconsistent with this hypothesis: the rate of gene substitutions was too high to occur in a genetically uniform population, and did not diminish over time. They hypothesized that the population remained genetically diverse, and that each new dominant could descend from any of the types present.

Tierra was used to test these hypotheses because it allowed them to monitor genealogies in full detail, which is not yet possible with real organisms. They seeded Tierra with a single minimal genotype, with parameters set so that organisms could only read, write, and execute instructions in their own genome. Experiments were run at three mutation rates: 0.002, 0.01, and 0.2 per genome per generation, which cover the range of natural spontaneous mutation rates. Relatively small populations were used, about 500 individuals, because of the need to record each genotype and its genealogical relations.

About 75% of the genotypes that arose during the experiments were stable replicators who bred true. The remainder consisted of inviable mutants (about 17%) and mutators, who consistently produced offspring unlike themselves. At low and intermediate mutation rates, they found that evolution followed the periodic selection hypothesis. With few exceptions each dominant was a direct descendant of the previous dominant, and most dominants achieved frequencies of 80% or higher in the population (Figure 8.2). But at high mutation rates (0.2/genome/generation) dominants generally "emerge from the fog of rare genotypes that collectively constitute a large fraction of the population" (Yedid and Bell 2001, p. 479).

Yedid and Bell (2001) interpreted these results as indicating that different patterns of adaptation may occur, depending on mutation rate. At lower mutation rates such as those believed to hold in microbes, the classical periodic selection hypothesis holds, with occasional selective sweeps as a new type evolves and replaces the current dominant. At higher genomic mutation rates, perhaps typical of multicellular eukaryotes, there are periods of dominance by a single dominant, but also long periods when multiple quasineutral types coexist, any of which could be the ancestor of the next dominant.

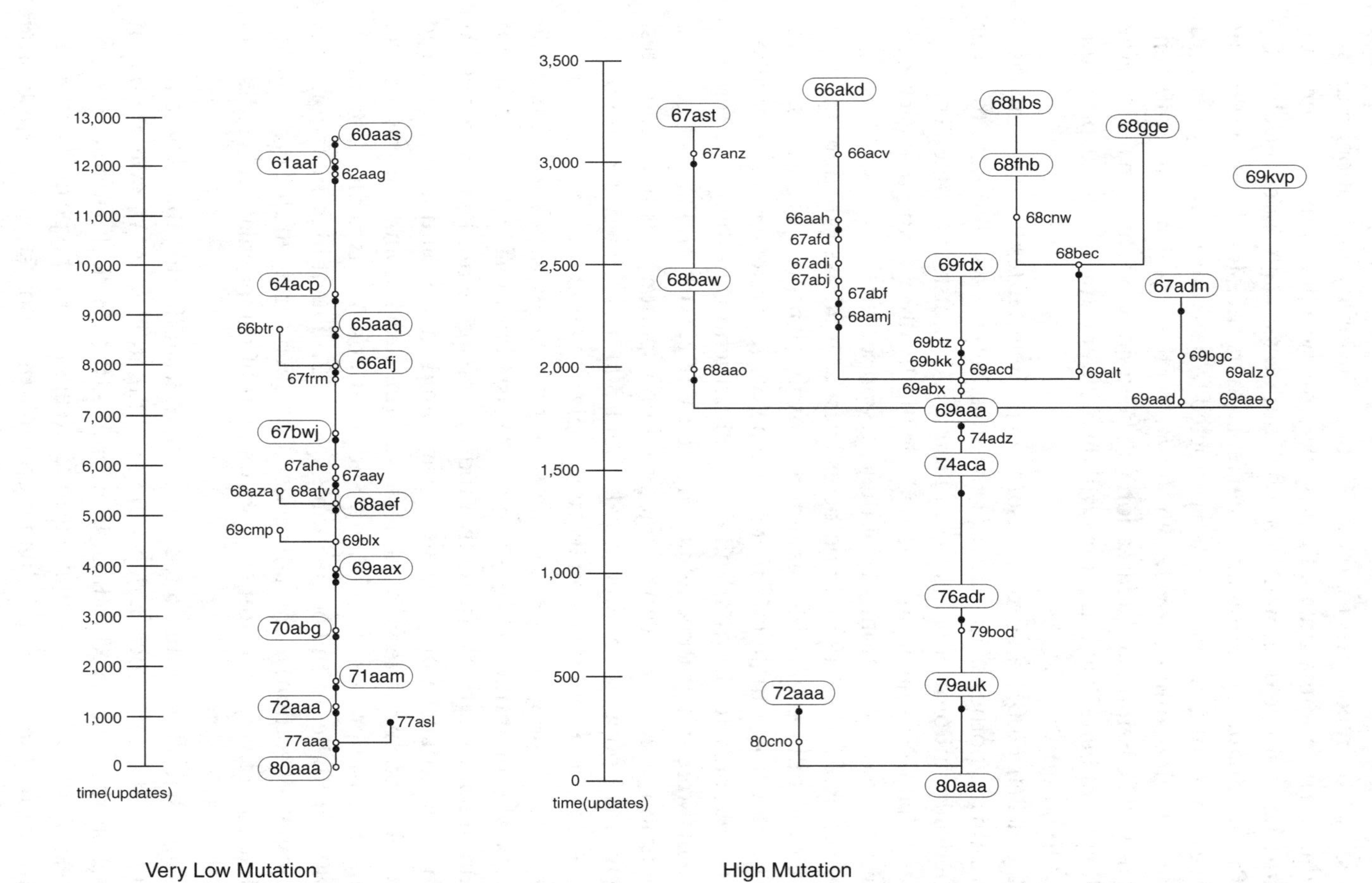

Figure 8.2 Genealogies of successive dominant genotypes under low and high mutation rates (from Yedid and Bell 2001). More common genotypes are coded by length and sequence of appearance.

8.2.3 Avida

The Avida program (Adami 2002; Ofria and Wilke 2004) is a modification of Tierra with two-dimensional space; Avidans "live" on a two-dimensional grid of cells with each individual occupying a single cell. Individuals interact through competition for space, which occurs at birth: when a parent reproduces, one of its neighbors is removed and replaced by the newborn. Reproduction is limited by CPU time as in Tierra, creating selection for short genomes. Individuals can also be given extra CPU time when they execute certain specific instructions, which allows selection for traits other than genome length. Avida uses a simpler genome than Tierra, consisting of a list of instructions such as "add," "copy," and "search in the forward direction." There are about 50 potential instructions, and the user specifies which of these are available in a run of the program.

Avida has been used to explore a number of issues in evolutionary biology, including effects of mutation rate (Wilke et al. 2001) and the processes whereby stepwise evolutionary change can produce complex adaptations (Lenski et al. 2003); see Wilke and Adami (2002) and O'Neill (2003) for reviews.

Chow et al. (2004) conducted Avida experiments on the relationship between productivity and species diversity in ecosystems. Species diversity is often (but not universally) observed to peak at intermediate levels of productivity, and a number of possible factors have been invoked to explain this pattern, including spatial heterogeneity and effects of predators. Chow et al. (2004) used Avida to test whether these factors are necessary to explain the observed pattern. They simulated a spatially homogeneous ecosystem without predators, in which individuals compete for a number of different "resources." Each resource was a CPU time reward for executing certain logical operations—an analog of obtaining food by carrying out a successful hunt, with different hunting behaviors required to obtain different types of food. Competition was modeled by having a finite supply rate of the reward for each operation, that was partially consumed each time an organism executed that operation. When a resource type becomes scarce because most of it has been consumed, the reward for the corresponding instruction goes down—"hunting" for that food item is less rewarding per unit of time.

In each experiment a grid of 300 sites was initiated with a founding genotype that could replicate but did not execute any of the rewarded operations, and evolution was allowed to proceed. Twenty-five replicates were run at each of seven values of the resource supply rate, which was interpreted as a proxy for productivity since it determined the rate at which individuals could reproduce. The results showed a clear pattern of maximum species diversity at intermediate productivity (Figure 8.3). Similar results were obtained with a generalist founder that could perform all rewarded operations. These results show that the observed qualitative relationship between diversity and productivity could result simply from evolutionary diversification in resource usage: the only requirement is a pool of different resources that organisms can exploit.

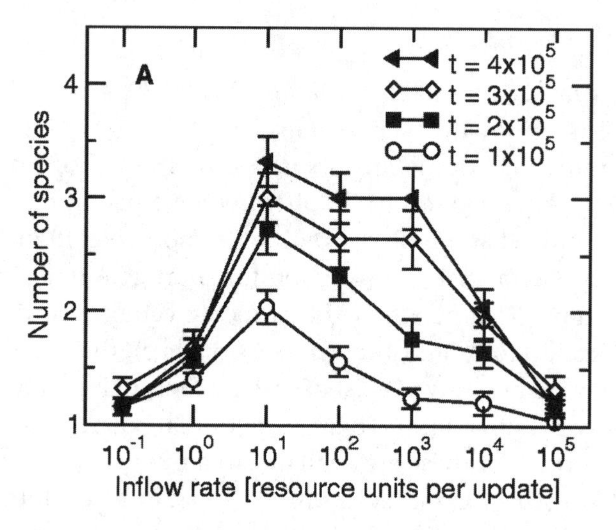

Figure 8.3 Relationship between resource supply rate and the number of distinct species descended from the ancestor, at different times in the simulation (time t is measured in "updates", corresponding roughly to the CPU time required for 30 instructions to be executed). From Chow et al. (2004)

The relevance of experiments in digital evolution depends on the analogy between digital and organic evolution. In some ways they are similar—being driven by mutation, selection, drift, and limited resources—but in other respects they can be quite different. Evolution in Tierra is driven by the benefits of a short genome. Yedid and Bell (2001) found that genome length most often changed via mutants that miscalculate their genome size and produce offspring shorter than themselves. Sometimes those offspring were still able to replicate and therefore had higher fitness than the ancestor. Consequently, adaptation in Tierra was dominated by processes very different from those typical of actual organisms.

Exercise 8.4. The online Supplementary Information for Waide et al. (1999) provides summary information on published studies of the relationship between productivity and species diversity. Considering only those studies showing a unimodal relationship between productivity and diversity, evaluate (and explain how you evaluated) the *quantitative* correspondence between Figure 8.3 and empirical studies. How should your findings affect our interpretation of the results in Chow et al. (2004)? What additional simulation experiments do your findings suggest?

Exercise 8.5. Read up on current theories for the origin of sexual reproduction, and conduct some experiments in Tierra or Avida to test one of them. Publish your results in *Science* or *Nature*, and send us a copy. If your paper is rejected, read up on current theories for the origin of cooperation and try again.

8.3 The Immune System and the Flu

The National Institutes of Health recommends yearly influenza vaccination, especially for certain groups. The influenza virus evolves rapidly, so each year a new vaccine is produced to provide protection against the strains predicted to be prevalent in the next flu season. Smith et al. (1999) studied how the efficacy of a vaccine depends on previous vaccinations. Historical studies of this question reached different conclusions as to whether repeated vaccination was effective, so Smith et al. (1999) developed an agent-based model to explore the efficacy of repeated vaccination.

Each individual has a vast number of B-lymphocytes, cells that produce antibodies that bind to viral and bacterial antigens and destroy them. Some antibodies are incorporated as receptors into the membranes of B-cells, and binding of these to antigen stimulates the proliferation of a clone of immune cells. Some of these progeny cells secrete antibodies while others are long-lived memory cells that allow for a faster and more vigorous response to similar antigens in the future. Vaccines stimulate the immune system to produce antibodies to antigens from disease organisms. Their effectiveness depends upon an individual's history of exposure to similar antigens.

Exposure to one strain of flu can confer partial or complete immunity to closely related strains. Also, vaccination against one strain affects the response to subsequent vaccination against similar strains. If an individual is vaccinated in two successive years with similar strains, the second vaccination produces a smaller immune response because the immune system already has produced circulating antibodies that bind to the antigens of the second vaccine. This may decrease the effectiveness of the second vaccination, especially against strains of flu that are more similar to the specific target of the second vaccine than the first.

The model of Smith et al. (1999) represented the specificity of each B-cell receptor, antibody, and antigen as a 20-letter word with a 4-letter alphabet. Thus, their model allows for 4^{20} different antibodies and antigens. They use the *Hamming distance* to define the specificity of antibody and antigen: the interactions between an antibody and an antigen are a function of the number of letters that are the same in the two. For example the words

ABCDABCDBCDADDCCABAB

ABCDABCDBCDADDCABBAB

have distance 2 because the letters in positions 16 and 17 differ. Cells and antigens whose distance is 7 or smaller were allowed to interact, the strength of the reaction decreasing with distance.

Consider a person who was vaccinated against strain 1 last year and strain 2 this year, but has just been infected with strain 3—because the circulating strain is not exactly the one predicted when the vaccine was produced. Is their immune response to strain 3 stronger or weaker as a result of the previous year's vaccination? Assume the distance from strain 1 to strain 2 is 2, and that the distance from strain 3 to strain 1 is 3. The distance from strain 3 to strain 2 can be as small as 1 or as large as 5. The immune system produced lots of antibody to strain 1 last year, some of which is still present in the body, and smaller amounts of antibody to strain 2 this year. If strain 3 is closer to strain 2 than to strain 1, then there may be a weaker response than if no vaccination occurred last year. On the other hand, if strain 3 is closer to strain 1, then there may be a stronger response than would have occurred with a single vaccination this year to strain 2.

The model studied the interactions between repeated vaccinations by simulating a collection of 10^7 B-cell clones. Rules were established for the proliferation and death of B-cells when challenged with antigen at different antigenic distances. Rules were also established for the proliferation of viruses in the system that are bound and unbound by antibody. Simulations were performed that corresponded to no vaccination, vaccination "last year" only, vaccination "this year" only, and vaccination in both years. In each case, the system was challenged with attack by a flu virus two months after the time of vaccination this year. Simulations were run for 450 days to observe the response of the system over two subsequent flu seasons.

Simulations of the model clearly showed both positive and negative interference effects of repeated vaccination. With repeated vaccination, the response of the system depended upon the antigenic distance of the virus to both the first and second vaccines. In all simulations, vaccination in the current year conferred greater immunity than no vaccination. However, in some cases there was greater immunity this year if no vaccination occurred last year.

The study implies a definite policy recommendation for the production of flu vaccines. Each year's vaccine targets strains that are predicted to be prevalent, but there is a range of uncertainty as to which strain will predominate. Within the range of the predicted strains, the authors recommend that vaccine be produced to those antigenically farthest from the vaccine of the previous year. This will produce the largest response of individuals who were vaccinated last year to vaccination this year.

8.4 What Can We Learn from Agent-Based Models?

The studies reviewed above illustrate that the things we can learn from complex, agent-based models are no different from what we can learn from any other kind of theoretical dynamic model. The models provide the link between process and pattern: how mechanistic assumptions about key variables and processes lead to predictions of observable phenomena. Yedid and Bell (2001) predicted how mutation rate affects the dynamics of adaptation; Chow et al. (2004) showed that evolutionary diversification in ecological niches could produce a hump-shaped relationship between productivity and diversity; Enquist and Niklas (2001) found that allometric resource allocation rules could generate observed size-frequency scalings in forests. In principle these are no different from (for example) our simple models for gene regulation networks, in which certain specific types of interactions were shown to induce certain kinds of temporal patterns in the state variables.

However, the unavoidable complexity of agent-based models can make it hard to know *why* the observed outcomes occurred, as critics of simulation models often complain. The Enquist and Niklas (2001) model reproduces the observed size-frequency scaling. But is that really due to their agent-level allocation rules, as they claim, or is it a consequence of competition for space and light that would hold for other allocation rules? If so, what is the set of allocation rules that would produce results consistent with the observed size-frequency scaling? Given the number of assumptions in a realistic agent-level model, it may not be feasible to just vary each assumption and see what happens. Four options for each of five assumptions yields 1024 alternative models, each needing to be explored through multiple simulations across a range of parameter values—and it is a rare agent-based model that only has five assumptions.

In the rest of this chapter, we consider two approaches for addressing this issue: structured computational methods for understanding the predictions of a complex model, and the formulation of simpler dynamic models where the link between assumptions and outcomes is more easily determined. Used in this way, analytic and computational models can be mutually reinforcing approaches to understanding a complex system, rather than polar alternatives.

In a sense these two approaches are analogous to doing a bifurcation analysis on a classical dynamical systems model with many parameters (Chapter 5). Bifurcation analysis identifies the special locations in parameter space where qualitative changes occur. The overall behavior of the model is then sketched out by locating the bifurcation curves or surfaces and identifying the qualitative changes that occur at each of them. For computational models we similarly want to start by figuring out where the action is happening, and then focus our attention there.

8.5 Sensitivity Analysis

The goal of sensitivity analysis is to evade the curse of dimensionality by identifying a subset of parameters and assumptions that most strongly affect a model's behavior, at much lower computational cost than a full exploration of parameter space. Ranking parameters by their impact on model predictions is also important information for planning real-world experiments, because it identifies which additional information will be most beneficial for narrowing down the range of model predictions.

The benchmark for computational cost is the exponential dependence on the number of parameters in a brute force approach: to check all possible combinations of k values for d parameters requires

$$k^d = e^{d\log(k)} \tag{8.1}$$

model runs (if a model is stochastic, then some quantities have to be estimated by doing repeated simulations at the same parameter values—for simplicity we will also refer to this as a "model run").

The simplest form of sensitivity analysis is a computational version of the local sensitivity analysis for eigenvalues of matrix models (Chapter 2). "Local" means that we only consider small perturbations from a reference set of parameter values. Let Y denote a quantity of interest calculated from the output of a model run, and $Y(p_i)$ its value as a function of the ith parameter p_i, with all other parameters held fixed at their reference value The sensitivity of output Y to parameter p_i is defined as the percentage change in Y relative to the percentage change in p_i, for a small change in p_i. Typically 5% or 10% changes are used with a centered difference estimate of the response. With a 5% change, the fractional change in Y is $(Y(1.05p_i) - Y(0.95p_i))/Y(p_i)$ and the fractional change in p_i is 0.1, so the sensitivity is computed as

$$s_i = \frac{Y(1.05p_i) - Y(0.95p_i)}{0.1 \times Y(p_i)}. \tag{8.2}$$

Note that [8.2] is what we called the elasticity in Chapter 2. Ecologists and economists use the term elasticity for [8.2] but in the simulation modeling literature it is called the sensitivity. You might as well get used to this—it's not going to change any time soon.

Local sensitivity analysis is computationally cheap, requiring $2d$ additional runs for a model with d parameters. Even with $k = 2$ (the absolute minimum) the computational cost relative to [8.1] is reduced over 50-fold with 10 parameters, and more than 25000-fold with 20 parameters. However it has some significant limitations:

1. It characterizes the model's response only near the reference parameter set. This has limited relevance in the typical situation where many or most parameter estimates have low precision.
2. It takes no account of interactions between parameters. If a small increase in parameter 1 has no direct effect on Y, but it makes Y much more sensitive to changes in parameter 2, local sensitivity analysis will incorrectly say that parameter 1 is unimportant.

A *global sensitivity analysis* removes these limitations by making simultaneous large changes in multiple parameters. The goal is to identify the most important parameters by doing model runs at a relatively small but well-chosen set of parameter combinations. The model is viewed as a "black box" that transforms a probability distribution of inputs—parameters, assumptions, and initial conditions—into a probability distribution of model outputs. The results from model runs are treated as experimental data and analyzed statistically to determine which inputs have the most effect on the output distribution. The results of this analysis are meaningful so long as the input-output relationship satisfies the assumptions of the statistical procedures used to analyze the "data."

Three main types of statistical procedures have been used:

1. *Response surface* methods posit that the relationship between parameters and model outputs Y can be approximated by an assumed functional form that is simple enough to be estimated well from a small number of model runs. Response surface methods are simple and easy to implement using a statistics package. However their reliability depends on how well the assumed functional form approximates the input-output relationship, which is difficult to assess.
2. *Correlation* methods estimate the strength and direction of the statistical association between each parameter and Y. Correlation coefficients can be misleading when the parameter-output relationship is nonmonotonic or strongly nonlinear. For example, the correlation between p and $Y(p) = 1 - p^2$ is exactly 0 if p is chosen at random from the interval $[-1, 1]$. However, these properties are easy to examine by plotting Y as a function of each parameter.
3. *Variance decomposition* methods decompose the variance in Y across model runs into contributions from each parameter, alone and in combination with other parameters. These methods are applicable in principle to any parameter-output relationship with finite variance of the output.

Selection of parameter combinations for a global sensitivity analysis is usually based on a set of *input distributions* $f_i(p_i)$ for each parameter p_i. Each f_i is a probability distribution representing the range and relatively likelihood of different values for the ith parameter. If parameters were estimated from data and interest focuses on parameters representing the experimental setting, a reasonable choice of f_i might be Gaussian distributions corresponding to confidence intervals on each parameter (see Chapter 9)—if $[a, b]$ is the 95% confidence interval, the corresponding Gaussian parameters are $\mu = (a + b)/2, \sigma = (b - a)/3.92$. But in general f_i can represent any parameter range over which you want to characterize the behavior of the model. If the model output of interest depends on the model's initial conditions, the initial value of each state variable can be treated as an additional parameter, or the analysis can be repeated for different sets of initial conditions.

8.5.1 Correlation Methods

Latin hypercube sampling (LHS, McKay et al. 1979) is way of sampling parameter space that for many purposes gives much more precise results than a random sample of the same size [Blower and Dowlatabadi (1994) review comparisons that have been made between LHS and other sampling schemes]. For LHS with sample size k, the range for each parameter is divided into k intervals having equal probability under the input distribution f_i. A value in each interval is selected, possibly at random. This assures that the full range of each parameter is represented in the sample. The outcomes from this process are assembled into a matrix P whose (i, j)th entry is the jth selected value of parameter i. For example, suppose there are two parameters a and b, and the $k = 3$ selected values are $a = \{1, 2, 3\}$, $b = \{10, 20, 30\}$. Then

$$\mathbf{P} = \begin{bmatrix} 1 & 2 & 3 \\ 10 & 20 & 30 \end{bmatrix}. \quad [8.3]$$

Each column of $\mathbf{P}$ is then a complete parameter vector for the model. These vectors are far from random: reading across each row the values are monotonically increasing. To destroy this correlation, each row of $\mathbf{P}$ is shuffled into random order. The columns of the resulting matrix $\tilde{\mathbf{P}}$ are then a set of k parameter vectors for the model, such that each selected value of each parameter occurs in exactly one of the parameter vectors, and there are no systematic cross-correlations between parameters. Two possible outcomes of row-shuffling [8.3] are

$$\tilde{\mathbf{P}}_1 = \begin{bmatrix} 3 & 1 & 2 \\ 20 & 30 & 10 \end{bmatrix}, \quad \tilde{\mathbf{P}}_2 = \begin{bmatrix} 2 & 3 & 1 \\ 30 & 10 & 20 \end{bmatrix}. \quad [8.4]$$

Matrix $\tilde{\mathbf{P}}_1$ says to do model runs with parameter vectors $(a, b) = (3, 20), (1, 30)$, and $(2, 10)$; $\tilde{\mathbf{P}}_2$ says to do runs with $(a, b) = (2, 30), (3, 10)$, and $(1, 20)$.

The model is run at all parameter vectors in one or more $\tilde{\mathbf{P}}$ matrices. Statistical measures of correlation are then used to summarize how strongly each parameter affects Y. Possible measures include the following:

1. The linear correlation coefficients between $p_{i,j}$ and Y_j, where $p_{i,j}$ is the value of p_i and Y_j is the value of Y for the jth model run. This measures the overall linear association between each parameter and Y.
2. The standardized linear regression coefficients. These measure the strength of each parameter-output association independent of the linear effects of other parameters. Their values can be obtained by doing multiple linear regression of Y on all parameters, with all variables scaled relative to their standard deviations.
3. Rank correlation coefficients. These can be used instead of linear correlations. The parameter values $p_{i,j}$ are replaced by their ranks $r_{i,j}$ across the parameter samples—the smallest value of $p_{1,j}$ is replaced by 1, the second smallest is replaced by 2, and so on. The same is done to the Y values, and then the linear correlation

	Optimistic			Pessimistic		
Parameter	Year 1	Year 5	Year 10	Year 1	Year 5	Year 10
Fraction of drug-sensitive cases treated	0.99	0.99	0.99	0.97	0.97	0.95
Transmission coefficient of drug-sensitive treated infection	−0.18	−0.65	−0.84	−0.10	−0.43	−0.60
Average survival time of drug-sensitive treated individuals	0.86	0.85	0.78	0.52	0.35	0.26
Increase in risky behavior				−0.86	−0.89	−0.90

Table 8.4 Sensitivity coefficients for key parameters in the Blower et al. (2000) model for antiretroviral treatment of HIV/AIDS, taken from Table 1 of that paper. The model output variable was the number of AIDS deaths averted by treatment, and the tabulated values are partial rank correlation coefficients between parameters and output where the correlation was above 0.5 in magnitude. "Optimistic" and "pessimistic" scenarios differ in their assumptions about the fraction of treated infections where resistance develops and the possible rate at which risky behaviors increase.

coefficient between rank values is computed. This statistic is called Spearman's ρ, and is available in most statistics package without the user having to compute the ranks.

To see why standardized regression coefficients are recommended, consider the "model output" $Y = p_1 + 2p_2$ with input distributions $f_1 =$ uniform on $[-1, 1]$ and $f_2 =$ uniform on $[-2, 2]$. Because p_2 has twice the impact per unit change and twice the range of variation, the answer we want for global sensitivity is that p_2 is four times as important as p_1. The linear regression coefficients are 1 and 2, but standardizing gives regression coefficients 1 and 4 (why?). The linear correlation coefficient includes this standardization automatically. Opinion is mixed about the merits of rank correlations. Some authors consider rank correlations more appropriate if parameter-output relationships are monotonic but strongly nonlinear. Others feel that transformation to ranks is too strong a distortion for the correlation coefficients to be meaningful.

Blower et al. (2000) used LHS with partial rank correlation coefficients to identify key parameters affecting the effectiveness of antiretroviral therapy for HIV/AIDS, using the model discussed in Chapter 6. The model has a total of 24 parameters, but four of these were identified as having particularly strong impacts on the numbers of AIDS deaths averted (Table 8.4). Note that the relative importance of parameters depends on the time horizon being considered, and also on whether the assumed range of possible parameter variation was "optimistic" or "pessimistic." The results from a global sensitivity analysis always depend on the input distributions that define what "global" means.

Exercise 8.6. Consider the "model output" $Y(p_1, p_2, p_3) = e^{p_1+2p_2+3p_3}$. Write a script that uses LHS in combination with linear correlation and also with rank correlation to do a global sensitivity analysis with the input distributions f_i all being uniform distributions on $[-1, 1]$. Parameter sampling with these f_i is relatively easy, because equiprobable intervals are equal in length. For example with $k = 5$ the 6 interval endpoints are $\{-1, -0.6, -0.2, 0.2, 0.6, 1.0\}$ and the selected values can be the midpoints $p_i = \{-0.8, -0.4, 0, 0.4, 0.8\}$. Write your script so that a user can choose the value of k by editing just the first line of the script.

8.5.2 Variance Decomposition

Variance decomposition methods for global sensitivity analysis are based on the *analysis of variance* decomposition from theoretical statistics. The model output of interest can be decomposed as

$$Y(p_1, p_2, \dots, p_d) = Y_0 + \sum_{i=1}^{d} Y_i(p_i) + \sum_{1 \leqslant i<j \leqslant d} Y_{i,j}(p_i, p_j) + \cdots + Y_{1,2,\dots,d}(p_1, p_2, \dots, p_d). \quad [8.5]$$

In statistical terminology, [8.5] decomposes Y into the "main" effect of each parameter, and interaction effects of higher and higher orders. If parameter values are chosen independently, this decomposition is unique under the constraints that the functions on the right-hand side of [8.5] all have expected value zero and are mutually uncorrelated. Let σ_i^2, $\sigma_{i,j}^2$, and so on denote the variances of the terms in [8.5]. Because the terms are uncorrelated, [8.5] implies a corresponding decomposition of the total variance in Y,

$$\sigma^2 = \sum_{i=1}^{n} \sigma_i^2 + \sum_{1 \leqslant i<j \leqslant d} \sigma_{i,j}^2 + \cdots + \sigma_{1,2,\dots,k}^2 \quad [8.6]$$

where σ^2 is the variance of $Y(p_1, p_2, \dots, p_d)$. Sobol' (1993) proposed sensitivity indices defined by scaling [8.6] relative to the total variance:

$$1 = \sum_{i=1}^{n} M_i + \sum_{1 \leqslant i<j \leqslant d} C_{i,j} + \cdots + C_{1,2,\dots,k}. \quad [8.7]$$

The "main" sensitivity index M_i is the fraction of the total variance in Y accounted for directly by parameter i. The individual C's are not particularly significant, but they can be combined into a measure of each parameter's total contribution to the variance,

$$T_i = M_i + \sum_{j=1}^{d} C_{i,j} + \sum_{1 \leq j<k \leq d} C_{i,j,k} + \cdots + C_{1,2,\dots,k}. \quad [8.8]$$

This "total" sensitivity index for parameter i is the sum of all terms in [8.7] that involve parameter i. Fortunately T_i can be estimated without estimating each individual term in [8.8]. For any i, we can split the parameter vector into p_i and

$p_{\sim i}$, the latter being a vector of all parameters besides p_i. Consider the quantity

$$Y(p_i, p_{\sim i}) - Y(p_i, p'_{\sim 1}) \qquad [8.9]$$

where p_i is a random draw from the input distribution on parameter i, and $p_{\sim i}, p'_{\sim i}$ are two independent random draws of all other parameters from their input distributions. Because both terms in [8.9] use the same value of p_i, their difference lacks the $Y_i(p_i)$ term in the decomposition [8.5], but there are two sets of all other terms. As a result,

$$\frac{1}{2}E[Y(p_i, p_{\sim i}) - Y(p_i, p'_{\sim i})]^2 = \sigma^2 - \sigma_i^2$$

and therefore

$$M_i = 1 - \frac{1}{2\sigma^2}E[Y(p_i, p_{\sim i}) - Y(p_i, p'_{\sim i})]^2 \qquad [8.10]$$

(Chan et al. 2000), where E denotes the expected value with respect to the parameter input distributions. Similarly in $Y(p_i, p_{\sim i}) - Y(p'_i, p_{\sim i})$ the main effects of all parameters other than p_i cancel out while any term including p_i appears twice, so

$$T_i = \frac{1}{2\sigma^2}E[Y(p_i, p_{\sim i}) - Y(p'_i, p_{\sim i})]^2 \qquad [8.11]$$

where p_i, p'_i are two independent random draws from the input distribution on parameter i (Chan et al. 2000).

Using these formulas the complete set of main and total sensitivity indices can be estimating efficiently using "winding stairs" parameter sampling (Jansen et al. 1994). The winding stairs method iteratively generates a matrix of parameter vectors and corresponding Y values with d columns. Let $p_{i,j}$ denote the jth sample of parameter i, drawn at random from the input distribution f_i. With $d = 3$ parameters in the model, the form of the matrix is

$$W = \begin{bmatrix} Y(p_{1,1}, p_{2,1}, p_{3,1}) & Y(p_{1,1}, p_{2,2}, p_{3,1}) & Y(p_{1,1}, p_{2,2}, p_{3,2}) \\ Y(p_{1,2}, p_{2,2}, p_{3,2}) & Y(p_{1,2}, p_{2,3}, p_{3,2}) & Y(p_{1,2}, p_{2,3}, p_{3,3}) \\ Y(p_{1,3}, p_{2,3}, p_{3,3}) & \cdots & \cdots \\ \cdots & \cdots & \cdots \end{bmatrix}. \qquad [8.12]$$

Construction of W starts at the top left with the first sampled value of each parameter. Then proceeding to the right in the top row, the second column "updates" the value of parameter 2, and the third column "updates" the value of the parameter 3, in each case leaving the previous values of the other parameters unchanged. Then dropping down to the next row, the entry in the first column updates the value of parameter 1. This process continues until r rows have been filled in.

The quantities [8.10] and [8.11] can all be estimated from the output matrix W. For example, to estimate M_1 we need to average squared differences of Y values

Parameter group	M_i	T_i
Parr per spawner	0.38	0.39
Parr to smolt survival	0.01	0.09
Smolt to adult survival	0.52	0.54
Proportional spawning at age 4	0.01	0.001

Table 8.5 Main and total Sobol' sensitivity indices for stage-specific parameter groups in a stochastic age-structured model for salmon populations, using parameter input distributions corresponding to the uncertainty in parameter estimates (Ellner and Fieberg 2003). The response variable Y is the predicted long-term population growth rate, estimated from 5000 model runs at each parameter set. The exact values of the indices must satisfy $M_i < T_i$, but this may not hold for estimated values that are computed from a finite sample of parameters. From Ellner and Fieberg (2003).

where only the value of p_1 is the same for both runs. The values in the first and last columns of W fit that description, so the estimate of M_1 is

$$\hat{M}_1 = 1 - \frac{1}{2r\hat{\sigma}^2} \sum_{j=1}^{r} (W_{j,1} - W_{j,d})^2.$$

For T_1 we need model runs where p_1 has changed but all other parameters are the same. Pairs like that occur in the construction of W every time one row is finished and the next row is started. The estimate is therefore

$$\hat{T}_1 = \frac{1}{2(r-1)\hat{\sigma}^2} \sum_{j=2}^{r} (W_{j,1} - W_{j-1,d})^2.$$

Similar choices of entries in W can be used to estimate the sensitivity indices for all parameters. The Y value pairs needed for estimating T_i are adjacent in the sequence used to compute W (only one parameter changes), and those for computing M_i are $d - 1$ steps apart (all parameters but one have changed).

Table 8.5 shows the main and total sensitivity indices for stage-specific parameter groups in a stage-structured model for salmon populations (Ellner and Fieberg 2003). The model response variable Y is the long-term population growth rate. The purpose of the sensitivity analysis was to identify which parameters contributed most to uncertainty in the predicted growth rate, so the input distributions reflected the ranges of uncertainty in parameter estimates. Because the parameters for each particular life stage were estimated from the same data, the parameter uncertainty distributions have within-stage correlations whereas the methods for computing sensitivity indices assume independence. However the parameters p_i can be vectors rather than single numbers, so we computed the sensitivity indices to the four independent groups of parameters characterizing different life stages. The results were striking: two of the four life stages account for virtually all of the uncertainty.

Winding stairs sampling makes the Sobol' sensitivity indices a general and easy to implement method for global sensitivity analysis. The principal limitation is the requirement that parameters be generated independently. For dimension reduction—finding parameters that have limited effect on the model—that requirement can be met by using independent random draws at each step of constructing the W matrix. However, for quantifying the impact of parameter uncertainty, correlated distributions of parameter uncertainty are inevitable unless each parameter comes from a separate data set.

Exercise 8.7. In Table 8.5 three of the life stages had $M_i \approx T_i$. What does that tell us about the impact of the corresponding parameter groups on population growth? For the fourth life stage (parr to smolt) T_i is larger than M_i by nearly an order of magnitude—what does that tell us?

Exercise 8.8. Consider again the "model output" $Y(p_1, p_2, p_3) = e^{p_1+2p_2+3p_3}$. Write a script that uses winding stairs to compute the main and total sensitivity indices for each parameter, with the input distributions f_i for all parameters being Gaussian with mean 0 and standard deviation 1. Write your script with so that a user can set the value of r (the number of rows in the W matrix) by editing only the first line. Report and interpret your results for $r = 1000$.

Exercise 8.9. Derive [8.10] and [8.11] using the analysis of variance decomposition [8.5] and the property that

$$\int Y_{i,\ldots,j,\ldots,k}(p_i, \ldots, p_j, \ldots, p_k) f_j(p_j) dp_j = 0$$

for any fixed values of the variables other than p_j (in words: the expectation of each $Y_{i,\ldots,j,\ldots,k}$ conditional on the values of any strict subset of its arguments is 0).

8.6 Simplifying Computational Models

A second way of trying to understand a complicated model is to try to find a simpler, approximating model that captures its essential features. This is sometimes done by trial and error, or on the basis of sensitivity analysis. However there are also some approaches that derive approximations directly from the model's dynamic equations.

8.6.1 Separation of Time Scales

Biological systems often include processes that happen on very different time scales. When this occurs, a model for the system can sometimes be simplified by assuming that the fast processes happen infinitely fast relative to the slow processes. We have seen examples of this in Chapters 1 (enzyme kinetics), 3

(Morris-Lecar model), and 4 (gene regulation). Recall the equations for mRNA m and repressor protein p in the repressilator model:

$$\dot{m}_i = -m_i + \frac{\alpha}{1+p_j^n} + \alpha_0$$
$$\dot{p}_i = -\beta(p_i - m_i). \qquad [8.13]$$

If β is very large, then p_i changes more rapidly than m_i, and we can imagine that the second equation in [8.13] goes all the way to its asymptotic state while m_i remains at its current value—implying that $p_i \approx m_i$ always holds. The model can therefore be reduced to the equations for mRNA:

$$\dot{m}_i = -m_i + \frac{\alpha}{1+m_j^n} + \alpha_0. \qquad [8.14]$$

When β is very large, the repressilator system is an example of a *fast-slow* or *singularly perturbed* vector field. Abstractly, we regard the system as having the form

$$\dot{\mathbf{x}} = \mathbf{f}(\mathbf{x}, \mathbf{y})$$
$$\epsilon\dot{\mathbf{y}} = \mathbf{g}(\mathbf{x}, \mathbf{y}), \qquad [8.15]$$

where we have split the state vector into the "slow variables" $\mathbf{x}$ and the "fast variables" $\mathbf{y}$. If ϵ is small, the fast variables may approach an equilibrium with $\mathbf{g}(\mathbf{x}, \mathbf{y}) = 0$ before $\mathbf{x}$ changes appreciably. The *singular limit* of the system [8.15] is then a system of *differential algebraic equations*

$$\dot{\mathbf{x}} = \mathbf{f}(\mathbf{x}, \mathbf{y})$$
$$0 = \mathbf{g}(\mathbf{x}, \mathbf{y}). \qquad [8.16]$$

If we can solve the equations $\mathbf{g}(\mathbf{x}, \mathbf{y}) = 0$ in the form $\mathbf{y} = \mathbf{h}(\mathbf{x})$, then we can reduce the system to $\dot{\mathbf{x}} = \mathbf{f}(\mathbf{x}, \mathbf{h}(\mathbf{x}))$. This is what we have done in each of the examples cited above. The following *quasi-steady-state hypotheses* must hold in order for this reduction to be possible in this explicit fashion:

1. There must be a clear separation of time scales for the variation of $\mathbf{x}$ and $\mathbf{y}$.
2. There must be a function $\mathbf{y} = \mathbf{h}(\mathbf{x})$ that we can compute so that equations $\mathbf{g}(\mathbf{x}, \mathbf{h}(\mathbf{x})) = 0$ for all $\mathbf{x}$ in the region of interest.
3. For each fixed x in the region of interest, the system $\dot{\mathbf{y}} = \mathbf{g}(\mathbf{x}, \mathbf{y})$ must have a unique stable equilibrium at $\mathbf{y} = \mathbf{h}(\mathbf{x})$.

When the fast equations do not have a unique stable equilibrium, a fast-slow system can have solutions that alternate between two different kinds of behavior:

- Periods when it behaves like the reduced slow system

  $$\dot{\mathbf{x}} = \mathbf{f}(\mathbf{x}, \mathbf{h}_i(\mathbf{x})), \qquad \mathbf{y} = \mathbf{h}_i(\mathbf{x}))$$

  where $\mathbf{h}_i(\mathbf{x})$ is one of the stable solutions of $\mathbf{g}(\mathbf{x}, \mathbf{y}) = 0$.

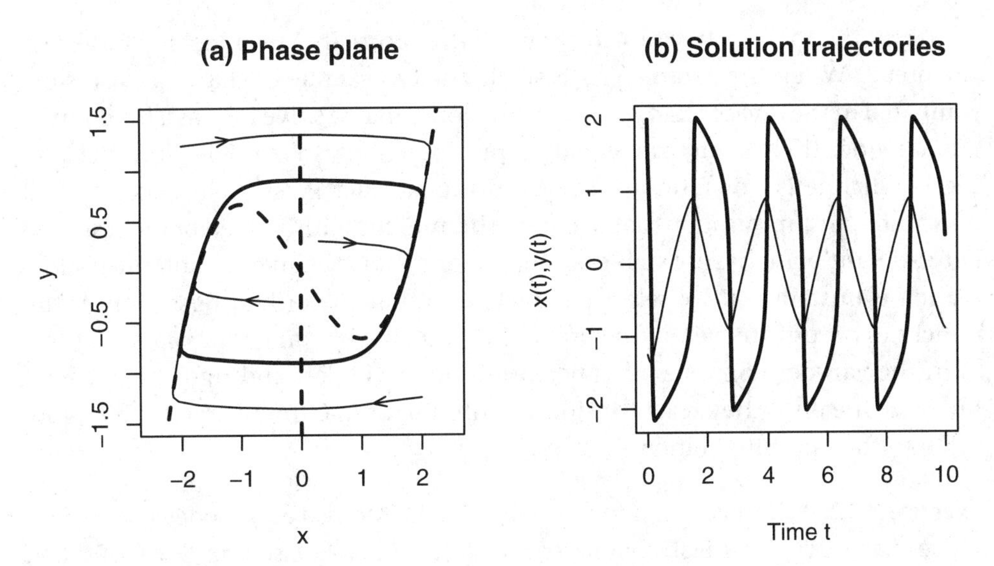

Figure 8.4 Relaxation oscillations in the van Der Pol model [8.17] with $C = 20$. In the phase portrait (a) the nullclines are drawn as dashed lines, and four solution trajectories as solid lines. All trajectories converge very quickly onto the limit cycle, drawn in bold. Panel (b) shows the state variables $x(t)$ (bold) and $y(t)$.

- Periods when the fast variables rapidly jump from one stable solution to another.

A simple example is given by the van der Pol system. Proposed as a model for heartbeat, the model can be rescaled into the form

$$\dot{x} = C(y - f(x)) \qquad \text{where } f(x) = -x + x^3/3$$
$$\dot{y} = -x. \tag{8.17}$$

The y-nullcline is the line $x = 0$. The x-nullcline is the curve $y = f(x) = -x + x^3/3$. The only equilibrium is their intersection at $(0, 0)$, which is locally unstable for any $C > 0$. When C is large this is a fast-slow system. Starting from any point above the x nullcline [i.e., $y > f(x)$)], $x(t)$ increases rapidly until the right branch of the nullcline $y = f(x)$ is reached (see Figure 8.4). The solution then remains on the nullcline, moving according to the slow equation. Eventually the solution "falls off" the bottom of the nullcline. At that point $y < f(x)$ so $x(t)$ decreases rapidly until the left branch of the nullcline is reached—and so on, leading to a strongly stable limit cycle.

The periodic orbits computed in Figure 8.4 are called *relaxation oscillations*. As $C \to \infty$ these orbits come closer and closer to closed curves formed from segments parallel to the x-axis that join the local minimum and maximum of the x-nullcline to the opposite branch of the nullcline, and the parts of the nullcline that connect these segments.

Exactly the same behavior can occur in the Morris-Lecar model as studied in Chapter 5. When the parameter ϕ is small, the typical rate of change of w is slow compared to the rate of change of v. So we would like to solve for v as a function of w and reduce the system to an equation just for w. But this can't be done because the v-nullcline is not monotonic. As in the van der Pol system it has a local maximum and minimum that separate the nullcline into three branches. The left and right branches are stable equilibria for the fast $\dot{v}$ equation and the middle branch is unstable. If we select parameters so that the only intersection of the nullclines occurs on the middle branch, we obtain the same type of oscillations as in the van der Pol system. Trajectories follow the left and right branches of the nullcline until they reach the local minimum or maximum, and then rapidly jump to the opposite branch.

Exercise 8.10. Plot nullclines and trajectories of the Morris-Lecar model in the (v, w) phase plane using the first parameter set of Table 5.1 but setting $\phi = 0.001$ and $i = 110$. In addition, plot solutions $v(t)$ and $w(t)$ as functions of time, observing the rapid jumps in v that alternate with periods of slower variation. What happens to w during the jumps in v? Note: be careful with the numerical integration and choice of time steps in doing this exercise, because the system is stiff as discussed in Section 5.7.

Exercise 8.11. Continuing the previous exercise, explore how the shape of the periodic orbits changes as $\phi \to 0$.

8.6.2 Simplifying Spatial Models

The formation of spatial patterns as a result of decisions and movements by individual organisms has recently been a major application area for agent-based models. Dieckmann et al. (2000), Parrish et al. (2002), and Chowdhury et al. (2004) summarize many such applications including bacterial colonies, ant trails, fish schools, and the walking paths followed by humans in crowds. Agent-based models are a natural starting point for phenomena that seem to arise from decisions by discrete agents responding to the behaviors of other discrete agents. Helbing et al. (2000) model escape panics, such as attempts to rapidly exit a room where a fire has started. Everyone's desire to exit rapidly leads to congestion and injuries at pileups that reduce the average exit rate. The model treats agents as circles of fixed diameter moving on a surface, that try to achieve a particular direction and velocity of movement but also try to avoid getting too close to other agents or the walls. Additional rules govern what happens when collisions occur. Simulations of the model replicate the counterproductive effects of panic, suggesting that the model could help design structures for rapid escape during emergencies. Paradoxically, a well-placed obstacle might actually increase the rate of escape during an emergency (Figure 8.5). The key features of the system—agents heading for an exit but bumping into each other and obstacles—would be

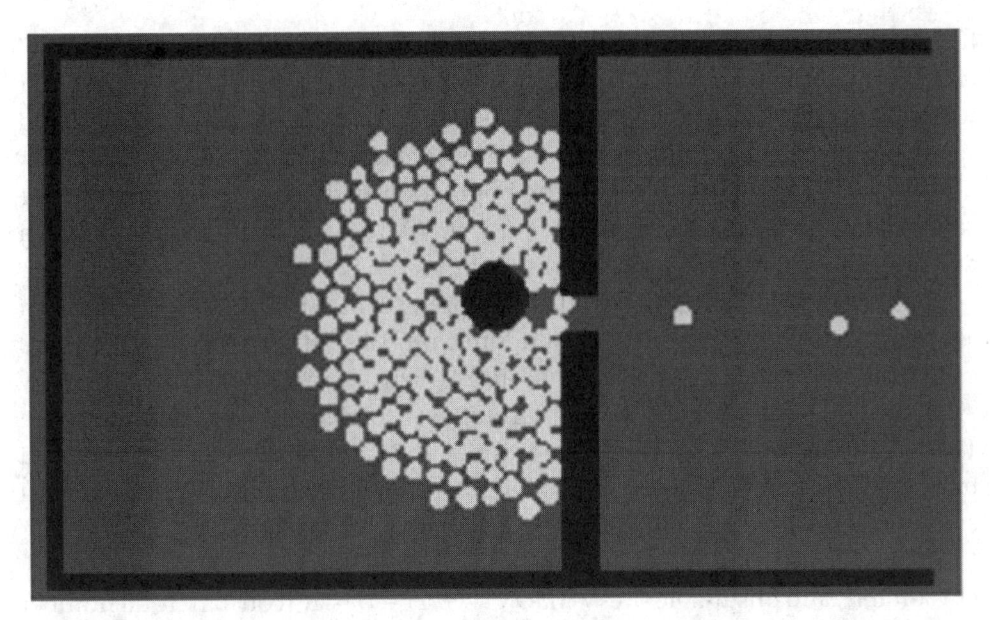

Figure 8.5 A simulation of the Helbing et al. (2000) model for exit panic with a column placed asymmetrically in front of the exit (from Bonabeau 2002). The presence of the column causes the agents to organize their movements in a way that decreases the number of injuries and increases the rate of exit.

more difficult to represent in a partial differential equation model for the "concentration" of pedestrians.

In this section we will briefly describe some methods that can sometimes lead to useful simplifications of spatial agent-based models: mean field equations, hydrodynamic limits, and moment equations. These methods are most easily introduced for spatial "lattice models" such as Avida. Space is represented by a regular discrete grid of sites—such as the points (x, y) in the plane where x and y are both integers. Each site is in one of a finite set of possible states. These models are an import from theoretical physics, where the agents might be elementary particles sitting at each vertex of a regular two- or three-dimensional lattice, with possible states "spin up" and "spin down."

For a simple biological example (based on Hiebeler 2000), consider a two-dimensional lattice such that each site can be suitable or unsuitable for occupancy (with this distinction being permanent), and each suitable site can be either empty or occupied by a single agent. We posit the following simple rules for what can happen between times t and $t + \Delta t$:

1. Each agent has probability $\mu \times \Delta t$ of dying.
2. Each agent has probability $b \times \Delta t$ of producing an offspring, which is dispersed at random into one of the four sites immediately adjacent to the parent (up, down, left, right). If the site is occupied or unsuitable, the offspring dies. If the cell is suitable and vacant, the offspring takes it and the cell is then occupied.

If all sites are suitable, these assumptions specify the classical "contact process" model for population spread. Hiebeler (2000) introduced unsuitable cells in order to model habitat loss in a fragmented landscape, and to ask how the amount and spatial arrangement of suitable habitat affects species persistence.

For what values of b and μ will the population persist? By rescaling time we can see that the answer will involve only the ratio b/μ, but beyond that it is a difficult question and the exact answer is not known.

One way of simplifying the model is to pretend that the sites around a given agent are drawn independently at random from the entire "landscape." This is called the mean field approximation, and is derived in general by asking how agents would behave if their local environments were statistically homogenized. Then assuming "many" agents, coin-tossing randomness is replaced by expected transition rates (as in Chapter 3). The result is typically a system of differential equations for the frequencies of agents in different possible states.

For the Hiebeler model let $O(t)$, $E(t)$, and U be the numbers of occupied, empty but suitable, and unsuitable sites with $O(t) + E(t) + U \equiv N$ (constant total number of sites). Since $O(t) + E(t) \equiv N - U$ we only need one state variable, $O(t)$. New occupied sites are created by births: offspring are created at rate $bO(t)$ per unit time, and the fraction that live to occupy a site is given by the fraction of sites that are suitable but unoccupied, $E(t)/N = (N - U - O(t))/N$. This calculation is where spatial structure is ignored: sites near a parent are more likely to be occupied than a "typical" site, so the fraction of offspring that survive is actually lower than $E(t)/N$. Occupied sites are vacated by deaths, which occur at total rate $\mu O(t)$ per unit time. Assuming enough sites that the coin-tossing randomness averages out, we then have

$$\dot{O} = bO(N - U - O)/N - \mu O$$

or equivalently

$$\dot{p} = bp(s - p) - \mu p \qquad [8.18]$$

where $p(t) = O(t)/N$ is the fraction of occupied sites and $s = (N - U)/N$ is the fraction of suitable sites. The condition for population persistence is easy to find by linear stability analysis of the equilibrium $p = 0$.

The mean field approximation does not totally ignore space. Agents are still viewed as interacting with their local neighbors, whose states are the outcome of random processes. These considerations can affect the expected rate at which sites change state. For example, suppose we assume that an agent's probability of death per unit time is $\mu(1 + C^2)$ where C is the number of occupied neighboring sites. If a given site has z neighboring sites, then in the mean field approximation C is a Binomial(z, p) random variable. The per agent death rate is then proportional to the expected value of $1 + C^2$, $E[1 + C^2] = 1 + (zp)^2 + zp(1 - p)$. The term $zp(1 - p)$ reflects the stochastic variation among the local neighborhoods of different agents. If we took the approximation one step further and pretended that each local neighborhood contained exactly $E[C] = zp$ occupied sites, this term would be eliminated. But because the mortality rate is a concave-up function of the number of occupied neighbor sites, the effect of local stochasticity is to increase the average mortality rate.

Sometimes the mean field approximation works—in the sense of preserving the model's qualitative properties—and sometimes it doesn't. We will see below that the Hiebeler model is a "doesn't," despite its simplicity. In contrast, Moorcroft et al. (2001) found that the mean field approximation for a very complicated and computationally intensive agent-based forest simulation model was very accurate, in the sense that it replicated the average trajectory of the model over many simulations. The mean field approximation in that case was a system of partial differential equations, describing the frequency distribution of trees classified by species, the amount of biomass in several compartments, and the time since a local disturbance of their site. The mean field equations also had to be solved numerically, but they could be scaled up to much larger land areas than could be simulated tree by tree. This allowed Hurtt et al. (2002) to use the mean field model for studying the carbon budget of the entire continental United States in response to past and projected future land use patterns.

For some types of models on a homogeneous lattice, Durrett and Levin (1994) conjectured that the model's behavior can often be predicted from the behavior of the mean-field approximate model:

- When the mean field model has a single fixed point with all state densities positive, the lattice model converges to a steady state with all state densities positive.
- When the mean field model has several fixed points, its long-term behavior depends on initial conditions. In contrast, the lattice model predictably converges to a steady state corresponding to one of the fixed points.
- When the mean field model has a limit cycle, the lattice model converges to a steady state with the same states present, and spatial average densities on intermediate spatial scales exhibit persistent cyclic oscillations.

Durrett (1999) reviews evidence in support of this conjecture, and some cases where mean field behavior is misleading. However, it is generally true (and proved for some models) that as the range of interagent interactions is increased, the model converges to the mean-field approximation. In the Hiebeler model this occurs as the range of offspring dispersal is increased so that offspring can land on any site within distance r of the parent, $r \gg 1$. When $r = \infty$ the site where an offspring lands really is a random draw from the entire landscape, so the model behaves just like the mean field approximation.

Exercise 8.12. Derive the complete mean field approximation to the Hiebeler model under the modified assumptions that in a short time interval of length Δt

(a) each agent has probability $\mu(1+\alpha_1 C+\alpha_2 C^2)\Delta t$ of dying, and

(b) each agent has probability $b(1+\beta_1 C+\beta_2 C^2)\Delta t$ of producing an offspring,

where C is the agent's number of occupied neighboring sites. Depending on their signs the α and β coefficients could model local competition or cooperation.

8.6.3 Improving the Mean Field Approximation

Mean-field equations obliterate the correlation between agents and their neighborhoods: a kindergarten class with one case of flu will soon have many, which limits the rate at which each infective child generates new cases. To retain these local correlations we need approximations that do not totally homogenize space.

The mean field approximation can be viewed as the limit of "mixing" the system more and more rapidly. Suppose that in each time interval of length $\Delta t \ll 1$ each pair of neighboring sites swap their contents with probability $m\Delta t$. Then if m is large enough the system approaches the mean-field property that each agent's neighbors are a random draw from the entire set of agents. Some local spatial structure can be preserved by simultaneously shrinking the size of lattice cells. As in the convergence of random walk to diffusion (Chapter 8), mixing and shrinking rates are linked so that agents' mean square displacement as a function of time approaches a limit. In physics this is called a "hydrodynamic limit." For some types of model the hydrodynamic limit is valid, meaning that an infinite-lattice model with rapid enough mixing has the same qualitative behavior. In many cases there is a simple recipe for the hydrodynamic limit: a mean field model of the form $du/dt = f(u)$ is replaced by a reaction-diffusion model of the form

$$\frac{\partial u}{\partial t} = f(u) + \frac{\partial^2 u}{\partial x^2} + \frac{\partial^2 u}{\partial y^2},$$

and similarly for models with more state variables (Durrett 1999).

Another way of improving the mean field is to expand the scale at which independence is assumed. Where mean field theory treats all sites as independent, *pair approximation* includes correlations between neighboring sites. Let ρ_i denote the fraction of sites in state i, and ρ_{ij} the fraction of neighboring site pairs in state (i, j). Assumptions of between-pair independence are then used to derive equations for the pair frequencies ρ_{ij}. In the Appendix to this chapter we show how approximate pair-frequency equations are derived for a simple model.

The pair approximation for the Hiebeler model is a system of three differential equations. There are nine different pair types, but only three state variables are needed because ρ_2 and ρ_{22} are constant by assumption, we have the symmetries $\rho_{ij} = \rho_{ji}$, and the pair frequencies sum to 1. Analysis of the pair approximation equations leads to the prediction that the fraction of suitable sites that are occupied depends on the conditional probability that the neighbor of a suitable site is unsuitable, not on the total fraction of suitable sites. The mean field approximation [8.18] makes exactly the opposite prediction. Simulations show that the pair approximation prediction is very close to the truth (Hiebeler 2000).

More accurate approximations can be obtained by using larger building blocks, but this is rarely done. A more important extension has been the use of similar ideas when interactions are defined by an interaction network rather than spatial proximity, for example, a network of social contacts allowing disease transmission (e.g., Rand 1999; van Baalen 2000; Eames and Keeling 2003).

Pair approximation is one example of *moment closure* methods, in which approximate equations are written for a limited number of statistics ("moments") characterizing the state of the system. For pair approximation these are the site-pair frequencies; in other cases the moments are means, variances, and covariances (in space and/or time) of agents of different types or agents in different states. These are more complicated than pair approximation but do not require artificially gridding up space into discrete cells. Bolker et al. (2000) review these methods and some of their applications.

8.7 Conclusions

Agent-based models are likely to remain an important approach in computational biology, in part because the agent-based paradigm has become a part of how we build models. It is increasingly common for classical differential equation models to be constructed by making assumptions at the level of agents, and then explicitly deriving the model as a mean-field or some other approximation to the "exact" agent-based model. This leads to deterministic models whose equations reflect the local stochastic nature of interactions among agents, which in the past typically would have been ignored.

We remind readers that this is the chapter where we have allowed ourselves to stray off the beaten path toward the "bleeding edge." It will probably become obsolete faster than the rest. We also caution readers that the sections on sensitivity analysis and model simplification are more a prescription for the future than a description of current practice. We believe that these methods have a lot of potential and deserve more widespread use, but it is too early to say if they will live up to their promise. In particular, it is not clear whether the methods to improve on mean-field approximations in spatial models will be useful for other kinds of agent-based simulations, or even for more realistic spatial models. The central message of those sections is that structured approaches, grounded in statistical and mathematical theory, can be enormously more efficient than trial and error for figuring out the behavior of a complicated computational model.

8.8 Appendix: Derivation of Pair Approximation

Consider the continuous-time Hiebeler model with (for simplicity) all sites suitable, and let states 0 and 1 indicate empty and occupied sites. This model is known as the "contact process." Each pair of neighboring cells is then in one of the four states $(0,0), (1,1), (1,0), (0,1)$. Although there are four types of site pairs, we only need two state variables because of the constraints

$$\sum_{i,j=0}^{1} \rho_{ij} = 1, \quad \rho_{10} = \rho_{01}. \tag{8.19}$$

It is convenient to use ρ_1 and ρ_{11} as the state variables. Because $\rho_{01} + \rho_{11} = \rho_1$ we have $\rho_{01} = \rho_{10} = \rho_1 - \rho_{11}$. Occupied sites are lost by deaths at total rate $\mu\rho_1$. They are gained by offspring birth and survival at total rate $b\rho_1 q_{0/1}$ where $q_{0/1}$ is the conditional probability that a randomly chosen neighbor to an occupied site is vacant. So ignoring the coin-tossing randomness, we have

$$\dot{\rho}_1 = b\rho_1 q_{0/1} - \mu\rho_1.$$

By the conditional probability formula $P(A|B) = P(A \cap B)/P(B)$ we have

$$q_{0/1} = \rho_{01}/\rho_1 = (\rho_1 - \rho_{11})/\rho_1$$

and therefore

$$\dot{\rho}_1 = b(\rho_1 - \rho_{11}) - \mu\rho_1. \tag{8.20}$$

Next we need the equation for ρ_{11}. A (1, 1) site pair is lost whenever one of the pair dies, so the total loss rate is $2\mu\rho_{11}$. (1, 1) pairs are gained when the vacant site in a (0, 1) or (1, 0) pair becomes occupied. Each occupied neighbor of the vacant site sends in offspring at rate $b/4$; one neighbor is occupied with probability 1, the other three with probability $q_{1/01}$, the conditional probability of state 1 in a randomly chosen neighbor of the 0 in a (0, 1) pair, $q_{1/01} = \rho_{101}/\rho_{01}$. The expected total rate of $(0, 1) \rightarrow (1, 1)$ transitions is therefore

$$\rho_{01}\frac{b}{4}(1 + 3\beta q_{1/01}).$$

$(1,0) \rightarrow (1, 1)$ transitions occur at the same rate so we have

$$\dot{\rho}_{11} = \rho_{01}\frac{b}{2}(1 + 3q_{1/01}) - 2\mu\rho_{11}. \tag{8.21}$$

The equation for ρ_{11} involves $q_{1/01} = \rho_{101}/\rho_{01}$, and the trio density ρ_{101} cannot be computed from our two state variables. This problem cannot be solved by adding more state variables, because the pattern continues—dynamic equations for site trio densities involve the densities of site quartets, and so on. At some point we have to somehow truncate this process. Pair approximation stops at site pairs by making the approximation $q_{1/01} = q_{1/0}$. The 0 site in a (0, 1) pair isn't any old vacant site—it's a vacant site with an occupied neighbor—but we ignore this and hope for the best. The total rate of $(0, 1) \to (1, 1)$ transitions is then approximated as

$$\rho_{01}\frac{b}{2}(1 + 3q_{1/0}).$$

The result, with a bit of algebra, is

$$\dot{\rho}_{11} = \frac{b}{2}(\rho_1 - \rho_{11})\left(1 + 3\frac{\rho_1 - \rho_{11}}{1 - \rho_1}\right) - 2\mu\rho_{11}. \qquad [8.22]$$

Equations [8.20] and [8.22] are now a closed system of equations that can be analyzed or solved numerically. For example, linear stability analysis of the equilibrium $(0, 0)$ gives population persistence for $b/\mu > 4/3$. The corresponding prediction of the mean-field model is persistence for $b/\mu > 1$. The actual critical value of b/μ (estimated from simulations) is about 1.65, so pair approximation has reduced by half the error of the mean-field approximation. But numerical accuracy is not really the point—you can do better by simulation. Pair and other moment approximations are valuable when they generate qualitative predictions that can be confirmed in simulations but would have been hard to discover inductively from simulations (e.g., Bolker and Pacala 1999; Keeling et al. 2000; Snyder and Chesson 2004).

8.9 References

Adami, C. 2002. Ab initio modeling of ecosystems with artificial life. Natural Resource Modeling 15: 133–145.

Blower, S. M., and H. Dowlatabadi. 1994. Sensitivity and uncertainty analysis of complex models of disease transmission: An HIV model, as an example. International Statistical Review 2: 229–243.

Blower, S. M., H. B. Gershengorn, and R. M. Grant. 2000. A tale of two futures: HIV and antiretroviral therapy in San Francisco. Science 287: 650–654.

Bolker, B. M., and S. W. Pacala. 1999. Spatial moment equations for plant competition: Understanding spatial strategies and the advantages of short dispersal. American Naturalist 153: 575–602.

Bolker, B. M., S. W. Pacala, and S. A. Levin. 2000. Moment methods for stochastic processes in continuous space and time. Pages 388–411 in Dieckmann et al. (2000).

Bonabeau, E. 2002. Agent-based modeling: Methods and techniques for simulating human systems. Proceedings of the National Academy of Sciences, U.S.A. 99, Suppl. 3: 7280–7287.

Brook, B. W., J. R. Cannon, R. C. Lacy, C. Mirande, and R. Frankham. 1999. Comparison of the population viability analysis packages GAPPS, INMAT, RAMAS, and VORTEX for the whooping crane (*Grus americana*). Animal Conservation 2: 23–31.

Chan, K., A. Saltelli, and S.Tarantola. 2000. Winding Stairs: A sampling tool to compute sensitivity indices. Statistics and Computing 10: 187–196.

Chow, S. S., C. O. Wilke, C. Ofria, R. E. Lenski, and C. Adami. 2004. Adaptive radiation from resource competition in digital organisms. Science 305: 84–86.

Chowdhury, D., K. Nishinari, and A. Schadschneider. 2004. Self-organized patterns and traffic flow in colonies of organisms: from bacteria and social insects to vertebrates. Phase Transitions 77: 601–624.

Cleavelend, S., G. R. Hess, A. P. Dobson, M. K. Laurenson, H. I. McCallum, M. G. Roberts, and R. Woodroffe. 2002. The role of pathogens in biological conservation. Pages 139–150 in Hudson et al. (2002).

DeAngelis, D. L., and L. J. Gross (eds.). 1992. *Individual-Based Models and Approaches in Ecology*. Chapman and Hall, New York.

DeAngelis, D. L., L. J. Gross, M. A. Huston, W. F. Wolff, D. M. Fleming, E. J. Comiskey, and S. M. Sylvester. 1998. Landscape modeling for everglades ecosystem restoration. Ecosystems 1: 64–75.

Dieckmann, U., R. Law, and J.A.J. Metz. 2000. *The Geometry of Ecological Interactions: Simplifying Spatial Complexity*. Cambridge University Press, Cambridge, U.K.

Durrett, R. 1999. Stochastic spatial models. SIAM Review 41: 677–718.

Durrett, R., and S. A. Levin. 1994. The importance of being discrete (and spatial). Theoretical Population Biology 35: 252–283.

Eames, K.T.D. and M. Keeling. 2003. Contact tracing and disease control. Proceedings of the Royal Society of London, Series B 270: 2565–2571.

Ellner, S. P., and John Fieberg. 2003. Using PVA for management despite uncertainty: Effects of habitat, hatcheries, and harvest on salmon. Ecology 84: 1359–1369.

Enquist, B. J., and K. J. Niklas. 2001. Invariant scaling relations across tree-dominated communities. Nature 410: 655–660.

Gould, S. J. 1990. *Wonderful Life: The Burgess Shale and The Nature of History*. W. W. Norton and Company, New York.

Harvell, C. D., K. Kim, J. M. Burkholder, R. R. Colwell, P. R. Epstein, J. Grimes, E. E. Hofmann, E. K. Lipp, A.D.M.E. Osterhaus, R. Overstreet, J. W. Porter, G. W. Smith, and G. R. Vasta. 1999. Emerging marine diseases: Climate links and anthropogenic factors. Science 285: 1505–1510.

Helbing, D., I. J. Farkas, and T. Vicsek. 2000. Simulating dynamical features of escape panic. Nature 407: 487–490.

Hess, G. 1996. Disease in metapopulation models: Implications for conservation. Ecology 77: 1617–1632.

Hiebeler, D. 2000. Populations on fragmented landscapes with spatially structured heterogeneities: Landscape generation and local dispersal. Ecology 81: 1629–1641.

Hudson, P. J., A. Rizzoli, B. T. Grenfell, H. Heesterbeek, and A. P. Dobson (eds.). 2002. *The Ecology of Wildlife Diseases*. Oxford University Press, Oxford, U.K.

Hurtt, G. C., S. W. Pacala, P. R. Moorcroft, J. P. Caspersen, E. Shevliakova, and B. Moore. 2002. Projecting the future of the US Carbon sink. Proceedings of the National Academy of Sciences U.S.A. 99: 1389–1394.

Huston, M., D. deAngelis, and W. Post. 1988. New computer models unify ecological theory. BioScience 38: 682–691.

Jansen, M.J.W., W.A.H. Rossing, and R. A. Daamen. 1994. Monte Carlo estimation of uncertainty contributions from several independent multivariate sources. Pages 334–343 in J. Gasman and G. van Straten (eds.), *Predictability and Nonlinear Modelling in Natural Sciences and Economics*. Kluwer Academic, Dordrecht.

Keeling, M. J., H. B. Wilson, and S. W. Pacala. 2000. Reinterpreting space, time lags, and functional responses in ecological models. Science 290: 1758–1761.

Lacy, R. C., M. Borbat, and J. P. Pollak. 2003. VORTEX: A stochastic simulation of the extinction process, version 9. Chicago Zoological Society, Brookfield, IL. Online at http://www.vortex9.org

Lenski, R. E., C. Ofria, R. T. Pennock, and C. Adami. 2003. The evolutionary origin of complex features. Nature 423: 139–144.

Lindenmayer, D. B., R. C. Lacy, and M. L. Pope. 2000. Testing a simulation model for population viability analysis. Ecological Applications 10: 580–597.

Lindenmayer, D. B., and R. C. Lacy. 2002. Small mammals, habitat patches and PVA models: a field test of model predictive ability. Biological Conservation 103: 247–265.

McKay, M. D. 1992. Latin Hypercube Sampling as a tool in uncertainty analysis of computer models. Pages 557–564 in J. J. Swain, D. Goldsman, R. C. Crain, and J. R. Wilson (eds.), Proceedings of the 1992 Winter Simulation Conference. Association for Computing Machinery, Arlington, VA. Online at portal.acm.org

McKay, M. D., W. J. Conover, and R. J. Beckman. 1979. A comparison of three methods for selecting values of input variables in the analysis of output from a computer code. Technometrics 22: 239–245.

Miller, P. S., and R. C. Lacy. 2003. VORTEX: A stochastic simulation of the extinction process, version 9.21 user's manual. Conservation Breeding Specialist Group (SSC/IUCN), Apple Valley, MN. Online at http://www.vortex9.org/v921manual.pdf

Moorcroft, P. R., G. C. Hurtt, and S. W. Pacala. 2001. A method for scaling vegetation dynamics: The ecosystem demography model (ED). Ecological Monographs 74: 557–586

Ofria, C., and C. O. Wilke. 2004. Avida: A software platform for research in computational evolutionary biology. Artificial Life 10: 191–229.

O'Neill, B. 2003. Digital evolution. PLOS Biology 1: 11–14.

Parrish, J. K., S. V. Viscido, and D. Grünbaum. 2002. Self-organized fish schools: An examination of emergent properties. Biological Bulletin 202: 296–305.

Peck, S. L. 2004. Simulation as experiment: A philosophical reassessment for biological modeling. Trends in Ecology and Evolution 19: 530–534.

Penn, A. M., W. B. Sherwin, G. Gordon, D. Lunney, A. Melzer, and R. C. Lacy. 2000. Demographic forecasting in koala conservation. Conservation Biology 14: 629–638.

Rand, D. A. 1999. Correlation equations for spatial ecologies. Pages 99–143 in J. McGlade (ed.) *Advanced Ecological Theory*. Blackwell Scientific Publishing, London.

Ray, T. S. 1991. An approach to the synthesis of life. Pages 371–408 in C. Langton, C. Taylor, J. D. Farmer, and S. Rasmussen (eds.), *Artificial Life II*. Santa Fe Institute Studies in the Sciences of Complexity vol. X. Addison-Wesley, Redwood City, CA.

Ray, T. S. 1992. Evolution, ecology and optimization of digital organisms. Santa Fe Institute Working Paper 92-08-042.

Rice, J. A., T. J. Miller, K. A. Rose, L. B. Crowder, E. A. Marschall, A. S. Trebitz, and D. L. DeAngelis. 1993. Growth rate variation and larval survival: Inferences from an individual-based size-dependent predation model. Canadian Journal of Fisheries and Aquatic Sciences 50: 133–142.

Smith, D. J., S. Forrest, D. H. Ackley, and A. S. Perelson. 1999. Variable efficacy of repeated annul unfluenza vaccination. Proceedings of the National Academy of Sciences USA 96: 14001–14006.

Snyder, R. E., and P. Chesson. 2004. How the spatial scales of dispersal, competition, and environmental heterogeneity interact to affect coexistence. American Naturalist 164: 633–650.

Sobol', I. M. 1993. Sensitivity analysis for nonlinear mathematical models. Mathematical Modeling and Computational Experiment 1: 407—414 [English translation from Matematicheskoe Modelirovanie 2: 112–118 (1990)].

van Balen, M. 2000. Pair approximations for different spatial geometries. Pages 359–387 in U. Dieckman, R. Law, and J.A.J. Metz (eds.), *The Geometry of Ecological Interations: Simplifying Spatial Complexity*. Cambridge University Press, Cambridge, U.K.

Waide, R. B., M. R. Willig, C. F. Steiner, G. Mittelbach, L. Gough, S. I. Dodson, G. P. Juday, and R. Parmenter. 1999. The relationship between productivity and species richness. Annual Review of Ecology and Systematics 30: 257–300.

Ward, J. R., and K. D. Lafferty. 2004. The elusive baseline of marine disease: Are diseases in ocean ecosystems increasing? PLOS Biology 2: 542–547.

Wilke, C. O., J. L. Wang, C. Ofria, R. E. Lenski, and Christoph Adami. 2001. Evolution of digital organisms at high mutation rates leads to survival of the flattest. Nature 412: 331–333.

Wilke, C. O., and C. Adami. 2002. The biology of digital organisms. Trends in Ecology and Evolution 17: 528–532.

Yedid, G., and G. Bell. 2001. Evolution in an electronic microcosm. American Naturalist 157: 465–487.

9 Building Dynamic Models

Modeling is often said to be more of an art than a science, but in some ways it is even more like a professional trade such as carpentry. Each modeler will approach a problem somewhat differently, influenced by their training, experience, and what happens to be fashionable at the moment. But given a set of specifications—what is it for, what data are available, and so on—two experienced modelers are likely to use similar methods and to produce functionally similar final products.

The goal of this chapter is to outline the modeling process and introduce some tools of the trade. Such generalities can get to be dry, but we hope that by now you are tempted to do some modeling on your own—or perhaps you are required to do some—so you will tolerate some words of advice.

Much of this chapter is concerned with connecting models and data—a subject called *statistics*. Nonetheless, contacts have been limited between the statistical research community and the scientific communities where dynamic models are used, such as economics and engineering—so limited that many other fields have independently developed statistical methods specific to their needs. Biologists seemingly have no such need—the specialty areas of biostatistics and statistical genetics are recognized in mainstream statistics—but there is very little on dynamic models in the mainstream statistics literature or curriculum. An online search in June 2005 using search engines at journal home pages, JSTOR (http://www.jstor.org), and the Current Index of Statistics (http://www.statindex.org) found two papers concerning a differential equation model in the last ten years of the *Journal of the American Statistical Association*, four in the last ten years of *Biometrics*, and five in the last ten years of *Biometrika*, which is under $\frac{1}{2}$% of the total for those journals over the same time period. Our bookshelves hold any number of statistics textbooks, elementary and advanced, with one or (usually) fewer applications to a differential equation model. The reverse is also true—one can get a Ph.D. in mathematical or computational biology without any formal training in statistics, so very few dynamic modelers are aware of how modern computer-intensive statistical methods can be applied to dynamic models.

This chapter reflects our belief that the next generation of dynamic modelers should not perpetuate the historical trend of treating statistical and dynamic modeling as two separate fields of knowledge, one of which can be left to somebody else. Connecting models with data is almost always the eventual goal. Taking the time to learn statistical theory and methods will make you a better modeler, and a more effective collaborator with experimental biologists. Reading this chapter is only a beginning.

9.1 Setting the Objective

Figure 9.1 outlines the steps involved in developing, evaluating, and refining a dynamic model. We will now proceed through them one by one.

The first, essential, and most frequently overlooked step in modeling is to decide exactly what the model is for. We cannot ask models to be literally true, but we can insist that they be useful, and usefulness is measured against your objectives and the value of those objectives.

One important aspect of setting objectives is to decide where they fall on the continuum between theoretical and practical modeling (Chapter 1). That is, will you use the model to help you *understand* the system and interpret observations of its behavior, or to *predict* the system, running either on its own or with outside interventions? Another important decision is how much numerical accuracy you need. Accurate prediction is often the primary goal in practical applications. But if theoretical understanding is the major goal, it may be good enough if the model gets the sign right or in some other way gives a reasonable *qualitative* match (treatment A really had a larger effect than treatment B; the system really does oscillate rather than settling down to an equilibrium; etc.).

The next step is to assess the feasibility of your goals. The most common constraints are time and data. Some pessimism about time requirements is usually a good idea, especially for beginners—such as students doing a term project. It is usually a good idea to start with a small project that can later be expanded to a more complete model, or a simple model to which more detail can be added later.

In contrast, assessment of whether the available data will meet your needs should be optimistic. Beginners frequently decide that a project cannot be done because some "crucial" piece of information is missing. But models often have several parameters or assumptions that have little or no impact on relevant aspects of model behavior. The only way to find out if the data you are missing are actually needed is to build the model, and then do a sensitivity analysis (Chapter 8) to find out which parts really matter. If you seem to have most of the data that you need, the odds are good that you or an experienced advisor can find some way of working around the gaps.

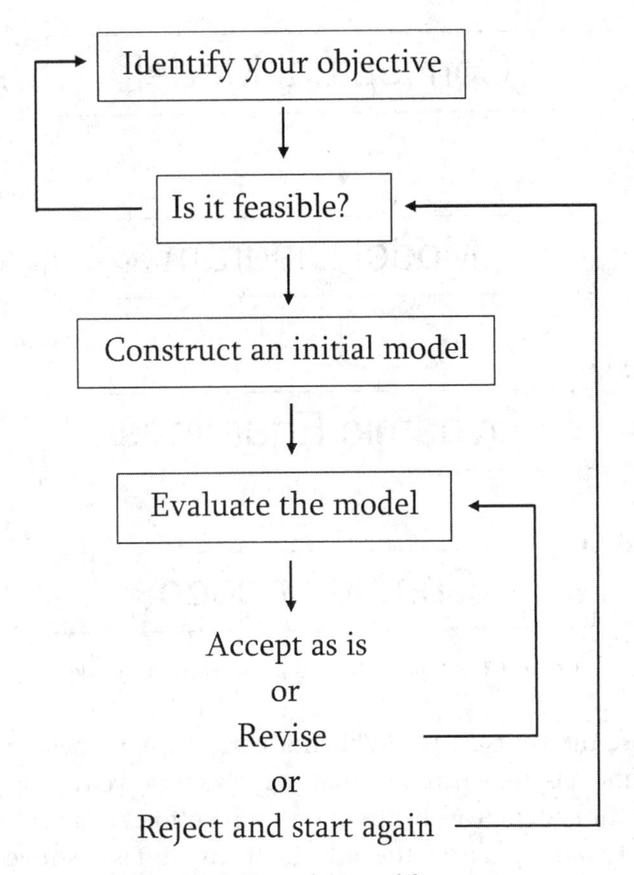

Figure 9.1 Outline of the modeling process.

9.2 Building an Initial Model

Figure 9.2 summarizes the steps in building a dynamic model. As our main example we will use continuous-time compartment models, because they are widely used and allow us to present the main ideas and methods with a minimum of terminology and notation.

Recall from Chapter 1 that the state variables of a compartment model are the amounts of a single kind of "stuff" in a number of locations or categories. The dynamic equations are

$$dx_i/dt = \sum_{\substack{j=0 \\ j \neq i}}^{n} \rho_{ij}(t) - \sum_{\substack{j=0 \\ j \neq i}}^{n} \rho_{ji}(t) \qquad [9.1]$$

where $x_i(t)$ is the amount in compartment i at time t, and $\rho_{ij}(t)$ is the flow rate from compartment j to compartment i at time t, with compartment 0 being the "outside"—the part of the world beyond the limits of the model. Note that there

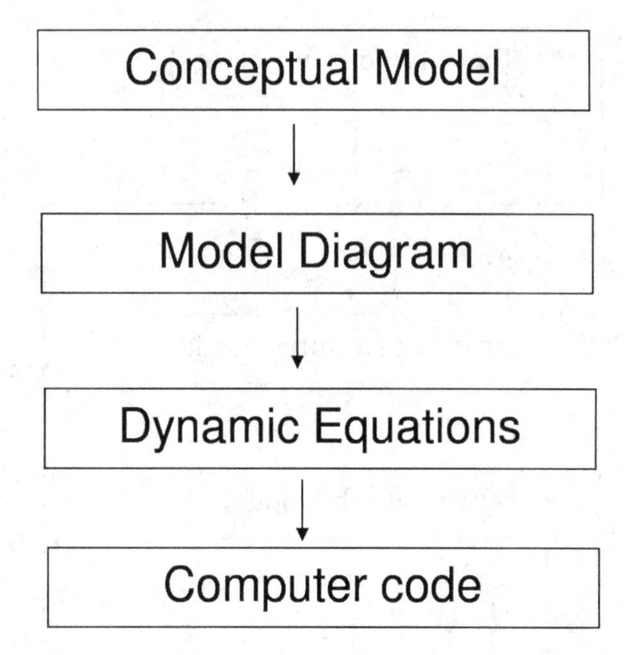

Figure 9.2 Outline of the steps in developing a model.

is no x_0 because the "outside world" is not part of the model. By convention, compartment models are written so that mass is conserved. If any new stuff is created within the system (e.g., births of new susceptibles in an epidemic model), it is represented as an input from the outside. If any stuff is destroyed (e.g., deaths of infected individuals), this is represented as a loss to the outside.

By using a deterministic model with continuous state variables, we are implicitly assuming that *many* individual units of the stuff are flowing through the system. Equation [9.1] cannot be applied to (say) five atoms of a trace metal moving through the body, since for each compartment we could only have $x_i = 1, 2, 3, 4$, or 5. Another assumption is that each compartment is *well mixed*, meaning that all individual units within the same compartment are identical, regardless of their past history or how long they have been in the compartment. Without this assumption, the x_i by themselves would not completely describe the state of the system. In the terminology of the last chapter, a compartment model is based on an agent-based model where each agent has a finite number of possible states. Equation [9.1] is then the mean field equation for the expected changes in the numbers of agents in each possible state.

9.2.1 Conceptual Model and Diagram

A model begins with your ideas about which variables and processes in the system are the most important. These may come from hard data and experimental evidence, or they may be hypotheses that are being entertained for the moment, in order to determine their consequences.

A useful first step in turning these concepts into a dynamic model is to represent the conceptual model as a diagram showing the state variables and processes. Compartment models can be depicted in a *compartment diagram* consisting of a labeled or numbered box for each compartment, and an arrow for each flow (each nonzero ρ_{ij}). As you draw the boxes and arrows, you are formalizing the conceptual model by choosing which components and processes are included, and which are outside the model—either ignored, or included in the external environment in which your model operates.

A good strategy for turning verbal concepts into a diagram is to start with the phenomena or data that figure explicitly in your objectives, and then work out from there until you are willing to declare everything else "outside."

- Quantities "inside" that will change over time as your model is run are your state variables.
- Quantities "outside" that change over time are called exogenous variables or forcing functions. They are taken as given, and only modeled descriptively (e.g., how temperature changes over a day or rainfall varies over the year; the rate at which bone marrow produces new T-cells that enter the blood stream).
- Quantities that do not change over time are called *parameters*.

Diagramming a model forces you to decide which processes require mechanistic description in order to accomplish your objectives. If it is enough for your purposes to know how a variable changes over time without knowing why, you can put it outside the model as an exogenous variable. More and more variables often move "outside" as a model is developed.

Another issue is choosing the level of detail. In a compartment model this is determined by the number of compartments, because units within a compartment are treated as identical, even though they usually are not. This is called *aggregation*, and causes *aggregation error*—treating things that really are different as if they were the same. For example,

- Species are combined into broad categories: birds, grasses, bacteria, etc.
- Physical location is often ignored: one bone versus another in the skeleton, location of a virus particle in the bloodstream, location of infected individuals.
- Compartments are often used to subdivide a continuum. Models of reproducing cell populations sometimes assume there are only a few discrete types of cells (e.g., distinguished by their stage of the cell cycle). Epidemic models often classify infected individuals as exposed (not contagious) versus infected (contagious), whereas there is actually a gradual transition. Space is often modeled as a grid of discrete cells, with variables allowed to differ between cells but not within them

(the reverse is also done, i.e., a tissue composed of discrete cells may be approximated as a uniform continuum for the sake of simplicity in theoretical models).

Useful models may have widely differing levels of aggregation, depending on the questions being asked, for example:

- *Carbon in humans.* A model of glucose metabolism in the human body (Cramp and Carson 1979) had nearly fifty different compartments, including eleven for different substances in the liver. Their "simplified" model recognized six different compartments for glucose in the body: (1) intestinal tract; (2) hepatic portal; (3) and (4) two forms in the liver; (5) blood; (6) "peripheral tissue." To model carbon flow in the Aleutian Island ecosystem, Hett and O'Neill (1974) used nine highly aggregated compartments, such as (1) atmosphere; (2) land plants; (3) man; (4) marine animals and zooplankton; and so on. The object was to see which pathways of carbon flow were most critical to the indigenous human population. This model ignores location, aggregates species, and instead of six or fifty compartments for the glucose in one human there is one compartment for all forms of carbon in all Aleut humans.
- *Carbon in soils.* The initial version of the CENTURY model for soil organic matter (SOM) used eight compartments as shown in Figure 9.3, but represented soil by a single layer (Parton et al. 1987, 1988). The current version (NREL 2001) allows for a vertically layered soil structure (so each SOM compartment is replicated multiple times), simulates C, N, P, and S dynamics, includes models for several different vegetation types (grassland/crop, forest or savanna), and can simulate agricultural management actions such as crop rotation, fertilization, and grazing. Currently, most global ecosystem models use soil carbon and nutrient modules that closely follow the basic CENTURY model (Bolker et al. 1998). In contrast, a global climate model developed by the U.K. government's Meteorological Office (Cox et al. 2000) uses just one compartment for the total amount of organic matter in a unit of area under a given type of vegetation (five vegetation types are recognized: broadleaf and coniferous trees, shrubs, C3 and C4 grasses). The overall model also includes oceanic and atmospheric components, and operates at the global scale by dividing the earth surface into a large number of grid cells. The model for vegetation/soil dynamics within each cell was kept simple so that all model components could be simulated simultaneously to represent dynamic climate-vegetation feedbacks (Cox 2001).

Overaggregation leads to errors if the units within a compartment are heterogenous in their behavior. Then no valid set of rate equations can be written, because a count of the total number in each compartment is not sufficient information for predicting what happens next. For other types of models, the equiv-

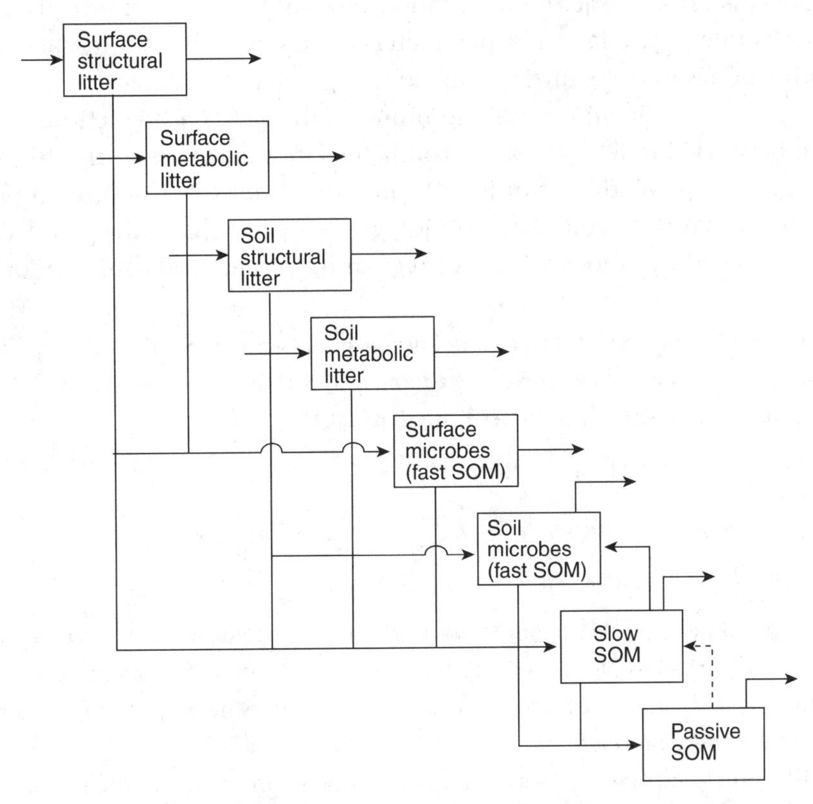

Figure 9.3 Diagram of the compartments for soil organic matter in the CENTURY model (from Bolker et al. 1998). The dotted arrow (from the passive to the slow pool of decomposing organic matter in the soil) represents a very small flow.

alent of aggregation is to ignore variables whose effect is assumed to be relatively unimportant—for example, quantities that really vary over time or space are assumed to be constant. In the fishpond model of Chapter 1, the state variables for phytoplankton represent the total abundance of several species—because one species always made up about 90% of the phytoplankton biomass, variation in which species made up the remaining 10% was simply ignored.

You can also get into trouble by including *too much* detail. Biologists often feel that adding more and more biological detail will make a model more accurate, but that is true only up to a point if parameters are estimated from data. More detail requires more parameters, so the number of observations going into each parameter goes down, and eventually all parameter estimates are unreliable. Conceptually,

$$\text{Prediction error} = \text{Model error} + \text{Parameter error}. \qquad [9.2]$$

(In practice there also are numerical errors in computing model solutions, but that is a separate issue.) Model error is error in predictions due to the fact that your model is not an exact representation of reality. Parameter error is error in predictions due to the fact that parameters values estimated from data are not the optimal ones for maximizing the model's prediction accuracy.

The best level of detail for making numerically accurate predictions strikes a balance between model error and parameter error. Consequently, the model with the lowest prediction error is often one that *deliberately* makes simplifying assumptions that contradict known biology. For example, Ludwig and Walters (1985) compared two models used for regulating commercial fishing effort:

1. The traditional simple Ricker model uses a single variable for the fish population, $B(t)$ = total fish biomass in year t. Changes in $B(t)$ are determined by the fishing effort in year t, $E(t)$, and the resulting harvest $H(t)$:

$$\begin{aligned} H(t) &= B(t)(1 - e^{-qE(t)}), \\ S(t) &= B(t) - H(t), \\ B(t+1) &= r(t)S(t)e^{\alpha - \beta S(t)}. \end{aligned} \qquad [9.3]$$

The first line of [9.3] specifies how increasing fishing effort $E(t)$ leads to a larger fraction of the biomass $B(t)$ being harvested. $S(t)$ is then the remaining unharvested "stock," which (in the last line) produces next year's population via survival and reproduction.

2. The "structured" model uses a full matrix population model for the stock. Only adults are harvested—which is true but omitted by the Ricker model—using the same equation for $H(t)$ as the Ricker model.

The manager's assumed goal is to choose $E(t)$ to maximize the long-run rate of economic return from the harvest. The available data in year t are the fishing effort and harvest in prior years, from which the parameters of each model need to be estimated. The fitted model is then used to set the fishing effort. Ludwig and Walters (1985) compared the two models by simulation, using fifty years of "data" generated by the structured model with reasonable parameter values. The structured model was exactly right, because it generated the data. Nonetheless, the Ricker model generally did as well *or better* at maximizing the total economic gain from the harvest, unless the simulated "data" included unrealistically large variability in fishing effort.

The details of this study are less important than the general message: by simulating on the computer the process of collecting data and estimating model parameters, you can explore whether your model has become too complex for the available data.

The effects of parameter error may be moot if numerical prediction accuracy is not important for your objectives. A theoretical model can add as many components as desired, to express the assumptions that the model is intended to embody. The danger in that situation is that the model becomes too complex to understand, which defeats its purpose.

Exercise 9.1. Find a recent (last five years) paper that uses a compartment model in your field of biology (or the one that interests you most). Draw a compartment diagram for the model, and identify (a) the state variables (b) the rate equation for all the flow rates (this is often challenging and may be impossible, using only what is contained in a published paper), and (c) any exogenous variables in the model.

Exercise 9.2. (a) Discuss the level of aggregation used in the paper you selected for the last exercise. In particular: which of the compartments seems most aggregated (composed of items that are really most heterogeneous)? (b) Draw the compartment diagram (boxes and arrows only) for a new model in which the compartment you identified in (a) has been disaggregated into two or more separate compartments.

Exercise 9.3. Find a recent paper in your field of biology that uses a dynamic model. List five errors in the model, where an "error" is an assumption of the model that is not literally and exactly true. For each error, state concisely the assumption made in the model, and the literal truth that the assumption contradicts.

Exercise 9.4. Choose a system in your area of biology that would be a suitable subject for modeling, in the sense that modeling would serve a useful scientific or practical purpose and the data needed are available. State the purpose of the model, and propose a tentative set of state variables for a model. Where does your model lie on the continuum from practical to theoretical discussed in Chapter 1?

9.3 Developing Equations for Process Rates

Having drawn a model diagram, we now need an equation for each process rate. To begin with the simplest case, consider a process rate ρ that depends on a single state variable x. For example, a flow rate ρ_{ij} in a compartment model may depend only on the amount in the compartment, x_j.

9.3.1 Linear Rates: When and Why?

The simplest possible rate equation is linear:

$$\rho(x) = ax \qquad [9.4a]$$

or more generally

$$\rho(x) = a(x - x_0). \qquad [9.4b]$$

The advantage of [9.4a] is that everything depends on the one parameter a. Therefore, "one point determines a line": given one simultaneous measurement of the

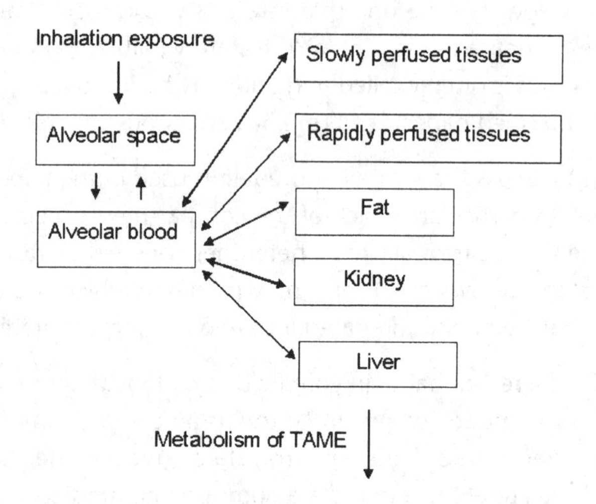

Figure 9.4 Compartment diagram of the model for TAME in the rat proposed by Collins et al. 1999. Double-headed arrows indicate a pair of flows, one in each direction, between two compartments.

state variable x and the flow rate ρ, one can estimate $\hat{a} = \rho/x$. In compartment models, a flow rate of this form,

$$\rho_{ij} = a_{ij}x_j,$$

is called *linear donor control.*

In compartment models, linearity occurs if the units in the compartment have no effect on each other. Consider for example the compartment model shown in Figure 9.4 for *tert*-amyl methyl ether (TAME) in laboratory rats (Collins et al. 1999). TAME is a gasoline additive that reduces carbon monoxide emissions, and the rat model was a step toward setting human exposure guidelines for TAME. Each flow between compartments was assumed to be linearly proportional to the concentration of TAME in the donor compartment (the input from inhalation exposure cannot be donor controlled because the donor—the air the rat is breathing—is "outside" the model, and the metabolism of TAME is assumed to follow Michaelis-Menten kinetics).

Suppose we start with q TAME molecules in the kidney, and imagine that they have been painted green; there will be some rate of flow of green TAME molecules from the kidney into the alveolar blood. Now let us add another q molecules to the kidney, painted red. Since the red and green molecules have no effect on each other (by assumption), the flow of red TAME molecules will be the same as the flow of green ones. So doubling the number of molecules doubles the flow rate.

Thus, one cause of linearity in compartment models is *dilution*: the stuff being tracked is so dilute that units never encounter each other. Conversely, if the units in a compartment are common enough that they interact, linear donor control may not be appropriate. The mechanism leading to nonlinearity may be direct interactions among units, or it could be indirect, meaning that the units affect some system property that in turn influences the other units. For example,

- substrate molecules binding to receptor sites may decrease the chance of other substrate molecules becoming bound
- when there are more prey available to a predator, each individual prey may be less likely to get eaten because the other prey contribute to satiating the predators, or because predators are busy hunting and consuming other prey

Exercise 9.5. Select one of the process rates from the model you proposed in Exercise 9.4 that could reasonably be modeled as linear, and explain why. Or, if there is no such process in your model, select one rate and explain why a linear rate equation is inappropriate.

9.3.2 Nonlinear Rates from "First Principles"

If enough is known about the mechanisms involved, that knowledge may imply a particular nonlinear functional form for rate equations, though not necessarily the numerical parameter values. For example,

- The law of mass action for chemical reactions states that reaction rates are proportional to the product of the concentrations of the reactants.
- Newton's laws of motion can be used to model animal locomotion as a set of rigid links connected at joints. The constraints due to the links make the equations nonlinear.
- Elasticity of heart tissue is coupled to its fluid motion in a nonlinear manner.
- The forces generated by muscle contraction depend nonlinearly on the length of the muscle.

Nonlinear functional forms also arise as consequences of biological assumptions about the system. In theoretical models rate equations are often derived as logical consequences of the model's biological hypotheses. For example,

- In the enzyme kinetics model in Chapter 1, the assumed underlying reaction scheme, combined with the law of mass action and assumptions about the relative time scales, implied the Michaelis-Menten form for the rate.

- In Ross's epidemic models the bilinear contact rate was derived as a consequence of assumptions about the transmission process.
- Membrane current through a population of channels depends upon the membrane potential and the open probability for individual channels in the population. The open probabilities are observed to vary in a way that must be modeled. In the Hodgkin-Huxley and Morris-Lecar models (Chapter 3), the open probability of each channel is expressed by gating variables, each of which is also a dynamic variable.

Exercise 9.6. Find a recent paper in your area of biology where one of the rate equations for a dynamic model was derived from first principles, or as a consequence of biological assumptions about the system, rather than on the basis of empirical data. State the assumptions or principles involved, and explain how those were used to derive the rate equation. Had you been the modeler, would you have made the same assumptions and used the same form of the rate equation?

9.3.3 Nonlinear Rates from Data: Fitting Parametric Models

In more practical models the rate equations are often estimated from data so that they quantitatively describe the system of interest. Continuing with the simplest case—a rate ρ depending on a single state variable x—estimating a rate equation from data means fitting a curve to a scatterplot of measurements $\{(x_i, \rho(x_i)), i = 1, 2, \ldots, N\}$. Two issues are involved: choosing a functional form, and estimating parameter values. We need to discuss the second question first—given a model, how do you estimate parameter values from data?—because the choice of functional form often comes down to seeing how well each of them can fit the data.

The data plotted in Figure 9.5a are the reproduction rate (estimated by the egg-to-adult ratio) in asexually reproducing rotifers *Brachionus calyciflorus* feeding on algae *Chlorella vulgaris* in experimental microcosms. These data come from some of the experiments reported by Fussmann et al. (2000). The variation in algal density results from the predator-prey limit cycles that occur in this system under suitable experimental conditions. Based on prior results with related organisms, a reasonable starting point is the Michaelis-Menten functional form

$$y = \frac{Vx}{K + x}. \qquad [9.5]$$

We want to find the values of V and K that give the best approximation to the data, according to some quantitative criterion. A common criterion is to minimize the sum of squared errors,

$$\text{SSE} = \sum_{i=1}^{N} \left(y_i - \frac{Vx_i}{K + x_i} \right)^2. \qquad [9.6]$$

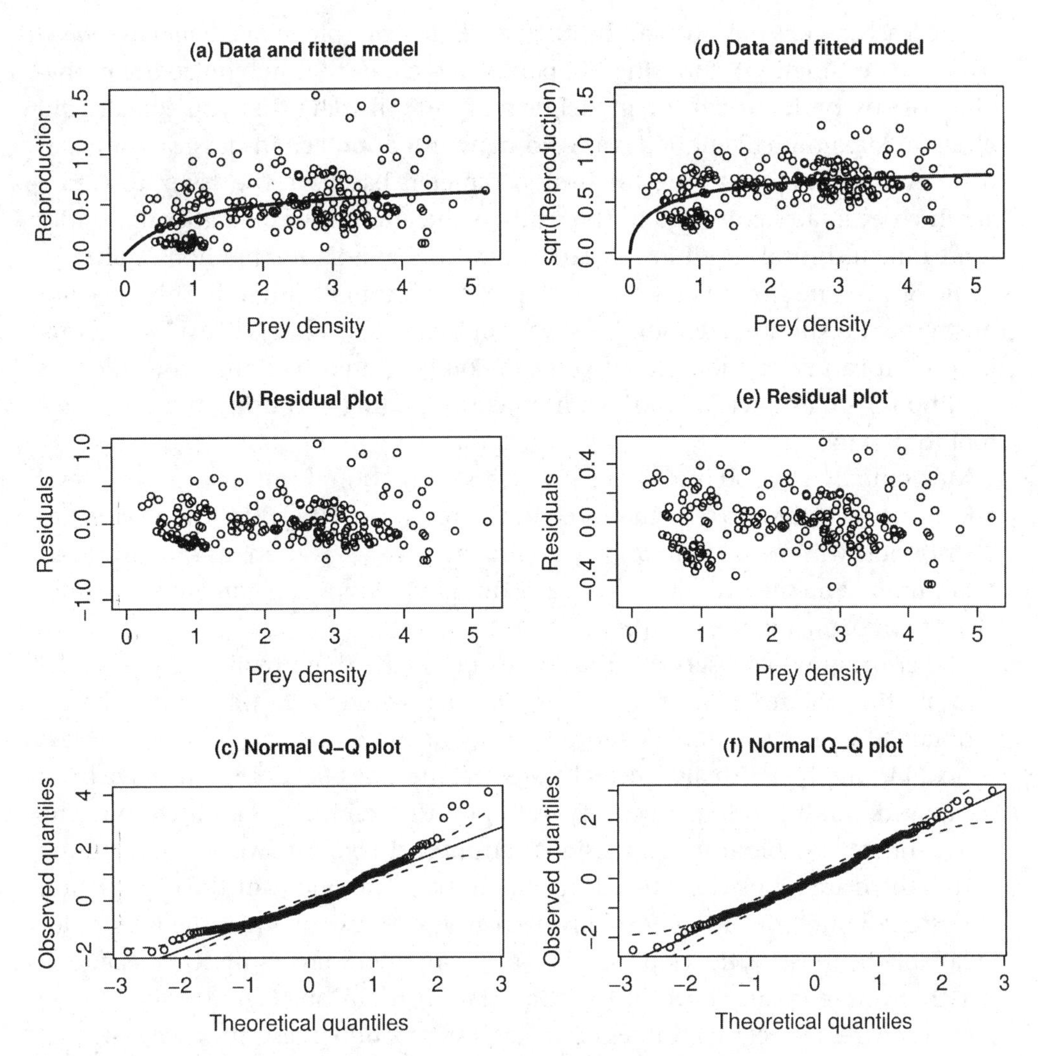

Figure 9.5 Fitting a parametric rate equation model. (a) The data and fitted curve (b) Residuals plotted against the independent variable (c) Quantile-quantile plot comparing the distribution of residuals to the Gaussian distribution assumed by the least-squares fitting criterion (d) Data and fitted curve on square root scale (e) Residuals from model fitted on square root scale (f) Quantile-quantile plot for square root scale residuals. The dashed curves in panels (c) and (f) are pointwise 95% confidence bands; the large number of points outside the confidence bands in panel (c) show that the residuals do not conform to a Gaussian distribution.

This is called *least squares*. Most statistics packages can do least squares fitting of nonlinear models such as this one; in a general programming language you can write a function to compute the SSE as a function of V and K, and then use a minimization algorithm to find the parameter values that minimize the SSE (the computer lab materials on this book's web page include examples of these). Either way, we find estimated values $\hat{V} = 0.72, \hat{K} = 0.95$ producing the curve drawn in Figure 9.5a.

The least squares method can be justified as an example of *maximum likelihood* parameter estimation: choosing the parameter values that maximize the probability of observing (under the model being fitted) the data that you actually did observe. Maximum likelihood is a "gold standard" approach in classical statistics. If we evaluate methods for estimating parameters based on the asymptotic rate at which estimates converge to the truth as the amount of data goes up, under some mild technical conditions it can be proved that no other method can have a convergence rate strictly better than that of maximum likelihood. This does not mean that maximum likelihood is always optimal. Rather, it is like a Swiss Army knife: if it can do the job (i.e., if you can compute and maximize your model's likelihood), you typically won't gain much by finding and using the exact best tool for the job.

Maximum likelihood leads to least squares estimation if the "errors" (the deviations between the model equation and the response data y_i) follow a Gaussian distribution with zero mean and constant variance. It is then a standard result that minimizing the sum of squares, as a function of model parameters, is equivalent to maximizing the likelihood.

The errors are not observable, but we can estimate them by fitting the model and plotting the residuals $e_i = y_i - \hat{V}x_i/(\hat{K} + x_i)$, where $\hat{V}, \hat{K}$ are the parameters estimated by least squares. Assumptions about the errors can now be checked by examining the residuals. Formal statistical methods for doing this have been developed, but it is often effective just to plot the residuals as a function of the independent variable and look for signs of trouble. In our case we see some (Figure 9.5b): there are no obvious trends in the mean or variance but the distribution is rather asymmetric. More formally, we can use a quantile-quantile (Q-Q) plot (available in most statistics packages, and computed here using R) to examine whether the residuals conform to a Gaussian distribution (Figure 9.5c). A Q-Q plot compares the relative values of the largest, second-largest, third-largest, etc., values in a data set, against the expected values of those quantities in a sample of the same size from the reference distribution. A perfect match between the data and the reference distribution results in the Q-Q plot being a straight line, and clearly here it isn't.

The model [9.5] seems to be good, but the residuals do not conform to the assumed error distribution. What do we do now? The simplest option is to just live with it. With non-Gaussian errors, least squares is no longer optimal but the estimates are still statistically acceptable; the same is true if the errors are Gaussian but their variance is nonconstant[1] (Gallant 1987, Chapters 1 and 2).

The next-simplest option is transformation: find a scale of measurement on which the error distribution conforms to the assumptions of least squares. The

[1]Technically, least squares estimates are still approximately Gaussian distributed, and still converge to the true values as sample size increases.

problem in Figure 9.5 is an asymmetric distribution with a long tail of large values. To pull in that tail we need a concave-down transformation such as log or square root. After some trial and error, square root transformation seems to do the trick. That is, the modified fitting criterion is

$$\mathrm{SSE2} = \sum_{i=1}^{N} \left(\sqrt{y_i} - \sqrt{\frac{Vx_i}{K + x_i}} \right)^2, \qquad [9.7]$$

which gives parameter estimates $\hat{V} = 0.70, \hat{K} = 1.1$. Finally, we look again at residuals to make sure that the transformation really has improved things, and the right-hand panels in Figure 9.5 confirm that it has. Instead of trial and error it is possible to estimate the power transformation that does the best job of producing Gaussian errors with constant variance; the procedures are described by Seber and Wild (1989, Chapter 2). For these data the result is $\hat{\beta} = 0.41$, not far from the trial-and-error result.

If transformation fails, then improving on least squares is more complicated and case specific, and may require either direct application of maximum likelihood or case-specific methods. A discussion of likelihood methods is not feasible here; Hilborn and Mangel (1997) give a very readable introduction.

Fitting nonlinear regression models by least squares or maximum likelihood is best done with a statistics program or package (*never* with spreadsheet programs, whose statistical functions are unreliable). Parameter estimates will generally be accompanied by a *confidence interval*—a range of possible parameter values that are credible, based on the data. By convention 95% confidence intervals are usually reported, or else a "standard error" σ for each parameter such that a range of $\pm 2\sigma$ centered at the estimated value is approximately a 95% confidence interval for large sample sizes. As Bayarri and Berger (2004) review, there are several different philosophical schools in statistics about how to define and compute confidence intervals, but all methods recommended for practical use have the same practical interpretation: in repeated applications to real data, the reported 95% confidence intervals obtained by a method should contain the true parameter value at least 95 times out of 100. Confidence intervals are useful when comparing model output against data—they limit how far parameters can be "adjusted" to get a good fit to data. They are also a good basis for setting ranges of parameter values to explore for sensitivity analysis (Chapter 8).

Exercise 9.7. Download the data from Figure 9.5 from this book's web site, and write a script to find least-squares parameter estimates for V and K on the untransformed scale (you should find that you duplicate the values above) and again using power transformation with $\beta = 0.41$. Do your parameter estimates for $\beta = 0.41$ indicate that trial-and-error choice of $\beta = 0.5$ was close enough?

Exercise 9.8. How does the choice of power transformation in the last exercise affect the 95% confidence intervals for the parameter estimates?

9.3.4 Nonlinear Rates from Data: Selecting a Parametric Model

We can now return to the task of selecting a functional form. The first step is to *plot your data.* If the data appear to be nonlinear, a useful next step is to see if some transformation straightens them out. For example, allometric relationships

$$y = ax^b \quad [9.8]$$

are pervasive in biology (see, e.g., Niklas 1994; West et al. 1997). Taking logarithms of both sides we have

$$\log y = \log a + b \log x, \quad [9.9]$$

a linear relationship between log-transformed variables. The Michaelis-Menten relationship [9.5] is linearized by taking the inverses of both sides:

$$\frac{1}{y} = \left(\frac{K}{V}\right)\frac{1}{x} + \frac{1}{V}. \quad [9.10]$$

The reason for trying this approach is that the eye is pretty good at telling if data are linear, but much poorer at telling the difference between one nonlinear curve and another.[2]

The next fallback is to "round up the usual suspects," a roster of conventional forms that modelers have used repeatedly to approximate nonlinear functional relationships. Figure 9.6 shows some of the most widely used forms. Replacing y by $y - y_0$ where y_0 is a constant produces a vertical shift of the curve by y_0; replacing x by $x - x_0$ produces a horizontal shift by amount x_0. Finally, if all else fails one can fall back on parametric *families* that allow you to add more and more parameters until the curve looks like your data. The most familiar are polynomials, $y = a_0 + a_1 x + a_2 x^2 + \cdots$.

The use of conventional functional forms is widespread, but we urge you to use them only as a last resort. The only thing special about them is their popularity. So before resorting to a conventional form, it is worthwhile thinking again if your knowledge of the underlying process, or reasonable assumptions about it, might suggest a rate equation with some mechanistic meaning. Or, if you have sufficient data, it might be preferable to use instead a nonparametric rate equation (as discussed below).

---

[2]Finding a linearizing transformation makes it tempting to fit the model by linear regression on the transformed scale. However, the transformation that produces linearity might not also produce errors that satisfy the assumptions for least-squares fitting. Fitting [9.5] by linear regression of $1/y$ on $1/x$ is often cited as an example of poor statistical practice. Small values of y and x transform into large values of their inverses, which are typically very inaccurate due to measurement errors even if the errors are small. When you fit on the transformed scale, the parameter estimates can be severely distorted in an attempt to fit those inaccurate values.

1. Concave up, increasing

$$y = ax^b \qquad a > 0, b > 1$$
$$y = ae^{bx} \qquad a > 0, b > 0$$
$$y = a + bx + cx^2 \qquad c > 0$$

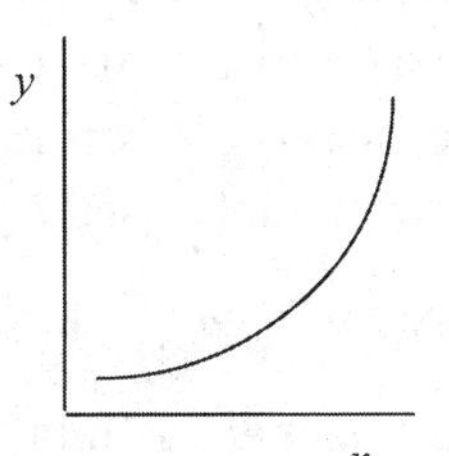

2. Concave down, increasing

$$y = ax^b \qquad a > 0, b < 1$$
$$y = ax/(b + x) \qquad a > 0, b > 0$$
$$y = a + bx + cx^2 \qquad b > 0, c < 0$$

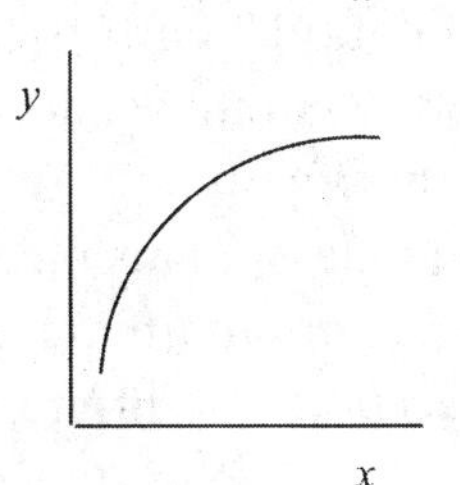

3. Concave up, decreasing

$$y = ax^{-b} \qquad a > 0, b > 0$$
$$y = ae^{-bx} \qquad a > 0, b > 0$$
$$y = a + bx + cx^2 \qquad b < 0, c < 0$$
$$y = A - \{\text{any from 2.}\}$$

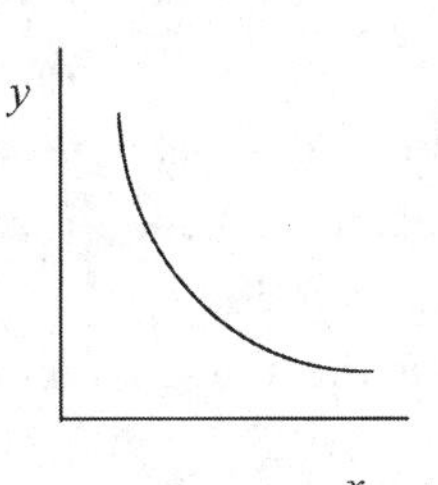

4. Concave down, decreasing

$$y = a + bx + cx^2 \quad c < 0$$
$$y = A - \{\text{any from 1.}\}$$

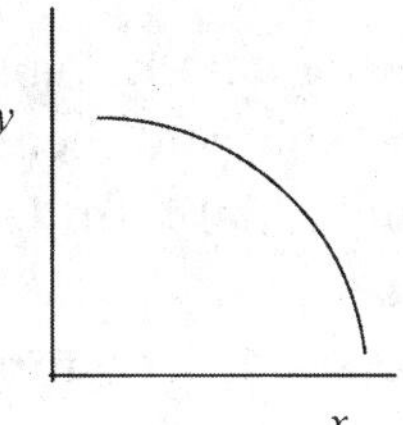

5. Sigmoid

$$y = \frac{1}{a + be^{-cx}} \qquad a, b, c > 0$$
$$y = ax^b/(c + x^b) \quad a, c, > 0, b > 1$$

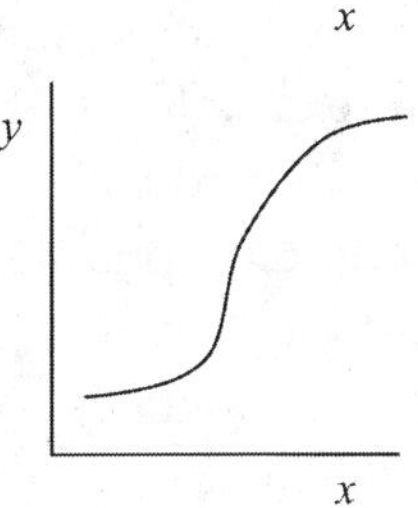

Figure 9.6 Some widely used equations forms to fit nonlinear functional relationships.

It is often difficult to identify one clearly best form for a rate equation just by visually comparing how well they fit the data (e.g., repeating something like Figure 9.5 for each candidate). Quantitative comparison can be based on values of the fitting criterion [e.g., [9.6] or [9.7]]. To fairly compare equations with different numbers of fitted parameters, those with more parameters need to be penalized—otherwise a cubic polynomial will always be better than a quadratic and worse than a quartic. Various criteria of this sort have been proposed, which attempt to pick the level of complexity that optimizes the ability to predict future observations. Two widely used criteria for least squares fitting are

$$\text{AIC} = N \log(\text{SSE}/N) + 2p, \qquad \text{BIC} = N \log(\text{SSE}/N) + p \log N,$$

where N is the number of data points, and p is the number of parameters in the model.[3] The model with the smallest value of AIC or BIC is preferred. For $N/p < 40$ a more accurate version of AIC is recommended, $\text{AIC}_c = \text{AIC} + 2p(p+1)/(N-p-1)$ (Burnham and Anderson 2002). If AIC and BIC disagree, they at least provide a range of plausible choices.

An alternative to approximate criteria such as AIC or BIC is to estimate directly the predictive accuracy of different functional forms, using a computational approach called *cross validation* (CV). For each data point (x_i, y_i), you

1. Form a reduced data set consisting of all other data points.
2. Fit the model to the reduced data set, obtaining estimated parameter vector $\hat{\theta}^{[-i]}$.
3. Generate a prediction of y_i using the reduced data set, $\hat{y}_i = f(x_i, \hat{\theta}^{[-i]})$.
4. Having done the above for all i, you compute the cross-validated prediction error $C = \sum_{i=1}^{N} (y_i - \hat{y}_i)^2$.

That is, you use the data in hand to simulate the process of fitting the model, predicting future data, and seeing how well you did. Repeating this process for each possible functional form lets you determine which of them gives the best predictive power.

Cross-validation penalizes unnecessarily complex models, because those models will "fit the noise" and therefore give bad predictions on actual data. Figure 9.7 shows an example using artificial data, generated by the linear model $y_i = 1 + x_i + 1.5e_i$ where the e_i are Gaussian distributed with mean $= 0$, variance $= 1$. Omitting one data point at a time, the linear model $y = a + bx$ and the quadratic model $y = a + bx + cx^2$ are both fitted by least squares to the remaining data. The quadratic model has an additional parameter so it always comes closer (on average) to the data used for fitting. However, the quadratic model may be further from the omitted data point—especially when the omitted data lets it "imagine" that there is some curvature present (the top left and bottom right panels). As a result, cross-validation selects the linear model: for the plotted data the cross-validated prediction errors are $C_{linear} = 46.6$, $C_{quadratic} = 160.1$. These data are only one example, but they illustrate what typically happens: repeating the same experiment 1000 times with different draws of the random errors e_i,

---

[3]Several other definitions of AIC and BIC are in circulation, differing from those given here by factors of 2 or N or by additive constants. All of these assign the same rank ordering to a set of models fitted to a given data set.

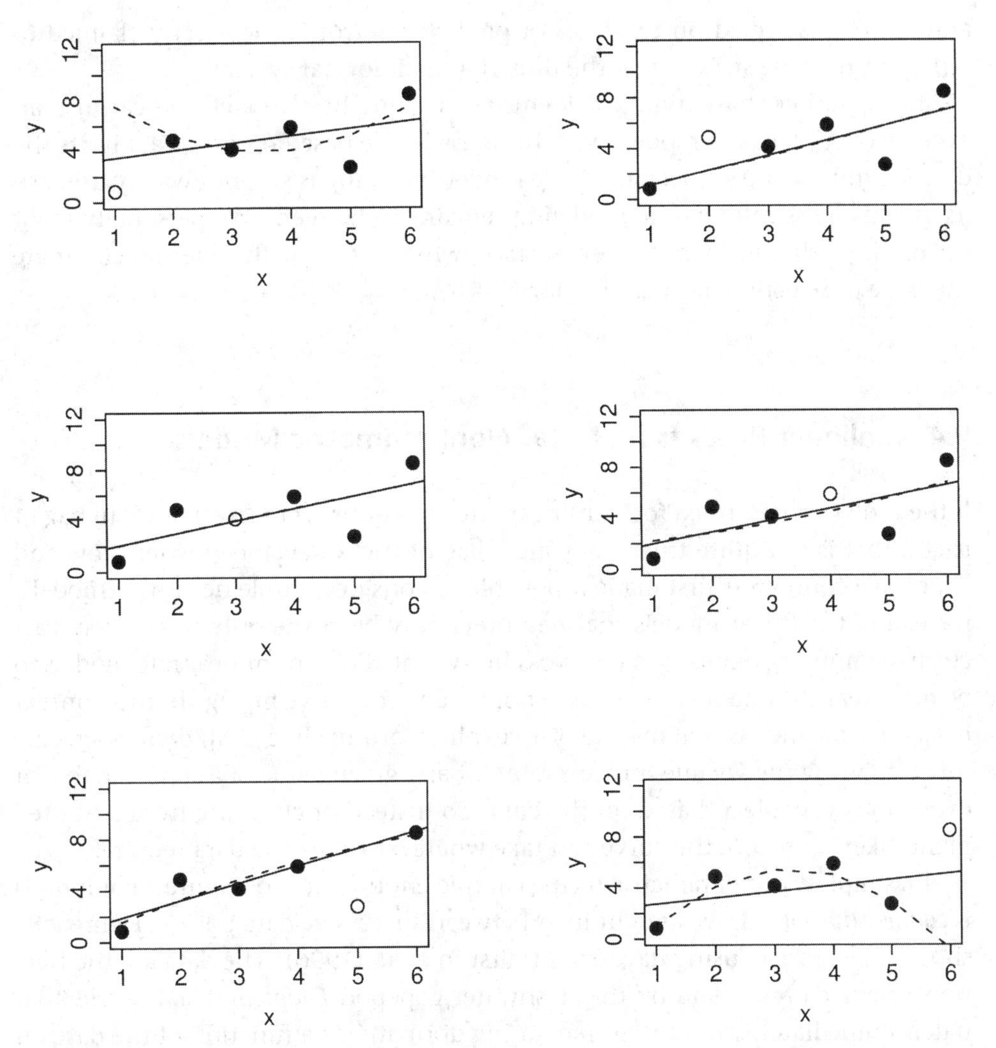

Figure 9.7 Cross validation for linear versus quadratic regression. In each panel, one of the 6 data points (shown as an open circle) is omitted from the data set, and the two models are fitted to the other data points. The solid line is the fitted linear model, the dashed line is the fitted quadratic.

the linear model was selected 78% of the time, which is pretty good for six noisy data points. Cross-validation is also not the last word in computer-intensive model selection; more sophisticated methods with better accuracy are available and development in this area is active (Efron 2004).

It is important to remember that cross-validation aims to find the rate equation that predicts best, given the data available. This is not the same as finding the "right" model, because of the tradeoff between model error and parameter error. If data are limited or noisy, any criterion based on prediction accuracy *should* select a model that is simpler than the truth. This distinction is often overlooked, and a model selected on the basis of prediction error is incorrectly claimed to represent the true structure of the underlying biological system.

Finally, rather than trying to identify a unique best model one can use an average over plausible models, weighting each one based on how well it fits the data (Burnham and Anderson 2002). Model averaging is still not widely used, so its effectiveness in dynamic modeling remains to be seen. We personally favor the methods discussed in the next section, which are more flexible and eliminate the issue of selecting a functional form.

9.4 Nonlinear Rates from Data: Nonparametric Models

If the tools just described for nonlinear rate equations seem like a random bag of tricks, that is not quite true. They are a bag of tricks developed when slow and expensive computers first made it possible to consider simple nonlinear models, instead of the linear models that had previously been the only possibility. Fast cheap computing opens up a new possibility that allows nonlinear rate models to be more realistic and less subjective: nonparametric curve fitting. In this context *nonparametric* means that instead of an explicit formula like [9.5], there is a recipe for computing the y value (process rate) for any given value of a state variable or exogenous variable x, based on the data. So instead of choosing from a limited menu like Figure 9.6, the curve can take whatever shape the data require.

The simplest nonparametric recipe (simple enough not to require a computer) is *connect the dots*: draw straight lines between successive data points. Figure 9.8a shows an example, using data from Hairston et al. (1996). The data are the fraction of egg clutches laid by the freshwater copepod *Diaptomus sanguineus* that hatch immediately rather than remaining dormant, as a function of the date on which the clutch was produced. The first and last dates plotted are field data. The intermediate dates are lab experiments, done in growth chambers set to mimic natural conditions on those dates (water temperature and photoperiod). Because only a few growth chambers were available, there are data for only a few dates, but each data point is reliable because a lot of copepods fit happily into one growth chamber—sample sizes are roughly 100–200. Because we trust each data point but know nothing about what happens in between them, "connect the dots" is reasonable. A more sophisticated version is to run a smooth curve through the data points, such as a *spline*. A spline is a polynomial on each interval between data points, with coefficients chosen so that polynomials for adjacent intervals join together smoothly. In this instance there was a "first principles" model based on population genetics theory (the light curve drawn in the figure), and connect-the-dots does a good job of approximating the theoretically derived function.

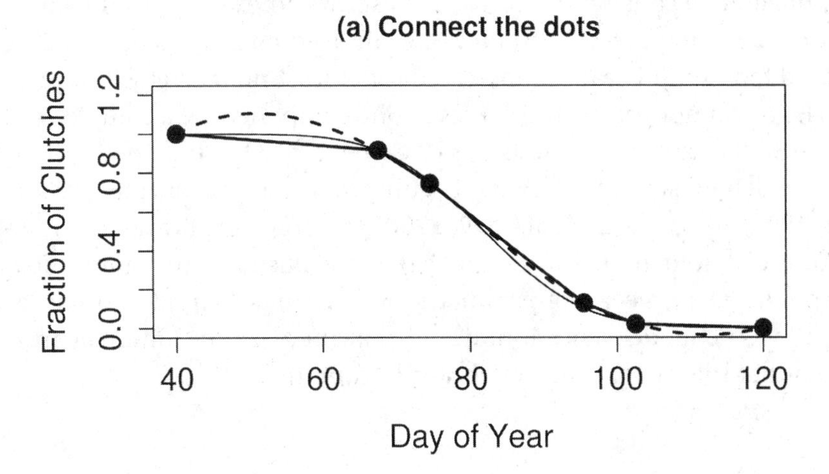

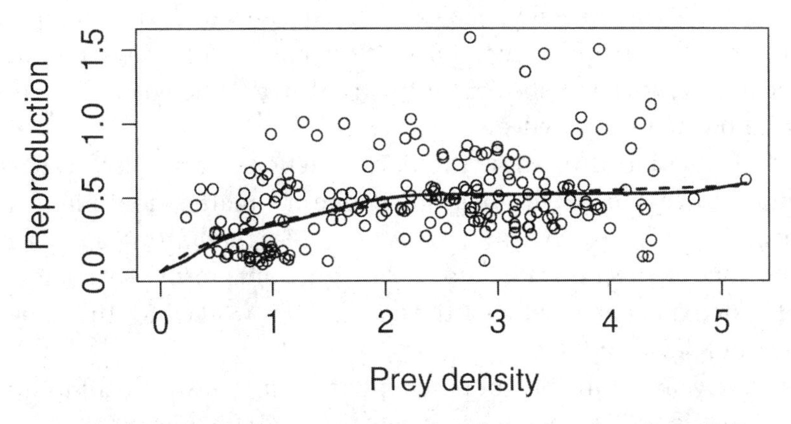

Figure 9.8 Examples of nonparametric rate equations based on data. (a) Two versions of "connect the dots." Heavy solid curve is linear interpolation, dashed curve is cubic spline interpolation. The light solid curve is a parametric model derived from population genetics theory for traits controlled by many genetic loci. (b) The solid curve is a nonparametric regression spline, fitted subject to the constraints of passing through $(0, 0)$ and being monotonically non-decreasing. For comparison, the dashed curve is the Michaelis-Menten model fitted to the same data ($\hat{V} = 0.7, \hat{K} = 1.1$).

The curves in Figure 9.8a go exactly through each data point—this is called *interpolation*. Interpolation no longer makes sense in a situation like Figure 9.5—a large number of imprecise data points. Nonparametric methods for this situation only became possible with sufficient computing power, so we are still in the stage where many different approaches are under parallel development, with new ideas appearing each year. We are partial to regression splines (Ruppert et al. 2003; Wood 2003) because they make it relatively easy to impose biologically meaningful qualitative constraints. Figure 9.8b shows an example (solid curve), with two constraints imposed: $y = 0$ for $x = 0$ (no feeding implies no breeding), and increased food supply leads to increased (or at least not decreased) reproduction rate. The use of nonparametric curves as one component of a complex statistical model is well established in statistics (Bickel et al. 1994), but applications to dynamic modeling have only recently begun; see Banks and Murphy (1989), Wood (1994, 1999, 2001), Ellner et al. (1998, 2002) for some applications. Another important use of nonparametric models is the one illustrated in Figure 9.8b: a close correspondence between nonparametric and parametric models supports use of that parametric model, since it indicates that there is no additional structure in the data that the parametric model cannot capture.

9.4.1 Multivariate Rate Equations

The curse of dimensionality (Chapter 1) afflicts even dynamic models if a rate equation depends on several variables. Half a dozen values can reveal the shape of a curve in one variable (Figure 9.8a), but would say little about a function of two variables and even less about a function of three. When data are sparse, some kind of simplification is needed.

The ideal situation is to know the right functional form. If you have a predetermined three-parameter equation, it doesn't matter how many variables are involved—fitting $\rho = ax + by + cz^2$ is no harder than fitting $\rho = a + bx + cx^2$, given the same number of data points. The functional form may come from first principles, or experience that a particular form has worked for the same process in models of related systems.

Alternatively, first principles or past experience may justify assumptions about the functional form. The two most widely used are the following.

1. Multiplication of rate limiting factors:

$$\rho(x_1, x_2, \ldots, x_m) = \rho_0 \rho_1(x_1) \rho_2(x_2) \cdots \rho_m(x_m). \qquad [9.11]$$

The assumption is that the relative effect of each variable x_i is independent of the other variables, analogous to a beam of light passing through a series of filters, each of which blocked a fixed fraction of the incoming photons. So for example, if a plant depends on N and P for growth, a given decrease in N availability causes the same percentage loss in growth rate, regardless of how much P is available. In statistics this would be called a generalized additive model (GAM) for $f = log\ \rho$,

$$f(x_1, x_2, \ldots, x_m) = f_0 + f_1(x_1) + f_2(x_2) + \cdots + f_m(x_m).$$

In R the MGCV package can be used to fit GAMs in which each f_j is a nonparametric regression spline.

x_1	2	3	4	5	6	1	1	1	1	1
x_2	2	2	2	2	2	1	2	3	4	5
ρ	10.0	10.7	9.6	8.3	5.4	16.6	8.3	4.2	3.6	6.2

Table 9.1 Artificial "data" for Exercise 9.9

2. Liebig's Law of the Minimum:

$$\rho(x_1, x_2, \ldots, x_m) = \rho_0 \times \min\{\rho_1(x_1), \rho_2(x_2), \ldots, \rho_m(x_m)\}. \quad [9.12]$$

Here we imagine a chain of subprocesses that are necessary for the overall process to occur, and the slowest of these acts as the rate-limiting step. Under this model, if a plant depends on N and P for growth and there is a severe N shortage, growth will be limited by lack of N and the (relatively) abundant P will not cause any additional decrease.

The advantage of these is that the multivariate model is reduced to a series of univariate models, each of which can be estimated using far less data. In addition, the individual functions can be based totally separate data sets, each involving variation in only one of the factors affecting the rate, and combined to model a situation where several factors are varying at once. The disadvantage of using these conventional forms is that you are skating on thin ice unless there are good reasons to believe that the chosen form is correct—such as results on similar systems, or evidence to support the mechanistic assumptions that lead to the particular form. Adopting a conventional assumption based on precedent should always be a last resort.

Finally, there are purely statistical approaches to dimension reduction, such as the generalized linear, neural network, and projection pursuit regression models (see, e.g., Venables and Ripley 2002). However, these have no clear biological interpretation, so choosing one of these over another is difficult unless data are so plentiful that dimension reduction is not necessary.

Exercise 9.9. For the "data" in Table 9.1, consider whether one of the models listed above is adequate.

(a) Which data points would you use to estimate ρ_1? Which would you use for ρ_2? Write a script to fit curves to these data (hint: the obvious works), and plot the data and estimated function curves.

(b) Combine your estimated ρ_1 and ρ_2 into an overall model, and evaluate how well it does by plotting observed versus predicted values.

Exercise 9.10. Propose a simple model for competition between two bacterial strains grown on an otherwise sterile "broth" of finely pulped vegetable (imagine cucumber in a blender on "high" for 10 minutes, diluted with water). The model is a step in identifying strains of bacteria that can be used to inoculate food products, so that if they are exposed to high temperatures and start to spoil, toxic bacteria will be outcompeted by harmless species. The salient biological facts are as follows:

- The experimental setup is closed culture (no inflow or outflow), kept well mixed at constant conditions.
- Substrate (the resources the bacteria need to survive, grow, reproduce) is available and not limiting.
- The main difference between strains is in their response to, and production rate of, lactic acid which they release into the culture medium. Lactic acid is a mildly toxic by-product of the metabolic processes that lead to growth and reproduction; as it accumulates the species have a decrease in their growth rates. On the time scale of interest, lactic acid does not degrade.

Your final set of equations can include parameters that would need to be estimated, and functions for process rates that would need to be determined experimentally.

9.5 Stochastic Models

Stochasticity can enter models in too many ways for us to consider or even enumerate here. But trying to do so would be pointless, because no fundamentally new issues arise. The only addition is the task of estimating the parameters that characterize the probability distributions for random components in the dynamic equations. To illustrate this point, we consider some types of stochastic models that we have seen in previous chapters.

9.5.1 Individual-Level Stochasticity

Models at the level of individual agents—for example, individual ion channels flipping between closed and open, or individual-based ecological models—are defined by the probabilities of different events rather than by process rates. Given the state of the system or of an individual agent, the model proceeds by asking what events could happen next, and what are their probabilities of occurring? For example, if a plant's size is x:

1. What is the probability that it will still be alive next year?
2. Live or die, how many seeds will it produce between now and next year, and how many of those will still be alive at the next census?
3. If it lives, how big will it be?

Questions of these types are the basis for any agent-based model. The agent's state is determined by a list of state variables—some discrete (alive or dead, healthy or infected, 0 or 1 for each bit in a Tierran organism, etc.), and some continuous. The model lists rules for how these change over time, and possibly for how they affect the numbers of agents that are created or destroyed. Typically the rules are stochastic, so the answer to "how big will it be?" is a probability distribution rather than a single number, with numerical values obtained from random number generators.

When the response variable is discrete—for example, live or die—the model requires a function $p(x)$ giving the probability of living as a function of agent state x. We can associate the outcomes with numerical scores $1 =$ live and $0 =$ die, but least squares fitting is still not appropriate because the distribution of outcomes is discrete and highly non-Gaussian. Fortunately the appropriate methods are so widely used that the basic models are included in most statistics packages. For binary choice (two possible outcomes) the most popular approach is a transformation method called *logistic regression*,

$$log\left(\frac{p}{1-p}\right) = f(x_1, x_2, \ldots, x_m). \qquad [9.13]$$

Here p is the probability of one of the possible outcomes, and $(x_1, x_2, \ldots, x_m)$ are the variables in the model that might affect the probability. Most statistics packages can at least do linear or polynomial logistic regression, and many allow f to be a nonparametric function. Much less is available prepackaged for more than two possible outcomes; most statistics packages can only fit logistic regression models in which the f for each outcome is linear. However, the likelihood function for more general logistic regression models is relatively simple, so these can be fitted via maximum likelihood.

In situations with multiple outcomes—such as offspring numbers in the list above—it is often more natural to use a parametric probability distribution to specify the outcome probabilities. For example, individual-based population models often assume that the number of offspring in a litter follows a Poisson distribution, whose mean depends on variables characterizing the individual (size, age, etc.) This is called *Poisson regression* and at least the linear model is available in many statistics packages. However it is again relatively easy to write down the likelihood function for other parametric probability distributions, for example, to modify the Poisson so that a litter must contain at least one offspring.

So fitting probability functions for an individual-level stochastic model is conceptually no different from fitting rate functions in a differential equation model, and the issues of model selection and complexity remain the same: linear or nonlinear? which nonlinear form? what form of multivariate response? and so on.

For continuous variables—such as individual size in the list above—the most common approach is a parametric probability distribution with parameters depending on agent and system state variables. This is like fitting a parametric process rate equation, except that functions are fitted for all parameters of the distribution, not just the mean. Figure 9.9 is an example, the size the following

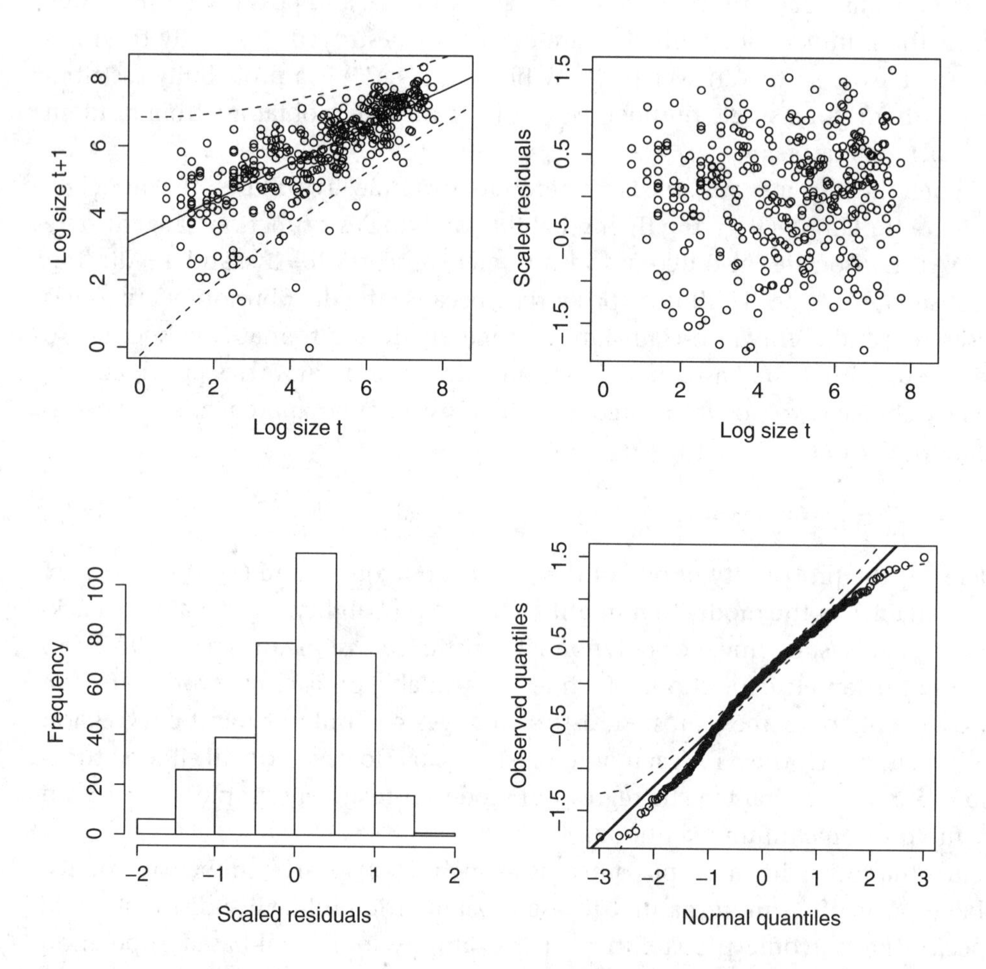

Figure 9.9 Fitting and testing a parametric model for the distribution of individual size in surviving plants of *Onopordum illyricum* (from Rees et al. 1999; data provided by Mark Rees). Panels (clockwise from top left) show the data from one study site with fitted mean and variance functions, the scaled residuals versus original size, Normal quantile-quantile plot and histogram of scaled residuals.

year of surviving plants in Illyrian thistle, *Onopordum illyricum* (Rees et al. 1999). The average log-transformed new size is a linear function of log-transformed current size, but the scatter about the regression line is not constant. Rees et al. (1999) proposed exponental size-dependence in growth variance. Letting x and y denote the current and next-year log transformed sizes, the fitted model for one of their study sites is a Gaussian distribution of y with mean and variance

$$\hat{y}(x) \equiv E(y|x) = 3.3 + 0.55x, \quad \sigma_y^2(x) \equiv \text{Var}(y|x) = 32.2e^{-0.58\hat{y}(x)}. \qquad [9.14]$$

Linear models such as [9.14] with nonconstant variance can be estimated by generalized least squares in many statistics packages.

The dashed lines in the scatterplot of the data (Figure 9.9) are curves based on the fitted variance function that should include 90% of all data. Clearly there are some problems: roughly 1% of the 720 plants in the data set had atypically poor growth, and these "outliers" have to be considered separately or ignored. Dropping the outliers, we can check the variance model in [9.14] by examining the scaled residuals $r_i = (y_i - \hat{y}(x_i))/\sigma_y(x_i)$. If the model is valid then their variance should be independent of x_i, which appears to be true, and their distribution should be approximately Gaussian.[4] The histogram and quantile-quantile plot for the scaled residuals indicate that the Gaussian assumption is not too bad (recall that exactly Gaussian scaled residuals would would give a perfectly straight quantile-quantile plot), but it is probably the least satisfactory aspect of this individual growth model.

There are also nonparametric estimates of probability distributions that can be used instead of a parametric distribution family (Venables and Ripley 2002, Chapter 5). For use in an agent-based simulation the simplest is to draw a scaled residual r at random from those computed from the data, and let the new size of a size-x individual be $\hat{y}(x) + \sigma_y(x)r$.

Exercise 9.11. The Poisson probability distribution with mean λ is given by $p(k) \equiv Pr(X = k) = e^{-\lambda}\lambda^k/k!$. The likelihood of obtaining a sample of values $x_1, x_2, \cdots, x_n$ is

$$L = \prod_{i=1}^{n} p(x_i).$$

(a) To see what the Poisson distribution is like, write a script file that generates 6 samples of size 50 from the Poisson distribution with mean 3, and produces a plot with a histogram of each sample. Then modify your script to do the same for samples of size 250 from the Poisson distribution with mean 20.

(b) Write a script file that computes $-\log(L)$ as a function of λ for the following data, and use it to find the maximum likelihood estimate of λ. Data values are $x = 1, 1, 1, 1, 2, 2, 2, 2, 2, 3, 4, 4, 4, 4, 5$.

9.5.2 Parameter Drift and Exogenous Shocks

"Parameter drift" means that the dynamics at any given moment follow a deterministic set of dynamic equations, but the parameters in the rate equations change randomly over time. The ideal situation is if the parameters themselves can be observed and measured. Then from multiple observations of the parameters, one can fit a probability distribution to their pattern of variation over time.

[4] More accurate residual scalings are possible and are recommended with small samples; see Davison and Hinkley 1997, section 6.2.

The situation is more difficult when the observable quantities are process rates that are jointly determined by state variables and parameter values. Then if the rate equation involves more than one parameter, the value of the process rate does not uniquely determine the parameter values that produced it. The most common way of dealing with this problem is to assume it out of existence, by adding a single parameter to represent the net effect of all stochastic variation in model parameters (and anything else), and then assuming that all other parameters are constant. That is, instead of a rate equation like

$$\rho(x,t) = \frac{V(t)x}{K(t)+x}$$

with stochastic variation in the parameters V and K, the rate equation might be

$$\rho(x,t) = \frac{Vx}{K+x} + e(t) \qquad \text{(additive noise)}$$
$$\rho(x,t) = \frac{Vx}{K+x}(1+e(t)) \quad \text{(multiplicative noise).} \qquad [9.15]$$

The terms with fixed parameter values are intepreted as the average process rate, and $e(t)$ produces random deviations from the average rate. In these equations, the randomness takes the form of external "shocks" to the system, rather than really specifying the mechanistic basis for fluctuations in process rates. The payoff is that fitting such models can be relatively straightforward. Consider for example Figure 9.5. We previously developed a model of the form

$$\sqrt{\rho(x)} = \sqrt{\frac{Vx}{K+x}} + e \qquad [9.16]$$

where the random "errors" e have a Gaussian distribution with 0 mean and constant variance. Ignoring the scatter about the fitted curve, we get a deterministic rate equation. But to take that variability into account in the model, the only additional task is to estimate their variance. As before, the residuals from the fit (Figure 9.5e) provide an estimate of the errors, and it is straightforward to show that for Gaussian-distributed errors, the maximum likelihood estimate of the error variance, $\hat{\sigma}_e^2$, is given by the mean of the squared residuals. For these data we have $\hat{\sigma}_e \doteq 0.18$ so the resulting stochastic rate model is

$$\rho(x,t) = \left(\sqrt{\frac{\hat{V}x}{\hat{K}+x}} + 0.18z(t)\right)^2 \qquad [9.17]$$

where the $z(t)$ are Gaussian random variables with mean 0, variance 1.

In fitting the model this way, we have assumed that the measured values of the rate are accurate—that the scatter about the fitted curve represents real variation in the rate, rather than errors in measurements. Otherwise, the fitted model overestimates the real variability in the rates. If there is measurement error, and if the measurement error variance is known, the model can be corrected to account for the component of the residual variance that is due to measurement errors. For example, if the measurement errors have constant known variance σ^2_{me}, we subtract this from the estimate of the total variance and have the estimate $\hat{\sigma}^2_e = \text{SSE2}/N - \sigma^2_{me}$.

9.6 Fitting Rate Equations by Calibration

So far we have assumed that data are available at the process level—simultaneous measurements of the process rate and of the state and exogenous variables thought to affect its value. When such data are not available, it may still be possible to fit the model by adjusting parameters so that model output—state variable trajectories—comes as close as possible in some sense to experimental measurements of state variable dynamics. This is called *calibration*. In some applications, the purpose of the model is to estimate otherwise unknowable parameters by calibration, such as the viral turnover rate in HIV (Chapter 6). Calibrating a model is then a way to connect the unobservable process rates and observable macroscopic quantities at the level of state variables, so that obtainable data can be used to estimate quantities that cannot be measured directly.

For example, consider fitting the so-called θ-logistic population model

$$dx/dt = rx(1 - (x/K)^\theta)$$

to a set of data on population growth over time (Figure 9.10). Although the model is a differential equation, we can still estimate parameters by least squares: find the values of r, K, θ and the initial value x_0, that minimize

$$\text{SSE} = \sum_{t=0}^{N} (\hat{x}(t) - \tilde{x}(t))^2, \qquad [9.18]$$

where $\hat{x}(t)$ are the data values, and $\tilde{x}(t)$ are solutions of the model. The starting value is treated as a parameter: were we to start model solutions at the first value in the data set, we would be giving that data point special treatment by insisting that the model match it exactly. Estimating parameters by minimizing SSE, or similar measures of the difference between model trajectories and data is called *path calibration* or *trajectory matching*.

In principle, path calibration of a deterministic model is just an example of nonlinear least squares, no different from nonlinear least squares fitting of [9.5], and it is commonly used for fitting dynamic models to data. Path calibration can also be used to estimate models with some nonparametric components (e.g., Banks and Murphy 1989; Banks et al. 1991; Wood 1994, 2001).

Path calibration does not make sense for stochastic models: two runs of the model yield different values of SSE, as would two replications of the experiment on the real system. Instead, we can fit the model by making model simulations

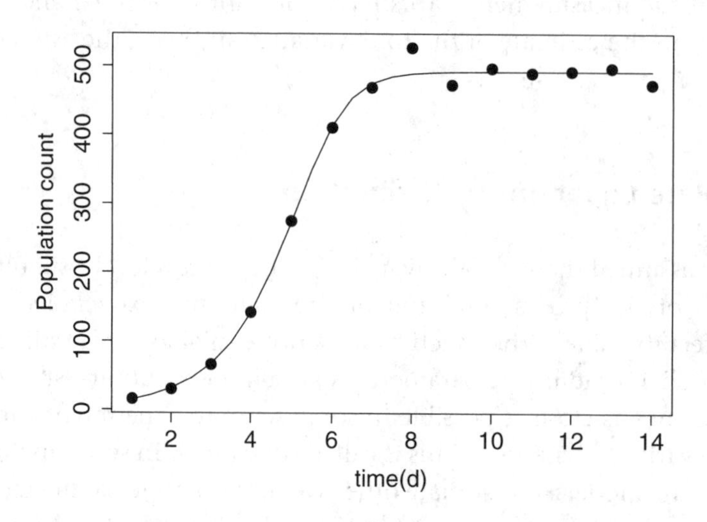

Figure 9.10 Results from calibration of the θ-logistic population model to data on population growth of *Paramecium aurelia* feeding on bacteria in Cerophyl growth medium; the data were digitized from Figure 2b in Veilleux (1976).

resemble the data in terms of statistical properties such as the mean, variance, and autocorrelation of state variable trajectories. Statisticians call this *moment calibration* or *method of moments*. It is widely used in economic modeling, because deliberate experiments are impossible and models have to be estimated and evaluated based on their predictions about observable variables such as stock prices, exchange rates, energy demand, etc. In the economics literature there is well-developed theory and numerous applications of moment calibration (e.g., Gourièroux and Montfort 1996), and biological applications are starting to appear (Ellner et al. 1995, 2002; Kendall et al. 1999; Turchin and Ellner 2000). However, *ad hoc* moment calibration is widespread, for instance, roughly estimating a few parameters by adjusting them until the model gives the right mean, variance, or range of state variables. More recently, general simulation-based methods are being developed for calibrating stochastic models by maximum likelihood or related Bayesian procedures, which can allow for both measurement errors in the data and inherently stochastic dynamics (e.g., Gelman et al. 2003; Calder et al. 2003; de Valpine 2004). This is a rapidly developing area, which will increase in importance as the necessary tools are incorporated into user-friendly statistics packages.

Exercise 9.12. The data plotted in Figure 9.10 are tabulated in Table 9.2. Are they fitted as well by the conventional logistic model in which $\theta = 1$? To answer this question, write scripts to do the following:

t	1	2	3	4	5	6	7	8	9	10	11	12	13	14
$x(t)$	15.58	30.04	66.05	141.6	274.6	410	468.8	526.4	472.5	496.6	489.5	492	496.8	473

Table 9.2 Experimental data on growth of *Paramecium aurelia* that are plotted in Figure 9.10, digitized from Figure 2b in Veilleux (1976)

Run	Period (days)	N_{max}/N_{min}
1	38.5 ± 1.5	36 ± 17
2	34.0 ± 1.5	77 ± 26
3	35.1 ± 0.4	240 ± 160

Table 9.3 Summary statistics for estimating the parameters P and δ of model (9.20) by moment calibration

(a) Path-calibrate by least squares the full model including θ and the reduced model with $\theta = 1$, and find which model would be selected by the AIC and BIC. Remember that the initial condition x_0 is included in the parameter count when fitting by path calibration.

(b) Path-calibrate both models repeatedly to reduced data sets that each omit one data point, and find which model would be selected by the cross-validation criterion.

Exercise 9.13. This exercise is based on Chapter 8 of Nisbet and Gurney (1982). The following simple model was proposed for oscillations in laboratory populations of sheep blowfly:

$$dN/dt = PN(t-\tau)e^{-N(t-\tau)/N_0} - \delta N(t) \qquad [9.19]$$

$N(t)$ is the number of sexually mature adults, and τ is the time required for a newly laid egg to develop into a sexually mature adult, about 14.8 days. Setting $x = N/N_0$ the model becomes

$$dx/dt = Px(t-\tau)e^{-x(t-\tau)} - \delta x(t). \qquad [9.20]$$

Write a script to estimate the values of P and δ by moment calibration, i.e., choosing their values to match as well as possible the estimates in Table 9.3 of the period of population oscillations, and the ratio between population densities at the peak and trough of the cycles. As conditions varied between experimental runs, parameters should be estimated separately for each of the three runs. Independent data suggest $\delta \approx 0.3/d$ and $100 < P\tau < 220$ for the first two experiments.

Note: your script can solve the model by representing it as a big density-dependent matrix model. Divide individuals into $m+1$ age classes $0-h, h-2h, 2h-3h, \ldots, (m-1)h-\tau$, and "older than 1" (adults) where $h = \tau/m$. One iteration of the matrix model then corresponds to h units of time. The number of births in a time step is $n_0(t+1) = hPn_{m+1}(t)e^{-n_{m+1}(t)}$, individuals younger than 1 have 100%

survival ($p_j = 1$ for $j = 0, 1, 2, \ldots, m - 1$), and adults have survival $e^{-\delta h}$. This method is computationally inefficient but makes it easy to add finite-population effects, for example, if the number of births in a time interval is a random variable with mean given by the expression above.

Exercise 9.14. Derive completely one of the process rate equations for the model from Exercise 9.4. That is, for one of the processes in the model, derive the equation which gives the rate as a function of the state variables and/or exogenous variables which affect it. This includes both the choice of functional form, and estimating numerical values for all parameters in the equation. (As noted above "rate" could also be an outcome probability or probability distribution of outcomes, depending on the kind of model.)

(a) Describe the process whose rate equation you are deriving.

(b) Explain the rationale for your choice of functional form for the equation. This could involve data; the hypothesis being modeled; previous experience of other modelers; etc.

(c) Estimate the parameters, explaining what you are doing and the nature and source of any data that you are using. If the available data do not permit an exact estimate, identify a range of plausible values that could be explored when running the model.

(d) Graph the "final product": the relationship between the process rate and one or two of the state or exogenous variables influencing it, in the rate equation.

9.7 Three Commandments for Modelers

The principles of model development can be summarized as three important rules:

1. Lie
2. Cheat
3. Steal

These require some elaboration.

Lie. A good model includes incorrect assumptions. Practical models have to be simple enough that the number of parameters does not outstrip the available data. Theoretical models have to be simple enough that you can figure out what they're doing and why. The real world, unfortunately, lacks these properties. So in order to be useful, a model must ignore some known biological details, and replace these with simpler assumptions that are literally false.

Cheat. More precisely, do things with data that would make a statistician nervous, such as using univariate data to fit a multivariate rate equation by multiplication of limiting factors or Liebig's law of the minimum, and choosing between those options based on your biological knowledge or intuition. Statisticians like to let data "speak for themselves." Modelers should do that when it is possible, but more often the data are only one input into decisions about model structure, the rest coming from the experience and subject-area knowledge of the scientists and modelers.

Steal. Take ideas from other modelers and models, regardless of discipline. Cutting-edge original science is often done with conventional kinds of models using conventional functional forms for rate equations—for example, compartment models abound in the study of HIV/AIDS. If somebody else has developed a sensible-looking model for a process that appears in your model, try it. If somebody else invested time and effort to estimate a parameter in a reasonable way, use it. Of course you need to be critical, and don't hesitate to throw out what you've stolen if it doesn't fit what you know about your system.

9.8 Evaluating a Model

Everybody knows what it means to evaluate a model: you plot model output and your data on the same graph, as in Figure 9.10, and if the points are near the line, the model is good. But there is more to it than that. Following Caswell (1976), we consider an extreme example to explain why.

Figure 9.11 shows a model where the points are quite near the line: a remarkably simple model for growth of the global human population developed by Von Foerster et al. (1960). Von Foerster et al. (1960) argued that the dominant force in human population expansion was our ability to cooperate in advancing our standard of living, thus increasing lifespan and offspring survival. Consequently, the net (birth-death) rate r should be modeled as an *increasing* function of population size N. They proposed the model $r = aN^{1/K}$, $a, K > 0$; hence $dN/dt = aN^{1+1/K}$. This model can be solved explicitly—the form of the solution is

$$N(t) = \alpha(\tau - t)^{-K}. \qquad [9.21]$$

Equation [9.21] fits the historical data remarkably well: it says that $log\, N(t)$ should be a linear function of $log(\tau - t)$, and for the data and solution in Figure 9.11 this linear regression explains over 97% of the variance in $log\, N(t)$.

So we have a good model, right? But look at what happens in [9.21] as $t \to \tau$: $N(t)$ becomes infinite. Von Foerster et al. (1961) called this *Doomsday* and estimated it to occur on Friday November 13, 2026: "Our great-great grandchildren will not starve to death, they will be squeezed to death" (von Foertster et al. 1960, p. 1295). With an additional forty years of data, the estimate of Dooms-

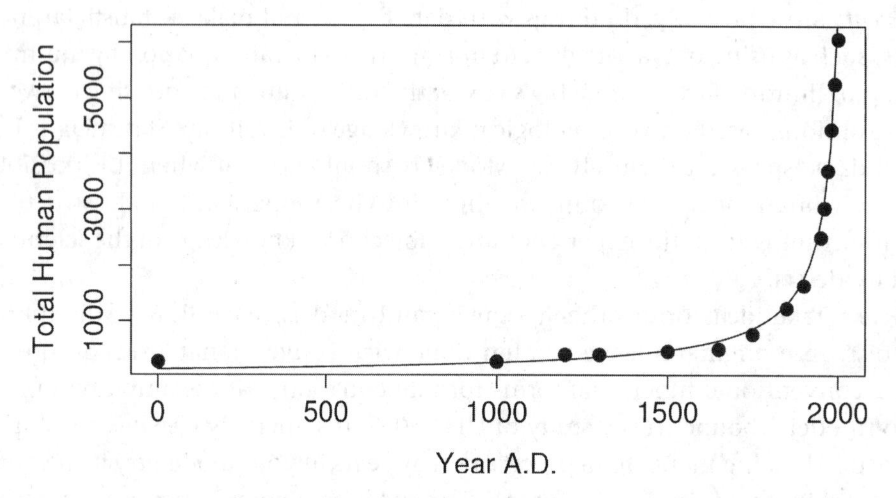

Figure 9.11 Solution of the Von Foerster et al. (1960) model (line) for the growth of total human population (solid dots, in millions), with parameters estimated by least squares on square-root scale. Population estimates were obtained from the International Programs Center of the U.S. Census Bureau (online: http://www.census.gov/ipc/www), accessed June 16, 2003. For 1950 and earlier, the value used was the median of tabulated historical estimates; for subsequent dates the Census Bureau estimate was used.

day is pushed back only to August 20, 2033—this stability of parameter estimates is what you would expect in a valid model. Of course Von Foerster et al. did not really believe in Doomsday: their actual conclusion was that current population trends could not be sustained, and that efforts should be made to promote reductions in family size.

This example illustrates the importance of evaluating models *relative to their purpose*. The Doomsday model's good fit to historical data justifies using it to predict future population growth under the assumption that current trends continue unabated—which was its purpose. But we can also interpret the model in a different way, as representing the mechanistic assumption that human population growth is driven by cooperative efforts that intensify as population grows. The model serves this purpose in a different way: by making impossible predictions, it shows that the model's assumptions are invalid or incomplete.

So let us try again: if your objectives are toward the practical end of the continuum and the model's main purpose is to make accurate predictions, then the model is good if model predictions are in accord with your data. Right?

Well, maybe. First, it would be more accurate to say that a model can be accepted if the data are within the range of possible outputs from the model; Waller et al. (2003) summarize methods for such comparisons and illustrate them on a model for the spatial spread of fox rabies. Also, was the model calibrated? If the goal is prediction, then Figures 9.10 and 9.11 seem to say that we have a good model. But the fit is good because we chose parameters to fit those particular data, so it does not say whether the same model could fit a second experiment, or whether another model might fit even better. If all available data were used to calibrate the model, then none is "left over" to test the model. Unless data are so plentiful that one can fit the model to part of the data and test it on the remainder (and be honest about it), successful calibration is only weak support for a model.

One more try: if your model lies toward the practical end of the continuum and its main purpose is to make accurate predictions, then the model is bad if model output is inconsistent with the data. Right?

Well, maybe. It could also mean that some of your parameters are poorly estimated, so it would be worth seeing if you can calibrate the model to fit the data. If not, you probably do have a bad model *for purposes of prediction*. If the model can be calibrated, you are back in the previous quandary that the model fits because you fiddled with parameters to make it fit.

In short: evaluating a model is hard. Fortunately it is easier, and usually more useful, to evaluate *several* models. The difficulties elaborated (and exaggerated) above stem from trying to benchmark a model against reality. No model can pass that test perfectly, so you're left trying to decide how close is "close enough". In contrast, when we compare how well two different models can account for the data, both are measured on the same scale.

9.8.1 Comparing Models

The appropriate criteria for comparing models are determined (as always) by your goals, so there is no "one size fits all" approach, or even an exhaustive list of possibilities.

At one extreme, if the only concern is prediction accuracy, then the methods for individual rate equations discussed in Section 9.3.4 apply equally to the model as a whole. In particular, one always has recourse to cross-validation, in principle. At the level of the entire model, it is often not feasible to automate the process of omitting individual data points and rebuilding the model. Instead, the data can be divided into several subsets, and subsets are omitted rather than single data points. Subsets can be chosen at random or in some systematic way. For example, if the models are fitted by calibrating to state variable trajectories, you might first calibrate to the first half of the data and see how well each model predicts the second half, then calibrate to the second half and try to predict the first.

Different approaches are needed if you are concerned with mechanistic validity. In some situations a structured cross-validation may be useful. For example, again assuming that you have data on several state variables, you might omit in turn the entire data series for one of the state variables, calibrate the model to the others, and see how well the model predicts the omitted state variable. A model that properly captures the causal relations among state variables would be expected to do better in such a test than one that has the wrong "wiring diagram."

More formal statistical approaches are possible if the models are *nested*, meaning that one of them (the *reduced model* $\mathbf{M}_R$) can be obtained by removing components of the other (the *full model* $\mathbf{M}_F$), or by constraining parameters of the full model to particular values. Comparison of these models addresses the scientific question of whether the additional features of the full model are actually playing a role in the system dynamics. Because $\mathbf{M}_F$ necessarily has more parameters (all those in $\mathbf{M}_R$ and then some), it always can fit the data better than $\mathbf{M}_R$. So in order to assay the actual relevance of the additional processes in $\mathbf{M}_F$, we have to determine whether the improvement in fit is too large to be solely a result of the additional parameters. We do this by adopting the *null hypothesis* that $\mathbf{M}_R$ is true, and computing the probability (under this assumption) of getting an improvement in fit as large, or larger, than the one that actually occurred. If this sounds familiar, it is: this is a standard statistical hypothesis test. However, in contrast to most descriptive statistical models, for dynamic models we typically have to use simulation methods to implement the test—a process called (somewhat confusingly) *parametric bootstrap*. In principle the recipe is simple:

1. Fit $\mathbf{M}_R$ and $\mathbf{M}_F$ to the data, and record for each a quantitative measure of how well they fit the data.
2. Use simulations of $\mathbf{M}_R$ to generate artificial data sets that mimic the real one (in terms of the amount, type, and accuracy of the data).
3. Fit $\mathbf{M}_R$ and $\mathbf{M}_F$ to each artificial data set, and record for each the same quantitative measure of how well they fit the data.
4. Compare the observed improvement (step 1) to those occurring in artifical data where $\mathbf{M}_R$ is really true. If the observed improvement is atypically large relative to those on the artificial data (e.g., larger than all but 5% or 1%), this implies that the improved fit of $\mathbf{M}_F$ is not just due it having additional parameters, hence the additional structure in $\mathbf{M}_F$ captures some features actually present in the data.

Note that a negative result is harder to interpret—as is true in general for statistical hypothesis tests. If the reduced model is not rejected, it may mean that the added ingredients in the full model are not present in the real system, or it may mean that the data set was too small or too inaccurate to demonstrate that those ingredients are present.

Pascual et al. (2000) used parametric bootstrap to test the hypothesis that cholera outbreaks are linked to climate variability associated with the El Niño Southern Oscillation (ENSO), mediated by increased sea-surface temperatures and higher numbers of zooplankton bearing the bacterium causing cholera. The data were monthly estimates of cholera incidence from a hospital in Bangladesh from 1980–1999, where a sample of incoming patients were tested for cholera. The reduced model included local seasonal climate variation as an exogenous variable affecting cholera dynamics; the full model added an index of ENSO based on sea-surface temperatures in the Pacific. The full and reduced models were compared based on their ability to predict cholera incidence two months into the future. One thousand artificial data sets were generated from the reduced model by adding randomly shuffled residuals to the reduced model's predictions of monthly cholera incidence. Fitting both models to these artificial data, the improvements in fit were all smaller than on the real data, providing very strong evidence of a link between ENSO and cholera.

Similarly, Turchin et al. (2003) used parametric bootstrap to compare models for population outbreaks of larch budmoth in Swiss forests. Budmoth outbreaks occur at regular intervals of 8–9 years, with roughly 100,000-fold variation between abundances at the peak and trough of the cycles. For the last several decades the experimental research has focused on the generally-accepted hypothesis that the cycles are driven by the interactions between budmoth and their food supply, larch needles. Turchin et al. (2003) also considered effects of parasitoids attacking the budmoth. Parametric bootstrap was used to compare models with and without the effects of parasitoids, and the results were decisive: the improvement in fit on the real data was larger than all but 0.3% of the improvements on the artificial data, demonstrating that the accepted hypothesis for the cycles was incorrect. A second reduced model with parasitoids but no effect of food supply was also rejected, hence it appears that both parasitoids and food supply affect budmoth dynamics.

As these examples illustrate, comparative evaluation of models is no different from comparative evaluation of two scientific hypotheses based on their ability to account for all the available evidence. It suffers from the same limitations—for example, there is broad scope for fiddling with hypotheses to explain away discrepancies, in the same way that there is broad scope for adjusting model parameters to fit whatever data are available. However, models add the potential for

- generating and testing predictions based on complex hypotheses involving the interactions of many components and processes, and
- making quantitative predictions, so that discrimination among hypotheses can be based on quantitative comparisons of how well each model can account for the available data.

That is, models used as quantitative expressions of contending hypotheses extend the scope of the scientific method. They let us evaluate ideas about an entire system, rather than about its individual parts one by one, and let us make those evaluations as rigorous as our knowledge and data allow.

Exercise 9.15. Produce two alternative versions of your model from Exercise 9.4, differing in some biologically meaningful assumption. Do an appropriate comparative evaluation to determine which model corresponds best with the actual behavior of the study system.

9.9 References

Banks, H. T., and K. A. Murphy. 1989. Estimation of nonlinearities in parabolic models for growth, predation, and dispersal of populations. Journal of Mathematical Analysis and Applications 141: 580–602.

Banks, H. T., L. W. Botsford, F. Kappel, and C. Wang. 1991. Estimation of growth and survival in size-structured cohort data—an application to larval striped bass (*Morone saxatilis*). Journal of Mathematical Biology 30: 125–150.

Bayarri, M. J., and J. O. Berger. 2004. The interplay of Bayesian and frequentist analysis. Statistical Science 19: 58–80.

Bickel, P. J, C.A.J. Klassen, Y. Ritov, and J. A. Wellner. 1993. *Efficient and Adaptive Estimation for Semiparametric Models*. Johns Hopkins University Press, Baltimore, MD.

Bolker, B. M., S. W. Pacala, and W. J. Parton, Jr. 1998. Linear analysis of soil decomposition: insights from the Century model. Ecological Applications 8: 425–439.

Burnham, K. P., and D. R. Anderson. 2002. *Model Selection and Multi-Model Inference*. Springer, New York.

Calder, C., M. Lavine, P. Muller, and J. S. Clark. 2003. Incorporating multiple sources of stochasticity into dynamic population models. Ecology 84: 1395–1402.

Caswell, H. 1976. The validation problem. Pages 313–325 in B. C. Patten (ed.), *Systems Analysis and Simulation in Ecology*. Academic Press, New York, Vol. 4.

Collins, A. S., S.C.J. Sumner, S. J. Borghoff, and M. A. Medinsky. 1999. A physiological model for *tert*-amyl methyl ether and *tert*-amyl alcohol: hypothesis testing of model structures. Toxicological Sciences 49: 15–28.

Colwell, R. 1996. Global climate and infectious disease: The cholera paradigm. Science 274: 2025–2031.

Cox, P., R. Betts, C. Jones, S. Spall, and I. Totterdell. 2000. Acceleration of global warming due to carbon-cycle feedbacks in a coupled climate model. Nature 408, 184–187.

Cox, P. 2001. Description of the "TRIFFID" Dynamic Global Vegetation Model. Hadley Centre Technical Note 24. Available at http://www.met-office.gov.uk/research/hadleycentre/pubs

Cramp, D. G., and E. R. Carson. 1979. Dynamics of blood glucose and its regulating hormones. In D. A. Linkens (ed.), *Biological Systems, Modeling and Control*. IEE, London.

Davison, A. C., and D. V. Hinkley. 1997. *Bootstrap Methods and Their Application*. Cambridge University Press, Cambridge, U.K.

de Valpine, P. 2004. Monte Carlo state-space likelihoods by weighted posterior kernel density estimation. Journal of the American Statistical Association 99: 523–536.

Efron, B. 2004. The estimation of prediction error: Covariance penalties and cross-validation. Journal of the American Statistical Association 99: 619–632.

Ellner, S., A. R. Gallant, and J. Theiler. 1995. Detecting nonlinearity and chaos in epidemic data. Pages 229–247 in D. Mollison (ed.), *Epidemic Models: Their Structure and Relation to Data*. Cambridge University Press, Cambridge, U.K.

Ellner, S. P., B. A. Bailey, G. V. Bobashev, A. R. Gallant, B. T. Grenfell, and D. W. Nychka. 1998. Noise and nonlinearity in measles epidemics: combining mechanistic and statistical approaches to population modeling. American Naturalist 151: 425–440.

Ellner, S. P., Y. Seifu, and Robert H. Smith. 2002. Fitting population models to time series data by gradient matching. Ecology 83: 2256–2270.

Fussman, G., S. P. Elner, K. W. Shertzer, and N. G. Hairston, Jr. 2000. Crossing the Hopf bifurcation in a live predator-prey system. Science 290: 1358–1360.

Gallant, A. R. 1987. *Nonlinear Statistical Models*. Wiley, New York.

Gelman, A., D. B. Rubin, J. B. Carlin, and H. S. Stern. 2003. *Bayesian Data Analysis*, 2nd edition. Chapman & Hall/CRC Press, Boca Raton, FL.

Gourièroux, C., and A. Montfort. 1996. *Simulation-based Econometric Inference*. Oxford University Press, Oxford, U.K.

Hairston, N. G., Jr., C. M. Kearns, and S. Ellner. 1996. Phenotypic variation in a zooplankton egg bank. Ecology 77: 2382–2392.

Hett, J. M., and R. J. O'Neill. 1974. Systems analysis of the Aleut ecosystem. Artic Anthropology 11: 31–40.

Hilborn, R., and M. Mangel. 1997. *The Ecological Detective: Confronting Models with Data*. Princeton University Press, Princeton, NJ.

Kendall, B. E., C. J. Briggs, W. W. Murdoch, P. Turchin, S. P. Ellner, E. McCauley, R. Nisbet, and S. N. Wood. 1999. Why do populations cycle? A synthesis of statistical and mechanistic modeling approaches. Ecology 80: 1789–1805.

Ludwig, D., and C. J. Walters. 1985. Are age-structured models appropriate for catch-effort data? Canadian Journal of Fisheries and Aquatic Science 42: 1066–1072.

Niklas, K. J. 1994. *Plant Allometry: the Scaling of Form and Process*. University of Chicago Press, Chicago.

Nisbet, R. M., and W.S.C. Gurney. 1982. *Modelling Fluctuating Populations*. Wiley, New York.

NREL (Natural Resource Ecology Laboratory). 2001. Century soil organic matter model version 5, user's guide and reference. Natural Resource Ecology Laboratory, Colorado State University, Fort Collins, CO. Online at http://www.nrel.colostate.edu/projects/century5

Parton, W., D. Schimel, C. Cole, and D. Ojima. 1987. Analysis of factors controlling soil organic levels of grasslands in the Great Plains. Soil Science Society of America Journal 51: 1173–1179.

Parton, W., J. Stewart, and C. Cole. 1988. Dynamics of C, N, P and S in grassland soils: A model. Biogeochemistry 5: 109–131.

Pascual, M., X. Rodo, S. P. Ellner, R. Colwell, and M. J. Bouma. 2000. Cholera dynamics and the El Niñ-Southern Oscillation. Science 289: 1766–1769.

Rees, M., A. Sheppard, D. briese, and M. Mangel. 1999. Evolution of size-dependent flowering in *Onopordum illyricum:* A quantitative assessment of the role of stochastic selection pressures. American Naturalist 154: 628–651.

Ruppert, D., M. P. Wand, and R. U. Carroll. 2003. *Semiparametric Regression*. Cambridge University Press, New York.

Seber, G.A.F., and C. J. Wild. 1989. *Nonlinear Regression*. Wiley, New York.

Turchin, P., and S. P. Ellner. 2000. Living on the edge of chaos: Population dynamics of Fennoscandian voles. Ecology 81: 3099–3116.

Turchin, P., S. N. Wood, S. P. Ellner, B. E. Kendall, W. W. Murdoch, A. Fischlin, J. Casas, E. McCauley, and C. J. Briggs. 2003. Dynamical effects of plant quality and parasitism on population cycles of larch budmoth. Ecology 84: 1207–1214.

Venables, W. N., and B. D. Ripley. 2002. *Modern Applied Statistics with S*, 4th edition. Springer, New York.

Veilleux, B. G. (1976). The analysis of a predatory interaction between *Didinium* and *Paramecium*. MS Thesis, University of Alberta.

Von Foerster, H., L. W. Amiot, and P. M. Mora. 1960. Doomsday: Friday, 13 November, A.D. 2026. Science 132: 1291–1295.

Waller, L. A., D. Smith, J. E. Childs, and L. A. Real. 2003. Monte Carlo assessments of goodness-of-fit for ecological simulation models. Ecological Modelling 164: 49–63.

West, G. B., J. H. Brown, and B. J. Enquist. 1997. A general model for the origin of allometric scaling laws in biology. Science 276: 122–126.

Wood, S. N. 1994. Obtaining birth and death rate patterns from structured population trajectories. Ecological Monographs 64: 23–44.

Wood, S. N. 1999. Semi-parametric population models. Aspects of Applied Biology 53: 41–50.

Wood, S. N. 2001. Partially specified ecological models. Ecological Monographs 71: 1–25.

Wood, S. N. 2003. Thin-plate regression splines. Journal of the Royal Statistical Society, Series B 65: 95–114.

Index